Electromagnetic Wave Absorption and Shielding Materials

This book reveals the latest research findings and innovations in electromagnetic wave absorption and shielding by exploring the design and application of absorbent materials, the optimization of shielding structures and the improvement of testing and evaluation methods.

From conductive materials to magnetic materials, and composite materials to nanomaterials, *Electromagnetic Wave Absorption and Shielding Materials* details the characteristics and advantages of various absorbent materials and explains their applications in electromagnetic wave absorption and shielding. It then introduces the different methods of electromagnetic shielding, including structural shielding and material shielding. The book also studies experimental and testing techniques, including measurement methods and evaluation criteria for electromagnetic wave absorption performance.

The book will be of interest to researchers and graduate students in electromagnetic compatibility, materials science and engineering.

Wei Lu earned his PhD from Tongji University in 2006. He is a full tenured professor at Tongji University. His current research focuses on electromagnetic wave shielding and absorption materials, and materials for soft electronics.

Hongtao Guan earned his PhD from Dalian University of Technology in 2006. He is currently a professor at Yunnan University. His recent research focuses on carbon materials, low-dimensional materials for electromagnetic shielding, absorption and electrochemical energy storage applications.

Electromagnetic Wave Absorption and Shielding Materials

Wei Lu and Hongtao Guan

CRC Press
Taylor & Francis Group
Boca Raton London New York

CRC Press is an imprint of the
Taylor & Francis Group, an **informa** business

Designed cover image: © Haojie Jiang

First edition published 2025
by CRC Press
2385 NW Executive Center Drive, Suite 320, Boca Raton, FL 33431

and by CRC Press
4 Park Square, Milton Park, Abingdon, Oxon, OX14 4RN

CRC Press is an imprint of Taylor & Francis Group, LLC

© 2025 Wei Lu and Hongtao Guan

ISBN: 978-1-032-78980-4 (hbk)
ISBN: 978-1-032-78981-1 (pbk)
ISBN: 978-1-003-49009-8 (ebk)

DOI: 10.1201/9781003490098

Typeset in Minion
by Newgen Publishing UK

Contents

CHAPTER 1

Introduction

SINCE THE ADVENT OF wireless communication technologies, a multitude of newly interconnected wireless devices, particularly those associated with the Internet of Things (IoTs), artificial intelligence (AI), wearable electronics and microwave sensors, have emerged. These innovations have transformed society and paved the way for an era of intelligence. While the rapid advancement of wireless technology has enhanced our quality of life, the widespread use of wireless electronics has also led to significant electromagnetic (EM) pollution, now recognized as the fourth major source of pollution after air, water and noise[1]. The detrimental effects of EM pollution primarily involve two aspects: (i) adverse impacts on human health; and (ii) interference with nearby electronic operations, resulting in signal loss or interruption, also known as electromagnetic interference (EMI)[2]. Compared to physical harm, EMI is becoming increasingly prominent, as it not only hampers the performance of advanced electronics and can lead to substantial economic losses, but also hinders the development of emerging fields. This is especially pertinent in the ongoing implementation of 5G mobile networks or the potential adoption of 6G in the future. Conflicts among these EM signals are intensifying due to the concentration of modern electronic devices' working frequencies within a narrow band (2–6.0 GHz)[3]. For instance, current wireless charging technology often experiences issues such as slow charging rates and poor energy efficiency, mainly due to signal interference from other appliances. Similarly, other promising applications such as self-driving cars, aircraft and telesurgery

DOI: 10.1201/9781003490098-1

remain significant challenges until the problems caused by EMI are addressed.

There are two potential approaches to mitigating EM pollution. The first involves exploring new wireless communication technologies that facilitate the directional transmission of EM waves. With such technology, the exchange of EM signals would be confined to the interaction between the transmitter and receiver, preventing devices from emitting EM waves into their surroundings. However, developing wireless communication technology capable of directional EM wave transmission poses a significant challenge, particularly for long-distance applications. The second approach is to investigate advanced EM materials that can interact with external EM waves. EM materials are commonly categorized as either EM absorbing or EM shielding materials, differing in their abilities to absorb (or dissipate) or reflect EM waves[4]. An ideal EM absorber should exhibit strong absorption (measured by reflection loss value (RL)) and effective absorption across a wide frequency range (f_E, representing the region with RL $\leq$ 10 dB). Additionally, low density and thin thickness are important considerations for selecting EM absorbers to minimize weight, especially for military aviation equipments[5]. On the other hand, EM shielding materials function by reflecting most EM waves to block EM radiation, based on the ultra-low skin-depth theory[6]. High-performance EM shielding materials are required to provide significant shielding effectiveness while maintaining an ultrathin profile. In contrast to EM shielding materials, EM absorbers absorb incident EM waves and convert the EM energy into heat, rather than reflecting it back directly[7].

Up until now, the development of EM absorption materials has progressed through two stages. In the first stage, the primary focus was on achieving excellent EM performance (referred to as first-generation EM absorption materials). However, these materials have faced significant limitations in practical applications due to sole emphasis on performance, without considering specific use-case scenarios. For instance, in the context of wireless charging, an ideal scenario would involve selectively receiving a specific EM signal for charging while being immune to interference from background EM waves[8]. To achieve this, the EM materials used should possess frequency-selective capabilities. This means that they should not only absorb unwanted EM waves but also facilitate efficient signal transmission between devices[9]. Consequently, it is necessary for these materials to exhibit very weak absorption values within specific frequency bands while demonstrating strong absorption in other bands. Another example pertains to the use of EM absorption materials in miniaturized electronic devices. In such cases, these materials must exhibit not only broad EM absorption capabilities but also excellent thermal resistance to prevent performance degradation as temperatures rise.

EMI shielding involves protecting a specific component from EM waves by using enclosures composed of electrically conductive and/or magnetic materials. With the increasing demand and utilization of electronic devices that rely on EM signals, it has become crucial to prevent potential interference from other devices. As a result, EMI protective elements are being increasingly employed to isolate electrical and electronic devices, various types of cables, and ensure radio frequency shielding protection in medical and laboratory equipments, among other applications. To achieve effective EMI protection, materials with electrical conductivity are required. Common EMI shielding materials typically consist of metal sheets, screens, or foams made from steel, copper, nickel, or aluminum alloys[10]. These materials possess high electrical conductivity and dielectric constants. However, protective systems based on metals have significant limitations, which restrict their applications. These drawbacks include high density, poor resistance to corrosion, expensive processing, and an EMI shielding mechanism primarily reliant on reflection. Consequently, they are not suitable for applications where EMI absorption is dominant, such as in stealth technology. Additionally, metal-based shielding systems can impact the functionality of other electronic circuits or components and may even cause damage.

While conductive polymers could potentially address some of these limitations, they often exhibit drawbacks such as low stability during processing, high cost, and/or limited thermal stability, and consequently, a restricted service temperature or overall poor mechanical performance. This has led to the consideration of polymer composites containing conductive carbon-based nanoparticles as possible alternatives. These composites merge the benefits of polymers with those offered by the addition of carbon-based nanoparticles, primarily electrical conductivity while maintaining a favorable balance of mechanical performance and thermal stability. Moreover, it has been demonstrated that under specific microstructural conditions, composites based on polymers with carbon nanoparticles may transition from the typical reflection mechanism observed in metals to a pure absorption or multiple reflection mechanism when shielding against EM radiation. This expanded applicability positions them as materials for advanced EMI shielding applications, including stealth technology. Notably, this is exemplified by foams derived from these polymer nanocomposites, which will be further addressed in the later part of this review.

It is important to highlight that the growing interest in carbon-based polymer nanocomposites for EMI shielding applications in recent years has been made possible primarily by the introduction of carbon-based nanoparticles, specifically carbon nanotubes (single and multiwall carbon nanotubes, SWNTs and MWNTs, respectively) and more recently, graphene-based nanoparticles (monolayer/bilayer graphene, graphene nanoplatelets, graphene oxide, reduced graphene oxide, etc.)[11]. Additionally, significant advancements have been achieved in the synthesis processes of these carbon-based nanoparticles, particularly in terms of production—a critical requirement for industrialization—and in controlling the characteristics of the synthesized nanoparticles, including crystalline characteristics, geometry, aspect ratio, surface modification and functionalization. Furthermore, progress has been made in the processes for integrating carbon-based nanoparticles into polymers.

The discovery of metal carbides, nitrides or carbon nitrides (MXenes), a class of 2D materials, can be attributed to the pioneering work of researchers from Drexel University in 2011 when they successfully synthesized $Ti_3C_2T_x$[12]. MXenes are derived from a group of materials called MAX phases, which consist of a combination of metal and a chemically-bonded carbon or nitrogen layer. The general formula for MAX phases is $M_{n+1}AX_n$, where M represents a transition metal, A denotes an A-group element (primarily groups 13 and 14 elements from the periodic table), X represents carbon and/or nitrogen, and n ranges from 1 to 4, indicating the number of M layers that separate the A layers[13,14]. One of the key advantages of MAX phases is their ability to amalgamate the desirable properties of both metals and ceramics. Metals possess excellent electrical and thermal conductivity and magnetic properties, while ceramics exhibit high elastic moduli, temperature resistance, and resistance to oxidation and corrosion[15]. By combining these characteristics, MXenes offer a unique blend of properties that make them highly attractive for various applications. MXenes possess unique properties that make them highly promising as EMI shielding materials. One of the key requirements for effective EMI shielding is electrical conductivity, which MXenes satisfy.

Additionally, they are corrosion-resistant, enabling their use in harsh environments without compromising their effectiveness as a shielding material. Moreover, MXenes are lightweight and flexible, making them particularly well-suited for use in applications where these characteristics are crucial, such as wearable devices or other applications where the EMI shield needs to be molded or shaped to fit a specific geometry. Their weight is also a critical factor, making MXenes a highly desirable choice for EMI shielding applications where reducing the overall weight of the device is a priority.

In this book, the electromagnetic absorption (EMA) and EMI shielding mechanism are firstly discussed; secondly, recent progress of EMA and EMI shielding material is systematically expatiated. Finally, we expect the prospective development and propose future research directions regarding EMA and EMI shielding materials. We really hope that this book will provide more ideas and inspiration for the exploitation of sustainable EMI shielding and EMA materials.

REFERENCES

1. Ma, Z., Kang, S., Ma, J., et al. (2020). Ultra Flexible and Mechanically Strong Double-Layered Aramid Nanofiber–Ti3C2Tx MXene/Silver Nanowire Nanocomposite Papers for High-Performance Electromagnetic Interference Shielding. *ACS Nano 14*: 8368–82.

2. Ma, Z., Xiang, X., Shao, L., et al. (2022). Multifunctional Wearable Silver Nanowire Decorated Leather Nanocomposites for Joule Heating, Electromagnetic Interference Shielding and Piezoresistive Sensing. *Angewandte Chemie International Edition 61*: e202200705.

3. Wang, L., Li, X., Shi, X., et al. (2021). Recent Progress of Microwave Absorption Microspheres by Magnetic–Dielectric Synergy. *Nanoscale 13*: 2136–56.

4. Lou, Z., Wang, Q., Kara, U.I., et al. (2021). Biomass-Derived Carbon Heterostructures Enable Environmentally Adaptive Wideband Electromagnetic Wave Absorbers. *Nano-Micro Letters 14*: 11.

5. Li, X., Li, M., Lu, X., et al. (2021). A sheath-core shaped ZrO2-SiC/SiO2 Fiber Felt with Continuously Distributed SiC for Broad-Band Electromagnetic Absorption. *Chemical Engineering Journal 419*: 129414.

6. Huang, M., Wang, L., Liu, Q., et al. (2022). Interface Compatibility Engineering of Multi-Shell Fe@C@TiO2@MoS2 Heterojunction Expanded Microwave Absorption Bandwidth. *Chemical Engineering Journal 429*: 132191.

7. Song, P., Ma, Z., Qiu, H., et al. (2022). High-Efficiency Electromagnetic Interference Shielding of rGO@FeNi/Epoxy Composites with Regular Honeycomb Structures. *Nano-Micro Letters 14*: 51.

8. Lv, H., Zhang, H., Zhao, J., et al. (2016). Achieving Excellent Bandwidth Absorption by a Mirror Growth Process of Magnetic Porous Polyhedron Structures. *Nano Research 9*: 1813–22.

9. He, P., Hou, Z.L., Cao, W.Q., et al. (2021). Rutile TiO_2 Nanorod with Anomalous Resonance for Charge Storage and Frequency Selective Absorption. *Ceramics International 47*: 2016–21.

10. Cheng, Z., Du, Z., Chen, H., et al. (2023). Metal-bonded Atomic Layers of Transition Metal Carbides (MXenes). *Adv Mater* e2302141.

11. Hu, P., Lyu, J., Fu, C., et al. (2019). Multifunctional Aramid Nanofiber/Carbon Nanotube Hybrid Aerogel Films. *ACS Nano 14*: 688–97.

12. Naguib, M., Kurtoglu, M., Presser, V., et al. (2011). Two-Dimensional Nanocrystals Produced by Exfoliation of Ti_3AlC_2. *Advanced Materials 23*: 4248–53.

13. Deysher, G., Shuck, C.E., Hantanasirisakul, K., et al. (2020). Synthesis of Mo_4VAlC_4 MAX Phase and Two-Dimensional Mo_4VC_4 MXene with Five Atomic Layers of Transition Metals. *ACS Nano 14*: 204–17.

14. Wyatt, B.C., Rosenkranz, A., Anasori, B. (2021). 2D MXenes: Tunable Mechanical and Tribological Properties. *Advanced Materials 33*: 2007973.

15. Gonzalez-Julian, J. (2021). Processing of MAX phases: From Synthesis to Applications. *Journal of the American Ceramic Society 104*: 659–90.

CHAPTER 2

Basic theories

INTRODUCTION

Electromagnetic compatibility (EMC) has become an increasingly important consideration in modern technology, as electronic devices and wireless communication systems continue to proliferate. Electromagnetic interference (EMI) can disrupt the operation of these systems—leading to malfunctions, data loss or even safety hazards. Therefore, effective electromagnetic absorption and shielding solutions are crucial for mitigating EMI and ensuring reliable system operation. To develop effective materials and structures for EMC applications, it is essential to understand the underlying electromagnetic absorption and shielding mechanisms. Electromagnetic absorption involves the dissipation of incident electromagnetic energy into heat, while electromagnetic shielding refers to the attenuation or blocking of electromagnetic waves (EMWs). Both absorption and shielding effectiveness are influenced by factors such as material composition, frequency range and the presence of conductive or magnetic components.

To evaluate the performance of materials and structures in terms of absorption and shielding, a range of testing methods are available. These methods include experimental measurements and numerical simulations, which allow for a comprehensive analysis of electromagnetic behavior. Each method has its own advantages and limitations, and the choice of method depends on factors such as the frequency range of interest, type of

DOI: 10.1201/9781003490098-2

material being evaluated and specific application requirements. This article aims to provide an overview of electromagnetic absorption and shielding mechanisms, as well as commonly used testing methods. By understanding these fundamental principles and testing techniques, researchers and engineers can develop optimized materials and structures for EMC applications, ultimately contributing to the advancement of technology in an increasingly interconnected world.

2.1 PRINCIPLE OF MICROWAVE ABSORPTION

EMWs absorbing materials or microwave absorbing materials (MAMs) are a category of materials to attenuate the microwave energy by absorbing and transferring it to heat or other sorts of energies. Microwave absorption can significantly reduce or eliminate the incident electromagnetic energy through energy conversion, and thus is the most effective technology for EMI protection[1,2]. The applications of MAMs in military can prevent the military equipment from being detected, and in architectural engineering and design can endow the buildings with EMI protection ability, thus protecting the devices and equipment inside the building from electromagnetic radiation and interference. To achieve a better absorption, there are two basic requirements that must be fulfilled. One is that the incident EMW should transmit into the interior of the material as much as possible, which is the so-called impedance matching; the other one is that the wave transmitting into the MAMs needs to be attenuated effectively, known as strong attenuation[3]. Both these two requirements have a close relationship with the electromagnetic parameters of the MAMs, viz, complex dielectric permittivity ($\varepsilon_r = \varepsilon' - j\varepsilon''$) and magnetic permeability ($\mu_r = \mu' - j\mu''$). To achieve a better impedance matching, the values of ε_r and μ_r need to be close to each other. To make a high absorption, the dielectric and magnetic loss tangents ($\tan\delta_e = \varepsilon''/\varepsilon'$, $\tan\delta_m = \mu''/\mu'$) are required to be high to attenuate the microwave effectively. Generally, the ε'' and μ'' stand for the loss and attenuation properties of the MAMs. Dielectric loss is usually caused by conduction, dipolar, polarization or associated relaxation, whereas magnetic loss is generally ascribed to resonance, hysteresis or eddy current loss[4].

Based on transmission line (TML) theory, the microwave absorption of a single layer homogeneous (SLH) absorber under a normal incidence is expressed in terms of reflection loss (RL) and can be obtained as[5]

$$RL = 20 \lg|\Gamma| = 20 \lg \left| \frac{Z_{\mathrm{in}} - Z_0}{Z_{\mathrm{in}} + Z_0} \right| \tag{2.1}$$

where Γ is the reflection coefficient, Z_0 is the characteristic impedance of free space with a value of $120\pi\ \Omega$, and Z_{in} is the input impedance of the absorber. Z_{in} can be obtained from the electromagnetic parameters through the following Eq. (2.2)[5,6].

$$Z_{\mathrm{in}} = Z_0 \sqrt{\frac{\mu_r}{\varepsilon_r}} \tanh\left(j \frac{2\pi f}{c} d \sqrt{\mu_r \varepsilon_r} \right) \tag{2.2}$$

where c is the wave velocity in free space and d is the MAM thickness.

Based on Eq. (2.1), when Z_{in} is designed to equal Z_0, there will be no reflection at the surface of the MAM, so all the incident waves will transmit into the material. This phenomenon is called ideal impedance matching. However, the transmitted wave will penetrate out from the back surface of the MAM and no microwave energy is attenuated under the condition $Z_{\mathrm{in}} = Z_0$. In this sense, impedance matching and high attenuation are contradictory for an SLH MAM. To achieve high microwave absorption, special structures must be constructed for MAMs.

2.1.1 Dielectric loss

According to electromagnetic absorption mechanisms, MAMs can be classified into mainly two types. One is dielectric loss material and the other is magnetic loss material. Dielectric loss material attenuates the incident microwave through polarization loss and conduction loss due to its high conductivity of the material[7]. When the material is under an applied electromagnetic field, the bound charges that exist in the material move locally and thus cause dipole moments, which lead to polarization and related relaxation. Generally, dielectric polarization mainly comes from atomic polarization, ion polarization, electron polarization, dipole polarization, and space charge polarization[8]. As a matter of fact, atomic polarization, ion polarization, and electron polarization play a negligible role in the microwave frequency range. Therefore, dipole polarization, interfacial polarization, conduction loss, and defect-induced polarization are the main contributing sources to EMW loss at gigahertz[9]. The polarization process has a close relationship with the dielectric constant and can tune the microwave absorption property of the material effectively[10].

Interfacial polarization is generally caused by the differences in surface potential, as well as electrical conductivities of the different components[11]. In this sense, porous or hollow materials are beneficial to enhance interface polarization. Dipole polarization mainly comes from the movement of dipoles under an alternating electromagnetic field, which is also known as orientation polarization[12]. Dipole/orientation polarization will result in frequency dispersion, which is crucial to widen the absorption bandwidth. Generally, dipole polarization can be achieved by anchoring sufficient polar molecules and functional groups on MAMs. However, it must be noted that not all dipole bonds contribute to polarization due to their poor orientation. For example, there exist abundant dipole bonds, such as –C–O, –C–N and –C=O, in amorphous materials, especially polymers or organics, however, they are weak in orientation force and thus contribute little to electromagnetic response[13]. Defect-induced polarization stems from the presence of defects in materials. Typically, the common defects in MAMs are brought from oxygen vacancies and heteroatom dopings. The oxygen vacancy can be regulated by various methods including thermal treatment, reduction processing and cation doping[14,15]. It is believed that the defects existing in carbon materials such as carbon nanotubes and reduced graphene oxides could induce significant dipole polarization. As to heteroatom doping, it is usually applied to carbon materials, such as graphene and biomass carbon. The foreign atoms with different electronegativity will change the distribution of the electron cloud of carbon and lead to the formation of electric dipoles in an electromagnetic field, thus contributing to EMW dissipation[9]. Conduction loss is related to the electrical conductivity of the material. Introducing high-conductive components into the MAMs can improve the conduction loss. For example, for polymer-based absorbing materials, conduction loss plays a crucial role in electromagnetic attenuation, since most of the fillers in polymer matrix are conductive components.

Dielectric loss in a MAM is a complicated process, and it generally can be expressed by Cole-Cole semicircle (ε'–ε'' plot) based on the Debye theory. Specifically, the frequency-dependent ε' and ε'' can be written as,

$$\varepsilon' = \varepsilon_{\infty} + \frac{\varepsilon_s - \varepsilon_{\infty}}{1 + \omega^2 \tau^2} \tag{2.3}$$

$$\varepsilon'' = \frac{\omega\tau(\varepsilon_s - \varepsilon_{\infty})}{1 + \omega^2 \tau^2} \tag{2.4}$$

where $\omega = 2\pi f$ stands for the angular frequency and τ refers to the polarization relaxation time. Thus, the Cole-Cole model could be built as:

$$\left(\varepsilon' - \frac{\varepsilon_s + \varepsilon_\infty}{2}\right)^2 + \left(\varepsilon''\right)^2 = \left(\frac{\varepsilon_s - \varepsilon_\infty}{2}\right)^2 \tag{2.5}$$

Each semicircle corresponds to one Debye relaxation. When there is a distorted Cole-Cole semicircle, it usually means that a complicated polarization mechanism is present. Furthermore, according to Debye's theory, considering the conduction loss, the frequency-dependent ε'' is composed of two parts[16].

$$\varepsilon''(\omega) = \varepsilon''_p + \varepsilon''_c = \left(\varepsilon_s - \varepsilon_\infty\right)\frac{\omega\tau}{1 + \omega^2\tau^2} + \frac{\sigma}{\varepsilon_0\omega} \tag{2.6}$$

where σ refers to the electrical conductivity. ε''_c and ε''_p represent conduction loss and polarization relaxation loss, respectively.

For a composite, when the filling ratio is much lower than the percolation threshold, the conductive network is not well constructed. In this case, the electron migration will dominate the conduction process and both the electrical conductivity and conduction loss will be rather low[17,18]. With the further increase of conductive fillings in the matrix, the conduction domination changes from electron migration to electron hoping after the conductive network is formed. The conduction loss will be enhanced accordingly in this situation.

2.1.2 Magnetic loss

Magnetic loss materials usually attenuate the microwave through hysteresis loss, natural resonance loss and domain wall displacement loss[19,20]. However, neither hysteresis loss nor domain wall loss is the main contributor to magnetic loss in the microwave region, since they mainly occur in a very low frequency range[21]. It is worth noting that in magnetic loss materials, eddy current loss has a great effect on the electromagnetic behavior in the microwave range[13,22,23]. Under an alternating field, electrons in the materials can respond to an external magnetic field and move across the material, so as to form a microcurrent. In this sense, eddy current loss is a common phenomenon and it is believed to be favorable to microwave attenuation in magnetic materials with low conductivity[24,25].

For a magnetic loss material, it should be noted that due to Snoek's limit, its magnetic permeability usually decreases dramatically at high frequencies, thus influencing its electromagnetic performance in GHz range. The cut-off frequency due to the Snoek's limit is demonstrated by the following relation[26],

$$\left(\mu_i - 1\right)f_0 = \frac{2\gamma M_s}{3\pi} \tag{2.7}$$

where μ_i is the initial permeability, γ is the material gyromagnetic ratio and f_0 is the cut-off frequency.

There are two possible approaches to improve the magnetic permeability at high frequencies. One is the modification of saturation magnetization, magnetocrystalline anisotropy constant, and internal strain through manipulating the structure and composition of the magnetic loss MAMs[27]. The other method is to alter the size and morphological structure of the particles[28]. Particles with nanosize, fiber shape, and high aspect ratio are effective in increasing the magnetic anisotropy in the material and thus can improve the magnetic permeability at high frequencies.

2.1.3 Impedance matching

Besides large electromagnetic attenuation properties, impedance matching is another important parameter that must be considered during a MAM design. Impedance matching means the difference between the input impedance of the MAM and that of the free space. It is generally characterized using impedance matching coefficient (z) and is expressed as $z = |Z_{in}/Z_0|$. The closer of z to unity, the better the impedance matching of the MAM, and the more incident microwave will enter the interior of the MAM. When $Z_{in} = Z_0$, i.e., $z = 1$, all the incident microwaves will propagate into the material without any surface reflection. This situation is the so-called perfect match. However, when $Z_{in} = Z_0$, the material will be electromagnetic transparent without any attenuation of the incident wave. So, better absorption and impedance matching is controversial for a homogeneous material, and impedance matching must be considered comprehensively when it is used for a MAM design[6]. Through constructing hierarchical or porous structures, an MAM can achieve great absorption and better impedance matching at the same time.

2.2 PRINCIPLE OF EMI SHIELDING

In the realm of EMI suppression and protection against electromagnetic radiation, electromagnetic shielding plays a significant role. This involves the utilization of conductive and/or magnetically permeable materials as shields to block the propagation of EMWs and weaken the electromagnetic field[29]. To grasp the intrinsic nature of EMI shielding, various shielding systems have been proposed, including the eddy role, electromagnetic field, and TML theories. Among these theories, the TML theory has gained widespread use due to its simplicity and convenient yet accurate calculations. It serves as a valuable tool for elucidating electromagnetic shielding mechanisms[30,31]. These mechanisms encompass three distinct processes: reflection, absorption, and multiple reflection[32]. When incident EMWs encounter the surface of a shielding material, only a portion is shown due to the mismatch between the shielding material and the surrounding air. The remaining unseen EMWs penetrate the shielding material, where they undergo numerous reflections and absorption. A small fraction of the EMWs manage to pass through the shielding material as transmitted waves[33].

The extent to which incident EMWs are attenuated is a crucial factor in assessing the shielding effectiveness of a material. To quantitatively express the shielding capacity of a material, EMI shielding efficacy (EMI SE, unit: dB) is employed[34]. It is calculated based on the ratio of the electric field strength, magnetic field strength or effect of EMWs before and after passing through the shielding material, as shown in Eq. (2.8)[35]:

$$SE = 20\log_{10}\left(\frac{E_t}{E_i}\right) = 20\log_{10}\left(\frac{H_t}{H_i}\right) = 20\log\left(\frac{P_t}{P_i}\right) \qquad (2.8)$$

In the equation, the subscripts i and t respectively denote incident and transmitted EMWs.

According to the Schelkunoff theory[36], the total EMI shielding efficacy (EMI SE) of a shielding material is determined by taking into account the overall reflections (SE_R), absorptions (SE_A), and multiple reflections (SE_M) of the EMWs[37].

$$SE_T = SE_R + SE_A + SE_M \qquad (2.9)$$

2.2.1 Refection loss (SE_R)

The reflections (SE_R) occur due to an impedance mismatching between the propagation medium of the incident EMWs and the surface of the shielding material. The magnitude of SE_R from the front to the back surface of the shield can be represented as follows:

$$SE_R = 20\log_{10}\left(\frac{Z+Z_0}{4ZZ_0}\right) = 39.5 + 10\log_{10}\frac{\sigma}{2f\Pi\mu} \tag{2.10}$$

Here, Z and Z_0 represent the impedances of the shielding material and air, while σ and μ denote the electrical conductivity and magnetic permeability. The symbol f stands for the frequency of the incident EMWs. As per the formula, SE_R is linked to the electrical conductivity, the magnetic permeability of the material, and the frequency of the EMWs.

2.2.2 Absorption loss (SE_A)

The SE_A primarily originates from the dielectric and magnetic losses of a material[38]. To achieve significant absorption loss, the following conditions are necessary: high electrical conductivity to enhance interactions between electrons and incident EMWs[39], and high dielectric and magnetic permeabilities to increase eddy current and magnetic hysteresis losses[40]. The absorbed electromagnetic energy is dissipated as heat. The absorption loss (SE_A) of a shielding material can be expressed as follows[41,42]:

$$SE_A = 20\left(\frac{d}{\delta}\right)\log_{10} e = 8.68\left(\frac{d}{\delta}\right) = 8.68d\frac{\sqrt{f\mu\sigma}}{2} \tag{2.11}$$

Here, d refers to the thickness of the shielding material, and δ represents the skin or penetration depth. In the case of conductive shields, the skin depth can be expressed as $\delta = \frac{1}{\alpha}(\pi f \sigma \mu)^{-1}$. Here, α means the EMWs attenuation constant of the shielding material and is expressed as: $\alpha = \omega\sqrt{\frac{\mu\varepsilon}{2}\left[\sqrt{1+(\frac{\sigma}{\omega\varepsilon})^2}-1\right]}$. Here, ω denotes the angular frequency ($2\pi f$), and ε represents the dielectric permittivity. Therefore, SE_A is dependent on the controllable thickness and electrical conductivity of the material, as well as its inherent electromagnetic properties.

2.2.3 Multiple reflections (SE_M)

The multiple reflections of EMWs between two boundaries of a thin shield lead to their attenuation. This process can be iterated until the electromagnetic energy is fully dissipated. The effectiveness of multiple reflections is calculated as follows[43,44]:

$$SE_M = 20\log_{10}\left(1 - e^{\frac{-2d}{\delta}}\right) = 20\log_{10}\left(1 - 10^{\frac{SE_A}{10}}\right) \tag{2.12}$$

The SE_M is primarily determined by the thickness of the material. This factor can be disregarded when the shield is thicker than the skin depth, or when the EMI shielding effectiveness exceeds 15 dB[45,46]. However, if the thickness is significantly thinner than the skin depth, various reflections must be taken into consideration when evaluating the shielding efficacy.

2.2.4 Internal scattering (internal multiple reflections)

Apart from the macroscopic reflections between the two boundaries of an EMI shield, there can also be reflections and scattering inside the micro-structure of the EMI shielding material. Such internal scattering, also known as internal multiple reflections, can further elongate the propagation pathway of EMWs, leading to increased absorption loss and overall shielding effectiveness. Thus, the microstructures that constitute a shielding material have a significant impact on EMI shielding.

2.2.5 Calculation of EMI shielding performance

For practical calculation of EMI SE, the calculation theory on the basis of either absorption or reflection loss is favored, depending on which property can be directly calculated from the scattering coefficients (S-parameters). Here, absorption loss and reflection loss are represented by $SE_{A'}$ and $SE_{R'}$, respectively, to distinguish them from concepts with the same name in the Schelkunof theory:

$$SE_{R'} = 10\log_{10}\left|\frac{1}{1 - S_{11}^2}\right| \tag{2.13}$$

$$SE_{A'} = 10\log_{10}\left|\frac{1 - S_{11}^2}{S_{21}^2}\right| \tag{2.14}$$

It is worth noting that macroscopic multiple reflection losses are encompassed by $SE_{A'}$ and $SE_{R'}$; thus, SE is defined as follows[47]:

$$SE = SE_{A'} + SE_{R'} \tag{2.15}$$

In addition, considering the density and thickness of the material, the practical shielding effectiveness of a material is defined by introducing the absolute shielding efficiency (SSE/t), which can be calculated as follows:

$$\frac{SSE}{t} = \frac{EMI\ SE}{\rho t} = \text{dB cm}^2\text{g}^{-1} \tag{2.16}$$

where ρ and t denote the density and thickness of the shielding material, respectively. A great SSE/t proportion is required for lightweight shielding usage.

2.3 METHODS FOR MEASURING PERFORMANCE

2.3.1 Measurement of microwave absorption performance

Typically, the electromagnetic (EM) absorption performance can be evaluated using either direct methods (such as the free-space method, also known as the Naval Research Laboratory (NRL)-Arc method) or indirect methods (e.g., coaxial-line and waveguide methods). Digital images of the equipment are shown in Fig. 2.1[48]. For powder-like materials, they need to be shaped into a specific form by uniformly mixing them with a molding matrix (e.g., resin and wax) in a specified weight ratio[49,50]. These molding matrices are non-magnetic and should exhibit extremely low dielectric loss, thereby demonstrating minimal EM absorption properties. Consequently, the EM absorption property measured for the matrix/sample originates solely from the sample itself.

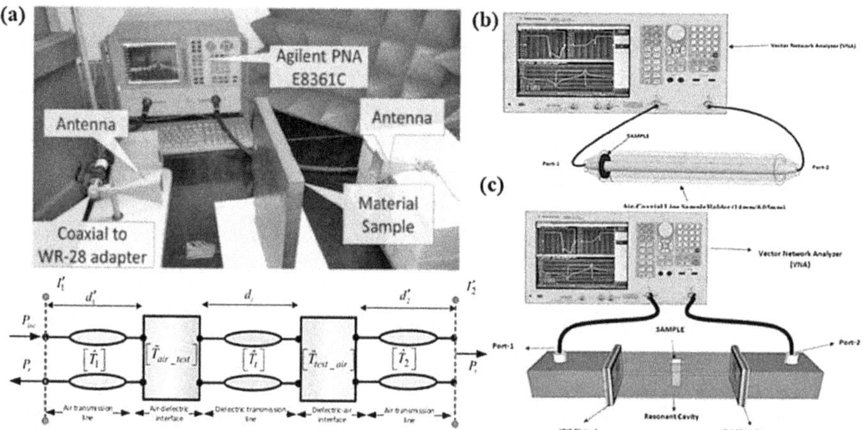

FIG. 2.1 Measurement of EM absorption performance: (a) free–space method (reproduced with permission from Ref.[48] Copyright (2010) Institutode Aeronautica. e–Espaco–IAE); (b) coaxial–TML method; (c) waveguide–method (reproduced with permission from Ref.[49] Copyright (2016) Institute of Electrical and Electronics Engineers Inc.).

2.3.2 Measurement of EMI shielding performance

EMI SE can be experimentally quantified through the use of widely employed network analyzers (NAs), which can be categorized into scalar NAs (SNAs) and vector NAs (VNAs). In practice, VNAs are typically preferred and more commonly utilized as SNAs are unable to measure complex signals, such as EM parameters (complex permittivity and complex permeability). Scattering parameters (S11 and S12, together with their reciprocals S22 and S21) are mathematically used to express both the incident and transmitted waves in a two-port VNA. These S parameters correspond to the reflection and transmission coefficients.

In general, VNAs allow for dynamic transmission, reception, and recording of EMI intensities across a broader frequency range. The most commonly used measurement techniques include transmission-reflection line configurations such as coaxial cables and waveguide techniques, as well as free space configurations like microwave anechoic chambers, NRL, and arch reflectivity system measurement instruments[51,52]. Fig. 2.2 provides a schematic representation of the major measurement techniques used for EMI SE evaluation of shielding materials.

For the coaxial line method, the sample being measured is fabricated in the form of a rectangular toroid shape, with an outer conductor diameter of 7 mm and an inner conductor diameter of 3.04 mm[53]. The key advantage of this method is that the same sample can be used to measure a wide range of frequencies. However, special attention must be given to carefully prepare thin samples without any air gaps that could introduce

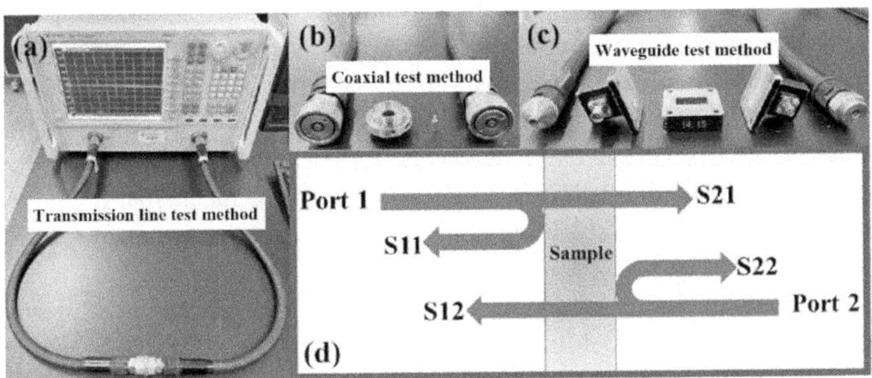

FIG. 2.2 (a) Vector network testing system; (b) coaxial test fixture; (c) waveguide test fixture; (d) S-parameter test principle.

measurement errors. In the waveguide method, a rectangular-shaped sample is required, and the size of the waveguide depends on the frequency being measured. It's important to note that the size of the waveguide decreases as the test frequency increases. The main advantage of this method is its simplicity in sample preparation. However, a limitation is that it has a relatively narrow frequency range, requiring multiple waveguides [54]. The free space method involves a contact measurement system consisting of two antennas placed on either side of the sample being measured. One significant advantage of this method is its ability to cover a wide frequency range while allowing manipulation of the incident angle of the EM waves during measurement. However, a drawback is that the size of the sample being measured needs to be larger compared to the other two methods[55].

CONCLUSION

Electromagnetic absorption and shielding mechanisms, along with their corresponding testing methods, are vital aspects in the field of EMC. Understanding these mechanisms and employing appropriate testing methods is crucial for developing effective materials and structures that can absorb and shield against electromagnetic energy. The mechanisms of electromagnetic absorption involve the conversion of incident electromagnetic energy into heat through various processes such as dielectric loss, magnetic loss and conductive loss. These mechanisms are influenced by factors such as material composition, frequency range and the presence of conductive or magnetic components. By comprehending these absorption mechanisms, researchers can design materials with optimized properties for specific EMC applications. Similarly, electromagnetic shielding mechanisms involve the attenuation or blocking of EMWs to prevent unwanted interference. Shielding effectiveness is influenced by factors such as material conductivity, permeability, thickness and the presence of conductive or magnetic layers. Understanding these mechanisms enables the design of effective shielding materials and structures for various EMC scenarios.

To evaluate the performance of materials and structures in terms of absorption and shielding, a range of testing methods are available. Experimental measurements, such as the TML method, cavity method and waveguide method, allow for direct assessment of absorption and shielding effectiveness. These methods provide valuable insights into the behavior of materials under specific electromagnetic conditions. In addition to experimental measurements, numerical simulations, such as finite element analysis (FEA) and boundary element method (BEM), enable virtual testing and prediction of absorption and shielding performance. These simulation techniques allow for a deeper understanding of the underlying physics and can aid in the optimization of material and structural designs.

By combining experimental measurements and numerical simulations, researchers and engineers can gain a comprehensive understanding of electromagnetic absorption and shielding mechanisms, as well as accurately evaluate the performance of materials and structures. This knowledge can guide the development of advanced materials and innovative designs to address the ever-increasing demands for EMC. In conclusion, a thorough understanding of electromagnetic absorption and shielding mechanisms, coupled with the appropriate testing methods, is essential for the development of effective EMC solutions. Continued research and advancements in these areas will contribute to the design of materials and structures that can efficiently absorb and shield against electromagnetic energy, ensuring the reliable operation of electronic systems in various industries.

REFERENCES

1. Raveendran, A., Sebastian, M.T., Raman, S. (2019). Applications of microwave materials: A review. *Journal of Electronic Materials 48*: 2601–34.
2. Xie, S., Ji, Z., Zhu, L., et al. (2020). Recent progress in electromagnetic wave absorption building materials. *Journal of Building Engineering 27*: 100963.
3. Pang, H., Duan, Y., Huang, L., et al. (2021). Research advances in composition, structure and mechanisms of microwave absorbing materials. *Composites Part B: Engineering 224*: 109173.
4. Kim, S.S., Jo, S.B., Gueon, K.I., et al. (1991). Complex permeability and permittivity and microwave absorption of ferrite-rubber composite in X-band frequencies. *IEEE Transactions on Magnetics 27*: 5462–64.
5. Suetake, Y.N.K. (1971). Application of ferrite to electromagnetic wave absorber and its characteristics. *IEEE Transactions on Microwave Theory and Techniques 19*: 65–72.
6. Duan, Y., Guan, H. (2016). *Microwave Absorbing Materials*; Pan Stanford; ISBN 9814745111
7. Quan, B., Liang, X., Ji, G., et al. (2017). Dielectric polarization in electromagnetic wave absorption: Review and perspective. *Journal of Alloys and Compounds 728*: 1065–75.

8. Ding, D., Wang, Y., Li, X., et al. (2017). Rational design of core-shell Co@C microspheres for high-performance microwave absorption. *Carbon* *111*: 722–32.

9. Qin, M., Zhang, L., and Wu, H. (2022). Dielectric loss mechanism in electromagnetic wave absorbing materials. *Advanced Science 9*: 2105553.

10. Liu, Q., Cao, Q., Bi, H., et al. (2015). CoNi@SiO2@TiO2 and CoNi@ Air@TiO2 microspheres with strong wideband microwave absorption. *Advanced Materials 28*: 486–90.

11. Li, Y., Wang, G., Gong, A., et al. (2022). High-performance ferroelectric electromagnetic attenuation materials with multiple polar units based on nanodomain engineering. *Small 18*: 2106302.

12. Kuriakose, M., Longuemart, S., Depriester, M., et al. (2014). Maxwell-Wagner-Sillars effects on the thermal-transport properties of polymer-dispersed liquid crystals. *Physical Review E 89*: 022511.

13. Lv H.L., Y.Z.H., Pan H. G., et al. (2022). Electromagnetic absorption materials: Current progress and new frontiers. *Progress in Materials Science 127*: 100946.

14. Su, Z., Zhang, W., Lu, J., et al. (2022). Oxygen-vacancy-rich Fe_3O_4/ carbon nanosheets enabling high-attenuation and broadband microwave absorption through the integration of interfacial polarization and charge-separation polarization. *Journal of Materials Chemistry A 10*: 8479–90.

15. Quan, B., Shi, W., Ong, S.J.H., et al. (2019). Defect engineering in two common types of dielectric materials for electromagnetic absorption applications. *Advanced Functional Materials 29*: 1901236.

16. Zhang, Y., Meng, H., Shi, Y., et al. (2020). TiN/Ni/C ternary composites with expanded heterogeneous interfaces for efficient microwave absorption. *Composites Part B: Engineering 193*: 108028.

17. Song, W.L., Cao, M.S., Hou, Z.L., et al. (2009). High dielectric loss and its monotonic dependence of conducting-dominated multiwalled carbon nanotubes/silica nanocomposite on temperature ranging from 373 to 873 K in X-band. *Applied Physics Letters 94*: 233110.

18. Wen, B., Cao, M.S., Hou, Z.L., et al. (2013). Temperature dependent microwave attenuation behavior for carbon-nanotube/silica composites. *Carbon 65*: 124–39.

19. Estevez, D., Qin, F.X., Quan, L., et al. (2018). Complementary design of nano-carbon/magnetic microwire hybrid fibers for tunable microwave absorption. *Carbon 132*: 486–94.

20. Wei, H., Zhang, Z., Hussain, G., et al. (2020). Techniques to enhance magnetic permeability in microwave absorbing materials. *Applied Materials Today 19*: 100596.

21. Green, M., Tian, L., Xiang, P., et al. (2018). Co_2P nanoparticles for microwave absorption. *Materials Today Nano 1*: 1–7.

22. Cao, M.S., Wang, X.X., Zhang, M., et al. (2019). Electromagnetic response and energy conversion for functions and devices in low-dimensional materials. *Advanced Functional Materials 29*: 1807398.

23. Cheng, J., Zhang, H., Ning, M., et al. (2022). Emerging materials and designs for low- and multi-band electromagnetic wave absorbers: The search for dielectric and magnetic synergy? *Advanced Functional Materials 32*: 2200123.

24. Li, X., Cui, E., Xiang, Z., et al. (2020). Fe@NPC@CF nanocomposites derived from Fe-MOFs/biomass cotton for lightweight and high-performance electromagnetic wave absorption applications. *Journal of Alloys and Compounds 819*: 152952.

25. Shen, Y., Wei, Y., Ma, J., et al. (2020). Self-cleaning functionalized FeNi/$NiFe_2O_4$/NiO/C nanofibers with enhanced microwave absorption performance. *Ceramics International 46*: 13397–406.

26. Snokek, J.L. (1947). Gyromagnetic resonance in ferrites. *Nature 159*: 90–90.

27. Ajia, S., Asa, H., Toyoda, Y., et al. (2022). Development of an alternative approach for electromagnetic wave absorbers using Fe–Cr–Co alloy powders. *Journal of Alloys and Compounds 903*: 163920.

28. Yang, P., Yu, M., Fu, J., et al. (2017). Synthesis and microwave absorption properties of hierarchical Fe micro-sphere assembly by nano-plates. *Journal of Alloys and Compounds 721*: 449–55.

29. Thomassin, J.M., Jérôme, C., Pardoen, T., et al. (2013). Polymer/carbon based composites as electromagnetic interference (EMI) shielding materials. *Materials Science and Engineering: R: Reports 74*: 211–32.

30. Yi, Y., Feng, S., Zhou, Z., et al. (2022). Wet mechanical grinding regulates the micro-nano interfaces and structure of MXene/PVA composite for enhanced mechanical properties and thermal conductivity. *Composites Part A: Applied Science and Manufacturing 163*: 107232.

31. Nazir, A., Yu, H., Wang, L., et al. (2018). Recent progress in the modification of carbon materials and their application in composites for electromagnetic interference shielding. *Journal of Materials Science 53*: 8699–719.

32. Iqbal, A., Sambyal, P., Koo, C.M. (2020). 2D MXenes for electromagnetic shielding: A Review. *Advanced Functional Materials 30*: 2000883.

33. Wang, C., Murugadoss, V., Kong, J., et al. (2018). Overview of carbon nanostructures and nanocomposites for electromagnetic wave shielding. *Carbon 140*: 696–733.

34. Zhang, D., Liang, S., Chai, J., et al. (2019). Highly effective shielding of electromagnetic waves in MoS_2 nanosheets synthesized by a hydrothermal method. *Journal of Physics and Chemistry of Solids 134*: 77–82.

35. Geetha, S., Satheesh Kumar, K.K., Rao, C.R.K., et al. (2009). EMI shielding: Methods and materials–A review. *Journal of Applied Polymer Science 112*: 2073–86.

36. Sankaran, S., Deshmukh, K., Ahamed, M.B., et al. (2018). Recent advances in electromagnetic interference shielding properties of metal and carbon filler reinforced flexible polymer composites: A review. *Composites Part A: Applied Science and Manufacturing 114*: 49–71.

37. Wang, H., Wang, G., Li, W., et al. (2012). A material with high electromagnetic radiation shielding effectiveness fabricated using multi-walled carbon nanotubes wrapped with poly(ether sulfone) in a poly(ether ether ketone) matrix. *Journal of Materials Chemistry 22*: 21232–37.

38. Zhang, D., Cheng, J., Yang, X., et al. (2014). Electromagnetic and microwave absorbing properties of magnetite nanoparticles decorated carbon nanotubes/polyaniline multiphase heterostructures. *Journal of Materials Science 49*: 7221–30.

39. Shahzad, F., Alhabeb, M., Hatter, C.B., et al. (2016). Electromagnetic interference shielding with 2D transition metal carbides (MXenes). *Science 353*: 1137–40.

40. Kumar, R., Choudhary, H.K., Pawar, S.P., et al. (2017). Carbon encapsulated nanoscale iron/iron-carbide/graphite particles for EMI shielding and microwave absorption. *Physical Chemistry Chemical Physics 19*: 23268–79.

41. Yun, T., Kim, H., Iqbal, A., et al. (2020). Electromagnetic shielding of monolayer MXene assemblies. *Advanced Materials 32*: 1906769.

42. Li, Y., Tian, X., Gao, S.-P., et al. (2020). Reversible crumpling of 2D titanium carbide (MXene) nanocoatings for stretchable electromagnetic shielding and wearable wireless communication. *Advanced Functional Materials 30*: 1907451.

43. Singh, B.P., Prabha, Saini, P., et al. (2011). Designing of multiwalled carbon nanotubes reinforced low density polyethylene nanocomposites for suppression of electromagnetic radiation. *Journal of Nanoparticle Research 13*: 7065–74.

44. Kim, B.R., Lee, H.K., Kim, E., et al. (2010). Intrinsic electromagnetic radiation shielding/absorbing characteristics of polyaniline-coated transparent thin films. *Synthetic Metals 160*: 1838–42.

45. Li, Y., Wang, B., Sui, X., et al. (2017). Facile synthesis of microfibrillated cellulose/organosilicon/polydopamine composite sponges with flame retardant properties. *Cellulose 24*: 3815–23.

46. Wang, B., Li, W., Deng, J. (2017). Chiral 3D porous hybrid foams constructed by graphene and helically substituted polyacetylene: preparation and application in enantioselective crystallization. *Journal of Materials Science 52*: 4575–86.

47. Peng, M., Qin, F. (2021). Clarification of basic concepts for electromagnetic interference shielding effectiveness. *Journal of Applied Physics* *130*: 225108.

48. Akhter, Z., Akhtar, M.J. (2016). Free-space time domain position insensitive technique for simultaneous measurement of complex permittivity and thickness of lossy dielectric samples. *IEEE Transactions on Instrumentation and Measurement 65*: 2394–405.

49. Zhang, Y., Ma, Z., Ruan, K., et al. (2022). Flexible $Ti_3C_2T_x$/(aramid nanofiber/PVA) composite films for superior electromagnetic interference shielding. *Research 2022*: 9780290.

50. Lv, H., Liang, X., Cheng, Y., et al. (2015). Coin-like α-Fe_2O_3@$CoFe_2O_4$ core–shell composites with excellent electromagnetic absorption performance. *ACS Applied Materials & Interfaces 7*: 4744–50.

51. Lv, H., Guo, Y., Yang, Z., et al. (2017). A brief introduction to the fabrication and synthesis of graphene based composites for the realization of electromagnetic absorbing materials. *Journal of Materials Chemistry C 5*: 491–512.

52. Qin, F.X., Peng, H.X., Pankratov, N., et al. (2010). Exceptional electromagnetic interference shielding properties of ferromagnetic microwires enabled polymer composites. *Journal of Applied Physics 108*: 044510.

53. Liu, J., Che, R., Chen, H., et al. (2012). Microwave absorption enhancement of multifunctional composite microspheres with spinel Fe_3O_4 cores and anatase TiO_2 shells. *Small 8*: 1214–21.

54. Bhattacharjee, Y., Arief, I., Bose, S. (2017). Recent trends in multi-layered architectures towards screening electromagnetic radiation: Challenges and perspectives. *Journal of Materials Chemistry C 5*: 7390–403.

55. Hayashida, K., Matsuoka, Y. (2015). Electromagnetic interference shielding properties of polymer-grafted carbon nanotube composites with high electrical resistance. *Carbon 85*: 363–71.

CHAPTER 3

Traditional electromagnetic wave (EMW) absorption materials

ELECTROMAGNETIC WAVE (EMW) ABSORBING materials, also known as microwave absorption materials (MAMs), refer to a class of materials that can absorb or significantly attenuate the EMW energy projected onto a surface as efficiently as possible. It converts electromagnetic (EM) energy into heat energy or other forms of energy (mechanical energy) through an absorbent, so as to achieve the purpose of absorbing EMW energy. Based on the EMW absorption theory, ideal EMW absorbing materials need to have a strong absorption intensity (RL ≤ −10 dB), wide effective absorption bandwidth (EAB), thin thickness, good chemical stability and low filling rate. Therefore, the key point of designing EMW absorption materials with high efficiency and strong absorption lies in combining the impedance matching and loss capacity of the materials with component manipulation and structural design.

Among the traditional absorbing materials, ferrite materials with a natural resonance absorption mechanism have been extensively studied because of their excellent absorption performance and low cost. The

DOI: 10.1201/9781003490098-3

application of ferrite in electromagnetic shielding and stealth technology is relatively mature[1,2]. However, due to the narrow resonance band, large matching thickness and high density, traditional ferrites are difficult to meet the requirements of low density, thin thickness, strong absorption capacity and wide bandwidth in current application scenarios. To better utilize the advantages of ferrites, it is necessary to further improve their composition and structure. Similarly, many other traditional EM absorption materials, such as carbonaceous materials, metals and their oxides, and conductive polymers, are limited in application due to their single loss mechanism[3,4]. Therefore, in order to better develop excellent EMW absorbing materials, it is necessary to introduce other components to construct a variety of composite structures and loss mechanisms to improve the EMW absorption performance.

3.1 PHASE COMPOSITION

With the development of science and technology, the components and phase composition of EMW absorbing materials were systematically investigated in order to achieve perfect impedance matching and attenuation capabilities. Graphene, as the most typical representative two-dimensional (2D) material, has only one atomic thickness (about 0.34 nm thick), and its carbon atoms are bonded in the form of covalent bonds in the plane to form a hexagonal honeycomb planar structure.

There are two structural designs for graphene, in which one is to modify the design itself, and the other is to build a composite material by introducing low dielectric properties or magnetic components[5]. Unmodified graphene has an ultra-high ε_r value, which is the main factor leading to the weak impedance matching and attenuation ability of graphene[6]. Therefore, in order to reduce the ε_r value, some researchers introduced heteroatoms (N, P, F, S, etc.) doping[7-10]. Notably, elementally doped graphene is divided into n-type and p-type. The n-type is mainly due to high ε_r values caused by enhanced conductivity and has little use. However, the p-type, on the other hand, reduces conductivity by enhancing the electron scattering effect. Therefore, the p-type graphene is a little more popular.

Furthermore, heteroatom doping is considered to be a reasonable method for constructing the electronic properties of graphene. When the carbon atoms in the graphene lattice are replaced by N atoms, the ideal sp^2 hybridization of the carbon atoms is destroyed, and the presence of N atoms with greater electronegativity contributes electrons to the conjugated system, resulting in increased dielectric loss. In the element doping chemical process, point defects or disordered structures can be formed, which can not only balance impedance matching, but also enhance polarization loss through unique electronic characteristics. For example, Liu et al. prepared N-doped graphene foams with open reticular structures by a hydrothermal reaction and then a freeze-drying method. When the filler loading is only 5 wt%, the RL_{min} can reach −53.9 dB at 3.5 mm and the EAB is up to 4.56 GHz at 2.0 mm. Meanwhile, three-dimensional (3D) ultralight N-doped graphene wax composites were fabricated via the vacuum infusion method, which showed a strong absorption performance[11]. It exhibited an RL_{min} of −53.25 dB at 11.0 GHz and an EAB up to 8.15 GHz, corresponding to a loading concentration of 3.6 wt%. Finally, the excellent EMW absorption performance is attributed to the fact that N heteroatoms are conducive to inducing the construction of open network walls and adjusting the electronic structure characteristics. As is known, pyrrole N and pyridine N can enhance dipole relaxation loss, and graphite N is beneficial for conduction loss.

Although the defect caused by the introduction of N, S, P and F atoms can reduce the ε_r value, the reason why these defects are concentrated is limited, making the ε_r value of graphene still too high, indicating poor impedance matching. In order to reduce the ε_r value more efficiently, the researchers tried to introduce some dielectric components such as polymers, SiO_2, NiO, and α-Fe_2O_3. Among them, NiO has a high dielectric constant, electronic properties, and chemical stability[12,13]. For instance, Xing et al. successfully prepared a novel flower-like NiO@graphene composite via hydrothermal reaction and an annealing process[14]. Data analysis illustrates that the flower-like NiO@graphene composite has excellent absorption performance and it has an RL_{min} of −59.6 dB at 14.16 GHz, when the filler loading is 25 wt%. The range of the EAB is from 12.48 GHz to 16.72 GHz at 1.7 mm. In addition, polyaniline (PANI) is also a dielectric material, featuring a high dielectric constant, chemical stability, affordability, lightweight construction, and simple synthesis, capable of promoting the absorption performance of MAMs[15]. For instance, sandwich-like graphene@NiO@PANI decorated with Ag particle was constructed in three steps, including hydrothermal reaction, calcination, in-situ polymerization and reduction of $AgNO_3$[16]. The graphene@NiO@PANI@Ag composite exhibited an RL_{min} of −37.5 dB at 13.4 GHz and an EAB of 4.9 GHz at 3.5 mm.

For an excellent EMW absorbing material, optimized impedance matching is indispensable, but the effective absorption area and thickness are also worth improving. Therefore, the component control of graphene-based composites is not limited to the introduction of heteroatom and the dielectric properties, but also the appropriate introduction of double magnetic loss mechanism and dielectric loss capacity of absorbing materials, such as magnetic oxides (ferrite, spinel). Finally, with the help of spinel ferrites, graphene-based/magnetic oxide absorbers are able to achieve a wide absorption region of 6–8 GHz, at a thickness of less than 2 mm[17,18].

3.2 HETERO-STRUCTURAL EMW ABSORPTION MATERIALS

Hetero-structured materials are a new class of materials consisting of heterogeneous regions with significantly different (>100%) mechanical or physical properties. The interactive coupling between these heterogeneous regions produces a synergistic effect in which the synthesis characteristics exceed the predictions of the mixing rules. Compared with homogeneous materials, heterostructure materials have superior functional and mechanical properties. Absorption performance is mainly determined by the contribution from individual components and devices and the interaction between interfaces. Hetero-structured composite materials are widely investigated as EMW absorbing materials because of their unique structure and chemical composition, and they have also been investigated in various fields such as battery, capacitance, adsorption, EM protection and electrocatalysis.

Heterostructure EMW absorbing materials can be classified according to interface components and interface structures. In more detail, heterogeneous structural materials can be divided into metal-metal, metal-polymer and inorganic semiconductor-metal oxide[19]. Zhang et al. enhanced the heterogeneous interface by constructing a 3D cross-network of intertwined one-dimensional (1D) heterogeneous component structures MXene fibers/CoNi/C and carbon nanotubes (CNTs)/CoNi, thereby improving the EMW absorption performance[20]. An RL_{min} of −51.6 dB was obtained at 1.6 mm and its EAB was 4.5 GHz. It is clear that MXene fibers@MOF-derived CNTs with double 1D heterostructures enable increased surface area, rich heterogeneous interfaces, and 3D conductive networks, and demonstrate excellent dielectric and magnetic loss capabilities.

3.2.1 Heterostructures with uniform interface

The structure of a heterogeneous interface can be divided into layer-stacking heterostructure and core-shell shaped heterostructure[19]. The layer-stacking heterostructures can be generated by chemical vapor deposition (CVD) or molecular epitaxial growth. Although this method enables the interface to be accurately controlled at the atomic scale, the low yield and time consumption limit its development. However, unlike layer-stacking heterostructures, core-shell shaped heterostructures have the characteristics of a simple preparation process and low cost. More importantly, core-shell-shaped structures have different electronegativity (η) due to differences in the materials of the core and shell composition. Therefore, dielectric losses rise due to interface relaxation losses. In order to better construct EMW absorption materials, so that the core-shell structure of different components makes maximum use of the interface effect, two conditions must be met. One is suitable impedance matching, for example, the input impedance of the absorbing material (Z_{in}) should be close to the impedance of free space (Z_0); the second is to maximize the use of interface effects.

Through the EMW absorption theory, it can be seen that the complex permittivity is controlled in the medium region to ensure suitable impedance matching. Therefore, in order to make Z_{in}/Z_0 as close as possible to 1 and meet the close-to theoretical impedance matching, it is necessary to select a magnetic component material with high complex permeability. However, maximizing the use of interface effects should make the electronegativity difference ($|\Delta\eta|$) between the core and the shell as large as possible[19]. To ensure better impedance matching, an equivalent dielectric strategy may be selected, like a core-shell shaped heterogeneous structure, because the complex permittivity of the core and shell is around 10–20. However, a material with a medium dielectric value makes it difficult to produce dielectric interface effects. Therefore, low utilization is negligible in many cases. As a result, in order to meet the dielectric difference strategy and excellent interface effect at the same time, the dielectric difference strategy has been presented. Since the two components exhibit significantly different composite permittivity, there is also a huge difference in their electronegativity (η), which is attributed to the loss behavior of interface relaxation[19].

According to the principle of dielectric difference, numerous researchers have designed core-shell heterostructure composites, which demonstrated excellent EMW absorption performance. For example, Fan et al. constructed core-shell structured SiO_2@MoS_2 nanospheres through hydrothermal method[21]. The RL_{min} value of SiO_2@MoS_2 nanospheres was −49.35 dB at 2.8 mm, and the EAB was up to 7.44 GHz. In addition, the ε'' values of SiO_2@MoS_2 are from 6.91 to 2.85, higher than pure MoS_2. Meng et al. designed a core-shell bilayer heterogeneous graphene-based aerogel microspheres (carbon@rGO/Fe_3O_4 AMs) via a coaxial electrospinning-freeze drying and calcination process (Fig. 3.1a)[22]. Its scanning electron microscopy (SEM) image in Fig. 3.1b shows a unique microchannel structure, which diverges gradually from the inside out. The carbon@rGO/Fe_3O_4 AMs-2 composites illustrated an RL_{min} of −61 dB at 2.5 mm with a filling ratio of 5 wt% and the largest EAB can reach up to 6.88 GHz with a

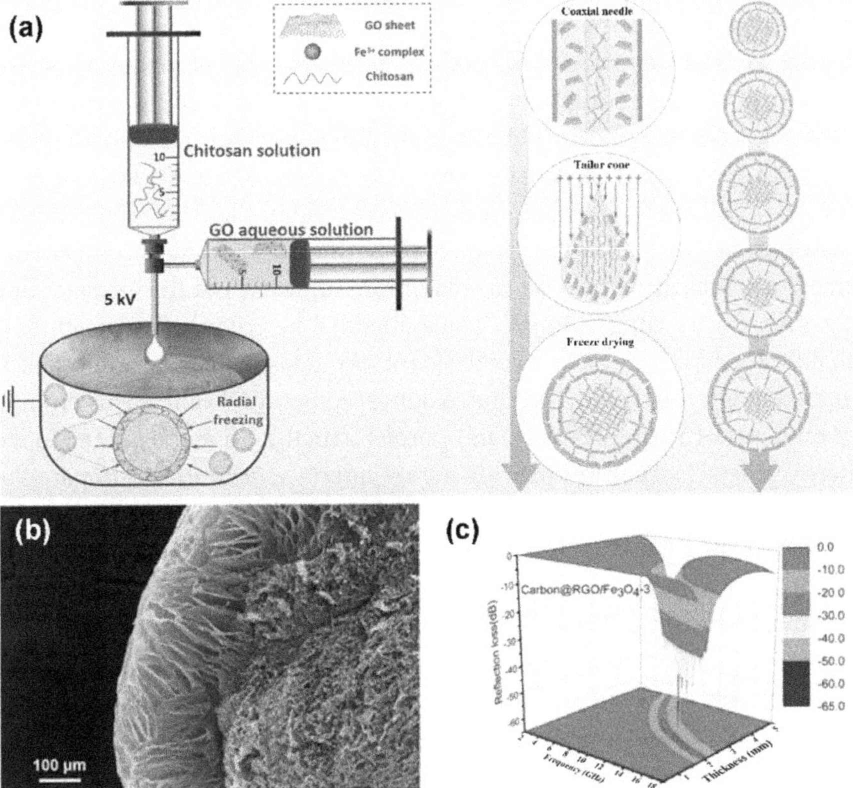

FIG. 3.1   (a) Schematic diagram of the synthesis of carbon@rGO/Fe_3O_4 AMs; (b) SEM image of partial enlarged cross-section of the as-prepared carbon@rGO/Fe_3O_4 AMs; (c) RL_{min} of Carbon@rGO/Fe_3O_4-3. (Reproduced with permission from Ref.[22] Copyright 2022, Elsevier.)

matching thickness of 2.6 mm. Meanwhile, when the shell layer thickness increased to 152 µm, an excellent MA performance was obtained with an RL_{min} of −51 dB at 9.28 GHz and 3.6 mm. Furthermore, when the layer thickness increased to 205 µm, the carbon@rGO/Fe$_3$O$_4$ AMs-3 displayed a higher RL_{min} value at 13.84 GHz, as shown in Fig. 3.1c. Finally, the core-shell bilayer structure facilitates heterogeneous interfacial polarization effects, resonant cavities, coordinating conductivity losses, and impedance matching.

3.2.2 Heterostructures with non-uniform interface

For heterogeneous structures with uniform interface shapes, it demonstrates controllable interfaces, components and EM coupling effects. The heterogeneous structure of the non-uniform interface can retain the original characteristics of the heterogeneous structure of the uniform interface, and can also have the ability of interface control. Commonly, a load-heterogeneous structure is formed by dispersing one component on the surface of another component. The synthesis of the loading structure is to create a host component first, and then the guest component continues to grow. However, the distribution of load heterogeneous structures is irregular. Therefore, they are prone to discontinuous and uneven interface distribution. Thus, the design of high-performance load heterogeneous structures mainly focuses on the maximum utilization of the interface and the selection of components[19]. The embedded heterogeneous structure is mainly based on the host structure, and the guest-assisted component is embedded in the host[23]. The choice of host structure should focus on high specific surface area, porosity, and porous structural materials. Embedded heterogeneous structures provide higher interface utilization compared to load heterogeneous structures. However, when the magnetic nanoparticles (NPs) are embedded in the carbon frame, it can strengthen the fixation degree of the heterogeneous interface, avoid its corrosion, and improve chemical stability.

Mesoporous carbon nanospheres are widely used in embedded heterostructures because of their controllable porosity, size and abundance in the dipole gene. However, it is well-known that mesoporous carbon has a low dielectric constant ($\varepsilon_r < 10$), so in order to better optimize impedance matching, some components with high dielectric constant need to be introduced, which just meets the principle of dielectric difference. For example, Wang et al. synthesized Fe-PBA@PDA@GO precursors by a solvothermal method. Subsequently, the core-shell heterostructured FeS$_2$/

Fe_7S_8@C@rGO composites were synthesized by vulcanizing the Fe-PBA@ PDA@GO[24]. The introduction of heteroatom S provides more defects and vacancies, and the S atoms with lower electronegativity was converted to metal sulfides, resulting in high dielectric constant and electrical conductivity. The core-shell FeS_2/Fe_7S_8@C microspheres were embedded in the crumpled reduced graphene oxide (rGO) layer. When the filler loading is 20 wt%, the FeS_2/Fe_7S_8@C@rGO composite displayed an RL_{min} of −62.7 dB and an EAB of 6.1 GHz at 2.2 mm.

Note that in 2023, Zhou et al. successfully prepared a 3D flower-like $Cu_{7.2}S_4$ with different pyrolysis temperatures embedded on rice husk-derived porous carbon (RPC) via carbonization-reaming-solvothermal method[25]. The RPC@$Cu_{7.2}S_4$-700 illustrated a strong RL_{min} of −53.02 dB at 2.0 mm and an EAB of 6.205 GHz. The carbon defects, Cu vacancies and O, S heteroatom doping in layered RPCs are important factors leading to dielectric losses. $Cu_{7.2}S_4$ with micron spherical structure and RPC in layered porous structure are helpful for optimizing impedance matching. Meanwhile, Xu et al. designed a Co NP-embedded biomass-derived porous carbon framework[26]. The embedded heterostructure is more conducive to comparing the utilization rate of the contact surface with the load heterogeneous structure. When the thickness is 2.3 mm, the EAB is up to 6.6 GHz. However, there are still some problems with embedded heterostructures, such as complex synthesis and how to control the growth of NPs in pores rather than attaching to their surfaces.

In the above discussion of the application of heterogeneous interfacial structure composites, it is clear that the interface effect has a significant relationship with dielectric loss and EMW absorption performance. Therefore, in the future, it is still necessary to explore some heterogeneous structures that can maximize the use of interface effect components.

3.3 EMW ABSORPTION MATERIALS WITH VARIOUS DIMENSIONS

3.3.1 Zero-dimensional materials

Zero-dimensional (0D) materials are materials in the 3D nanoscale range (1−100 nm) or composed of them as basic units, which is about equivalent to the scale of 10−100 atoms closely packed together. There are various types of 0D nanostructure units, commonly are NPs (1−100 nm), ultrafine particles (< 0.1 μm), ultrafine powder (< 10 μm), artificial atoms, quantum dots (QDs) (2−10 nm), atomic clusters and nanoclusters (< 2 nm), with the difference in the size range. Below, we will discuss the application of some typical 0D materials in EMW absorption.

3.3.1.1 Zero-dimensional carbon spheres (CSs)

CSs, as one of the most basic 0D carbon materials, have dielectric loss characteristics and unique structural properties, which provide a basic template and way for the preparation of high-performance EMW absorption materials. CSs exhibit typical spherical morphologies and have excellent dispersibility in a wide range of solvents and resins. Amorphous CSs produce good impedance matching in free space and absorbing medium. In the application of EMW absorption, however, the nanostructure design and composition regulation of CSs is mainly constructing cavity structure, porous structure, embedding magnetic particles (Fe, Co, Ni, Mn, etc.), introducing heteroatoms (N, P, S, etc.) doping, and depositing multi-component materials to improve the absorption performance[27]. CSs feature regular geometry, large specific surface area, narrow particle size distribution, and a simple synthesis process, which is of great significance in the design of carbon-based materials.

Although some CSs have certain absorption properties, they still have narrow EABs and needs high filling degrees, which is precisely due to the weak dielectric properties provided by conductive losses and inferior impedance matching[28]. Therefore, to improve the utilization rate and absorption performance of CSs, some researchers have carried out structural design and developed multi-component composites of CSs. For instance, Wang et al. successfully synthesized watermelon-derived micro carbonaceous spheres by one-pot solvothermal method. The organic components of the initial material were removed by high-temperature calcination, and the pre-prepared micro carbonaceous spheres had excellent microwave absorption (MA) performance at Ku band[29]. Thus, through construction of nanostructures and synergistic effects between different components, CS composites can have better EMW absorption performance than pure CSs.

From a part of the studies, CSs can greatly improve impedance matching by high-temperature pyrolysis, thereby improving absorption performance. However, some researchers have designed CSs by introducing heteroatom (N, P, S) and magnetic metal particles (Fe, Co, Ni, Mn) and achieved excellent absorption performance. Recently, Chen et al.[30] introduced magnetic Fe and Fe_3C (Fe–Fe_3C) NPs into N-doped carbon spheres (NCSs) and adjusted the content of Fe NPs by dilute hydrochloric acid treatment. The prepared NCSs/Fe–Fe$_3$C composites had excellent EMW absorption performance, and the optimal reflection loss value reached −63.77 dB at 9.92 GHz. Moreover, the EAB reached 4.88 GHz at a matching thickness of 2.9 mm. The eminent absorption performance can be attributed to the synergistic effect of conductivity, polarization and magnetic losses generated by different components of the composite. Moreover, the change of Fe content can adjust the dielectric permeability characteristics of NCS/Fe–Fe$_3$C composites, and finally balance the impedance matching and attenuation coefficient.

In 2021, Song et al.[31] successfully prepared an ultrafine zinc oxide (ZnO) NPs loaded on a 3D ordered mesoporous CS (ZnO/OMCS), using silica inverse opal as the raw material and phenolic resol precursor as the carbon source (Fig. 3.2a). The synthesized lightweight ZnO/OMCS nanocomposites demonstrated a highly ordered 3D CS array and dispersed imperceptible ZnO NPs on the mesoporous cell walls of CSs. ZnO/OMCS-30 nanocomposites with a theoretical content of 30 wt% ZnO NPs demonstrated a −39.3 dB MA capacity at 10.4 GHz with a small thickness of 2 mm and a wide EAB of 9.1 GHz (Figs. 3.2(c and d)). It is a remarkable fact that ordered porous carbon can solve the problem of nonuniform ZnO distribution, thereby further effectively enhancing the interfacial polarization. However, ZnO nanocomposites can effectively increase interfacial polarization, which is very significant in microwave attenuation (Fig. 3.2e). This is because it can be improved by enhancing the interface between different dielectrics in materials.

For the internal hollow structure, impedance matching can be promoted, and the thinner shell thickness can facilitate the entry of EMWs. In addition, multiple reflections and scattering within the MAMs will facilitate the loss of the incident EMWs, refer to transmission line theory[31]. In early 2011, the core-shell Fe$_3$O$_4$ poly(3,4-ethylenedioxythiophere) microspheres were constructed by a two-step method in polyvinyl alcohol and p-toluenesulfonic (ρ-TSA) acid, which exhibited excellent microwave absorbing properties in the layer thickness range of 3–4 mm and a minimum reflection loss (RL_{min}) of about −30 dB at 9.5 GHz with a layer thickness of 4 mm.

3.3.1.2 QDs composites

In 1981, a new concept of nanoscale conductor-emitting particles was discovered, which was called "QDs". QDs are aggregates of atoms and

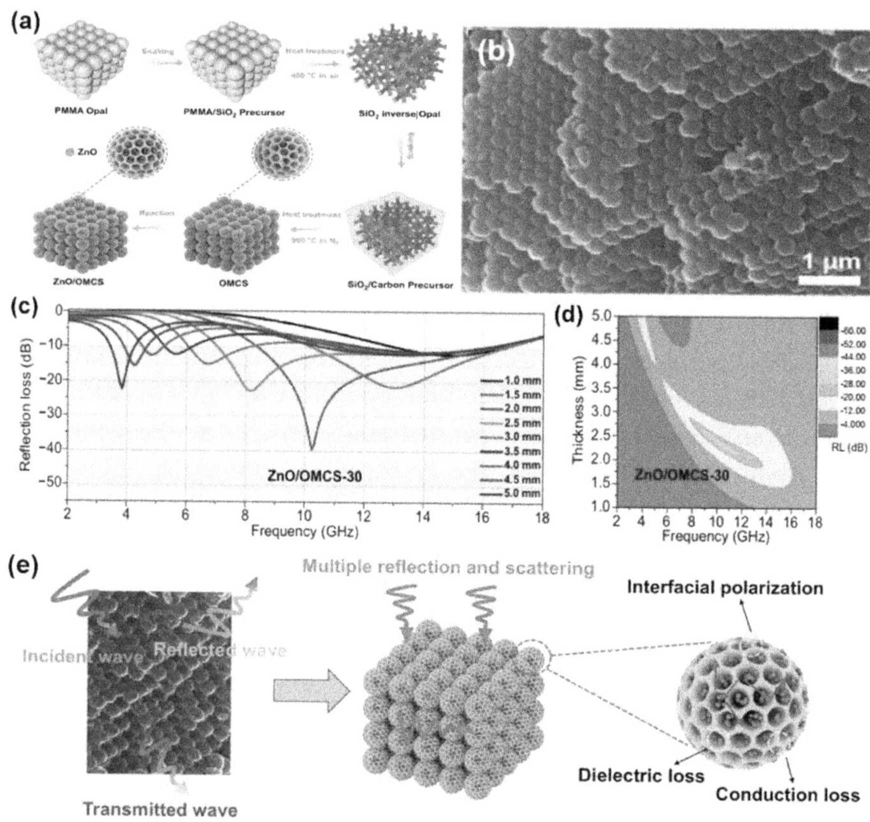

FIG. 3.2   (a and b) Schematic illustration of the synthesis of ZnO/OMCS and its SEM image; (c and d) reflection loss and the Delta value maps; (e) schematic illustration of the microwave absorbing mechanism of ZnO/OMCS. (Reproduced with permission from Ref.[31] Copyright 2021, Springer.)

molecules at the nanoscale, mainly homogeneous or core-shell NPs composed of II–IV B (CdSe, CdTe, CdS, ZnSe) and III–V A (InP, InAs) group elements, also known as semiconductor nanocrystals. QDs are generally spherical crystals with a diameter of 1–10 nm and are nanomaterials composed of a limited number of atoms that are quasi-0D, and the size of three dimensions is in the order of nanometers. The basic properties of QDs have multiple effects such as quantum surface effect, size effect, quantum confinement effect and macroscopic quantum tunneling effect. QDs can exhibit numerous optical and physical properties different from macroscopic substances, so they have miscellaneous application prospects in optics, electricity, magnetic media, photocatalysis[32], medicine, life sciences, functional materials, and other fields. Herein, the application of QDs in EMW absorption mainly comes from relevant research in recent years.

In 2017, Wu et al. aimed to load magnetic QDs on monodisperse amorphous CSs to achieve broadband EMW absorbers. Thus, the authors successfully prepared phenolic resin sphere (RS)/Fe-glycolate by thermal decomposition, and converted RS and Fe-glycolate into amorphous spherical Fe_3O_4/Fe QDs through a high-temperature carbonization process, respectively[33]. In 2022, He et al. used DC arc plasma technology to prepare porous carbon materials embedded in adjustable QDs for MAMs. The transition metal oxide MnO and magnetic Fe were coupled to the carbon matrix to construct heterogeneous core-shell composites. Nanoscale MnO not only produces more dipole polarization and relaxation, but also regulates the catalytic action of magnetic Fe[34]. In 2021, Tang et al. synthesized a hexagonal boron nitride nanosheet (BNN)-supported graphene QDs (GQDs) composite by a solvothermal method. The author, for the first time, used GQDs as microwave absorbents, and explored the relationship between the concentration of different GQDs and MA performance in GQDs/BNNs composites (defined as S9, BNNs and GQDs molar ratio = 1:9). The results show that the S9 composite has excellent MA performance, with RL_{min} values as low as −59.9 dB at 2.8 mm in the 2−18 GHz band[35].

As a ternary transition metal oxide, $ZnFe_2O_4$ has a relatively low complex permittivity and magnetic permeability, which limits its application in the field of EMW absorption. However, $ZnFe_2O_4$ can be used as a soft magnetic material with high magnetic loss ability, and it can be compounded with other dielectric materials to optimize impedance matching and explore loss mechanisms. Meanwhile, $ZnFe_2O_4$ NPs have small size effects and semiconducting properties with significant quantum effects such as electron and electron-hole confinement. However, this QD effect has great significance and value in the field of dielectric microwaves. For example, Wu et al.[36] synthesized amorphous carbon-coated $ZnFe_2O_4$ QDs by electrostatic assembly technology. $ZnFe_2O_4$ QDs demonstrated a reflection loss of −40.68 dB at 11.44 GHz and a thickness of 2.5 mm and an EAB of 3.66 GHz (9.87−13.52 GHz).

In 2023, Zhou et al. fabricated carbon foam embedded with magnetic Co quantum dots (Co-QDs/CF) via a two-step method concluding hydrothermal treatment and carbonization process, which displayed an EAB of 8.74 GHz at 2.5 mm in the 2−18 GHz with a filling rate of 20 wt%, as shown in Fig. 3.3a[37]. Co-QDs with 0D ultra-small size are distributed on carbon foam as magnetic components, which obtain considerable high-frequency magnetic responsiveness at lower densities, and the surface effect of Co-QDs NPs enhances the switching coupling of medium- and high-frequency frequency bands, thereby suppressing the eddy current effect. However, the contribution of exchange resonance to magnetic loss replaces eddy current loss, and the attenuation of the eddy current effect reduces the reflection of EMWs on the surface of Co-QDs/CF materials (Fig. 3.3b)[37]. As can be seen from Fig. 3.3b, forming a unique honeycomb structure can optimize impedance matching, improve carrier mobility, and provide more scattering paths.

In fact, the quantum surface effect is that when the number of atoms in QDs is reduced, the particle size will decrease and the specific surface area will increase. Although composites with QDs have excellent EMW absorption properties, the influence of size and shape on the EMW mechanism is still unclear.

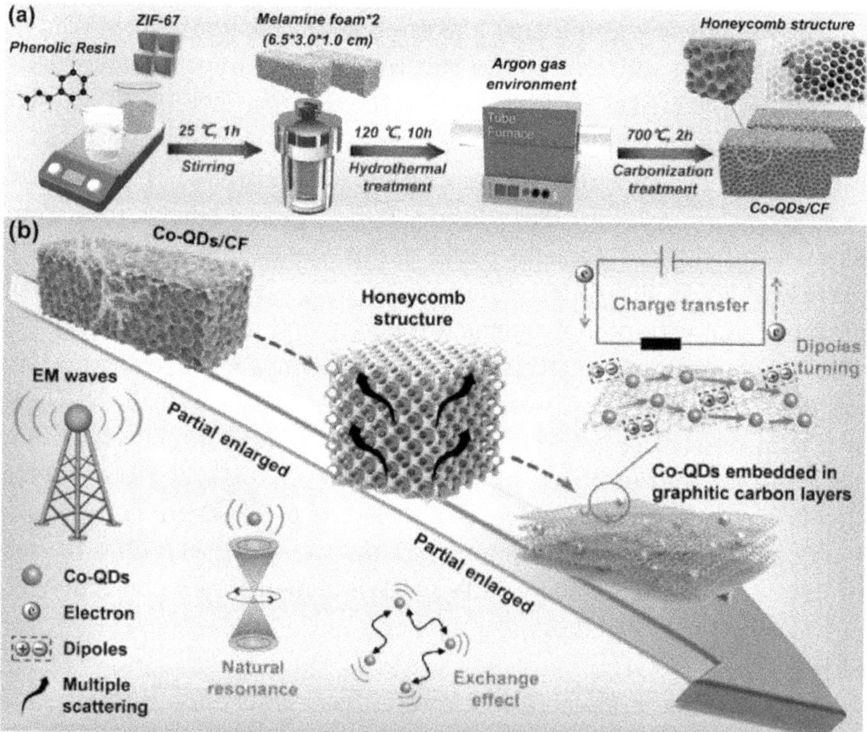

FIG. 3.3 The synthesis process (a) and schematic diagram of the electromagnetic wave absorption mechanism (b) of the honeycomb-like 3D Co-QDs/CF composites. (Reproduced with permission from Ref.[37] Copyright 2021, Elsevier.)

3.3.2 One-dimensional materials

The definition of 1D materials is that electrons only move freely in a non-nanoscale direction such as nanowires, nanofibers, nanobelts, nanoribbons, nanorods, and nanotubes. The movement of electrons in it is left in only one direction. CNTs, as a unique 1D nanostructure, have low density, corking electrical conductivity, chemical stability, high mechanical strength and large length-to-diameter ratio, and are widely used in various fields such as EMW absorption, EM shielding, supercapacitors, batteries and electrocatalysis[38–40]. Importantly, CNTs are able to provide large interfacial polarization regions and high charge transfer rates along the axial direction[41,42]. However, high conductivity, poor dispersion and magnetic loss property affect impedance matching of EMW absorption. Therefore, single CNTs are hindered in applications of EMW absorption[43,44]. Thus, numerous researchers have tried to compound CNTs with other materials to ameliorate their problems, such as optimizing impedance matching, and thus improving the attenuation ability of EMW.

3.3.2.1 CNTs-based composites as EMWs

3.3.2.1.1 CNT/magnetic component composites   As is well known, one-component CNTs must have a magnetic loss mechanism and good dispersion to achieve eminent absorption capacity. However, many researchers have improved the permeability and interfacial interaction by combining CNTs with magnetic loss mechanism materials such as Fe[45], Co[46], Ni[47], Ag[48], alloys[49,50], Fe_3O_4[51,52], MFe_2O_4 (Zn, Co, Mn, Ni), $NiCo_2O_4$, $MnFe_2O_4$, and $BaFe_{12}O_{19}$. As an example, Liu et al.[53] synthesized Co/CNT nanocomposites via deposition of Co NPs on the surface with a hydrothermal method followed by a novel carbon reduction route. The Co/CNT composites demonstrated a strong RL_{min} of −36.5 dB at 4.1 GHz with a low filling content of 20 wt%. The synergistic effect between the EM loss provided by Co NPs and the dielectric loss provided by CNTs is an important factor in optimizing impedance matching. Meanwhile, the CNTs/Ni hybrid was fabricated through catalytic carbonization with recycled waste masks[54]. The CNTs/Ni hybrids exhibited superior absorption performance, that is, an RL_{min} of −56.3 dB and an EAB of 4.3 GHz at 2.0 mm, which outperforms than other C/Ni hybrids. Importantly, element doping can introduce numerous point defects in the host material, and is also an effective way to improve the dielectric properties of EMW-absorbing materials. According to Debye's theory, under an EM field, these defects would act as dipoles to improve the dielectric properties of the material[55].

To investigate the EMW absorption performance of N-doped carbon materials, it is necessary to introduce a nitrogen source. There are various nitrogen sources such as melamine, dicyandiamide and nitrogen-containing organic ligands. Furthermore, a series of M@NCNTs (M:Fe/Co/Ni) composites were prepared via a modified non-poisonous pyrolysis method employing metal chlorate and melamine, and the Fe@NCNTs samples exhibited splendid EMW absorption performance. The RL_{min} and EAB of Fe@NCNTs are −30.43 dB at 3.2 mm with 10 wt% loading and 5.7 GHz, respectively[56]. Compared with monometals, bimetals are more conducive to excellent EMW absorption because they can induce electron transfer and enhance spin polarization during alloying[57]. Until now, numerous studies have confirmed this. For instance, FeNi-CNTs and FeNi-CNPs (carbon nanoparticles) were successfully fabricated and their MA performances were discussed[58]. The FeNi-CNTs composite had a doughty dielectric loss than FeNi-CNPs and displayed an RL_{min} of −32.2 dB at 2.8 mm with an EAB of 8.0 GHz. In another example, Song et al. constructed a FeCo@C-CNTs core-shell nanostructure with CNTs (10−15 nm) coated by diminutive FeCo@C NPs with sizes of 5 nm via a metal-organic chemical vapor deposition (MOCVD)[59]. The CNT-NP has a splendid EMW absorption performance, i.e., an RL_{min} of −79.2 dB and an absorption bandwidth of 6.3 GHz at 2.0 mm.

In addition, Ferrites (MFe_2O_4) have been extensively studied due to their magnetic properties, remarkable thermochemical stability, low cost, and diverse structure[60–63]. However, the small dielectric loss, narrow absorption bandwidth, and high density of ferrites greatly affects their application in the field of EMW absorption[64,65]. In order to advance the EMW absorption performance more effectively, some researchers took full advantage of interface engineering to strengthen the interface polarization and constructed mixed-dimensional CNT composites. In 2023, Peng et al. designed a mixed-dimensional (CNTs)@Fe_3C@Fe_3O_4 composite, which is mainly composed of 3D CF, 1D CNTs and 0D Fe_3C@Fe_3O_4 NPs through a plain CVD method[52]. Moreover, the effects of different contents of CNT and pyrolysis time on absorption performance were explored. The results show that CF/CNTs@Fe_3C@Fe_3O_4 samples have an RL_{min} of −44.48 dB at 1.68 mm, and an EAB of up to 5.0 GHz at 1.81 mm with the increase of CNT content and pyrolysis time. When the pyrolysis time is increased from 30 min to 60 min, the $\tan\delta_\varepsilon$ values of the sample are enhanced, which indicates that the dielectric loss capacity can be enhanced. The increased content of CNTs can also enhance conductivity and polarization loss capacities[52]. Zeng et al. fabricated a 3D hierarchical urchin-like Fe_3O_4/CNTs by using a hydrothermal method. The urchin-like Fe_3O_4/5CNTs architecture shows an RL_{min} of −56.8 dB at 11.12 GHz with 2.15 mm[51]. Moreover, the impedance matching of Fe_3O_4/5CNTs is superior to the other two Fe_3O_4/3CNTs and Fe_3O_4/7CNTs composites.

In numerous ferrite materials, not only Fe_3O_4 can be combined with CNTs as a source of magnetic loss, but ferrite (MFe_2O_4) can also be. Moreover, studies have shown that it is possible to reduce its size to the nanoscale to improve its absorption performance. In addition, designing layered nanostructures, constructing heterogeneous interfaces, and combining with other loss mechanism materials can optimize the absorption properties of the magnetite-based MAMs. For example, in Zhang's work,[66] $CoFe_2O_4$/CNTs were prepared by a solvothermal reaction and exhibited an RL_{min} of −37.39 dB at 1.7 mm and an EAB of 5.2 GHz.

3.3.2.1.2 CNTs/polymer composites   The research design of CNTs is not limited to magnetic metals, alloys, spinel, etc. CNTs can also be dispersed in a polymer matrix to form CNT-polymer composites. However, the easy agglomeration of CNTs remains as the biggest difficulty. Therefore, different decentralized technologies are crucial for the formation of CNT-polymer composites. There are numerous CNTs dispersion technologies such as mechanical mixing, solvent spraying, chemical functionalization and spark plasma sintering[67]. Mechanical mixing is the most extensively used dispersion technology, but the resulting CNTs have a large number of cracks, thus it is only used for industrial large-scale applications at short notice[68,69].

Chemical functionalization can more effectively create uniform dispersion compared to mechanical mixing and solvent spraying[70,71]. In the process, the groups on the surface of nanotubes are chemically modified to accelerate their dispersion in dissolution and polymers. A variety of CNTs-dispersed polymers have been extensively studied such as PANI/CNTs, polypyrrole (PPy)/CNTs and poly(vinylidene fluoride) PVDF/CNTs. Excellent MA can be achieved by using interface modulation-induced interfacial polarization. For instance, in 2019, Zhou et al. prepared a CNTs@PANI hybrid absorber with an interface modulation strategy by optimizing the CNT nanocore structure[72]. When the filler loading was only 10 wt.% and the coating thickness was 2.4 mm, the CNTs@PANI-4h exhibited an RL_{min} of −45.7 dB at 12.0 GHz and an EAB ranging from 10.2 to 14.8 GHz.

3.3.2.2 1D silicon carbide (SiC)-based composites as EMWs

SiC is a covalently bonded IV–IV compound with alternating layers of Si and C, in which each Si (or C) atom is surrounded by four C (or Si) atoms in a strong tetrahedral sp^3 hybrid bond [73]. SiC crystal structure is composed of a tight packing of two layers of Si and C atoms. Different stacking sequences for Si–C bilayers can cause crystal stack failures and layer arrangement changes. Diverse permutations result in diverse SiC configuration types, known as polytypes. Different SiC has significant differences in conductivity, band gap, dielectric responsiveness, and shape[74]. In the past few decades, 1D SiC crystals with a variety of structures such as nanowires/nanorods[75], nanoneedles[76], nanotubes[77], nanoribbons[78], layered nanostructures[79], beaded nanochains[80], bamboo[81], twin and double crystal structures[82], crystalline nanostructures[83], nanosprings[84], core-shell structures[85-87], and nanoarrays[88,74] have been studied. Moreover, SiC has been widely used in harsh environments such as high pressure, strong corrosion, and high temperature, which makes it a very attractive dielectric loss material in high-temperature applications. However, SiC is a dielectric loss material with poor impedance matching and low conductivity. Moreover, its EAB lies in the high-frequency region.

Researchers have studied the influence crystal length and temperature on the EMW absorption performance of SiC. Some studies have shown that as the temperature increases, the dielectric loss of SiC will increase, and the dielectric constant will also increase, which will affect the impedance matching ability[89,90]. For SiC with different lengths, the longer the length, the higher the dielectric loss capacity[91,92]. However, the conductivity loss of SiC at room temperature is particularly small, and dipole polarization loss cannot be achieved. Therefore, in order to utilize the advantages of SiC and make up for its shortcomings, some researchers have proposed the following ideas: (1) Modifying SiC nanowires (NWs) with doping and structure designing, to introduce defects or improve the interfacial polarization of materials. SiC can be doped with non-metallic elements such as N, C and B. Due to their different atomic radii, the atoms entering the SiC NW introduce a large number of defects, thereby improving the EMW absorption performance. (2) Compounding SiC with magnetic loss materials to optimize their impedance matching capabilities. (3) Improving the conductivity of SiC NWs by aligning them with conductive substrate materials (such as carbon materials and conductive polymers). (4) Designing multiple loss mechanisms to enhance EMW absorption.

For example, Zhang et al. successfully fabricated three kinds of SiC_{nw} with different morphology, i.e., linear, bamboo-shaped and worm-like, by a simple carbothermal reduction and then attaching them to carbon fibers through vapor-liquid-solid (VLS) mechanisms[93]. The bamboo-shaped SiC_{nw} felt displayed an RL_{min} of −44.3 dB at 3.85 GHz. In another instance, Wang et al. designed a magnetic/dielectric multi-component Fe_3Si/SiC composite by an in-situ carbothermal reduction strategy[94]. When the filler loading is 10 wt% and the thermal reduction temperature is 1400 °C, Fe_3Si/SiC composites with a double loss mechanism demonstrated that the minimum RL and EAB can reach −61.67 dB and 6.96 GHz, respectively.

This hetero-magnetic and dielectric multi-component composite material can form multi-polarization, magnetic loss, and dielectric loss effects, which is conducive to the development of high-performance EMW absorbing materials[95]. However, the application of carbon and magnetic materials at high temperatures is limited. Therefore, it is necessary to rationally construct interfaces and pores, such as porous structures, core-shell structures, layered, and 3D networks to improve impedance matching. Multifunctional SiC aerogels were synthesized by freezing casting technology, followed by carbonation and calcination processes. They were composed of 1D SiC nanofiber, in situ generated SiC NWs, and 2D SiC nanosheet (Fig. 3.4a)[96]. The obtained SiC aerogel-2 showed excellent absorption performance with an RL_{min} of −61.6 dB and a maximum EAB of 9.82 GHz (Fig. 3.4b). In conclusion, SiC aerogels with large pores allow more incident waves to enter the materials from the surface, thus optimizing their impedance matching and consuming EMW energy efficiently.

In conclusion, SiC-based materials have made some research progress in the field of EMW absorption, but SiC still has the limitations of narrow absorption bandwidth and large thickness when used at high temperatures. At room temperature, the conductive loss of SiC is small, and it is difficult to achieve strong dipole polarization loss. At medium or high temperatures, the dielectric loss increases, but the real part of the dielectric constant also increases, affecting the impedance matching ability. Therefore, in order to

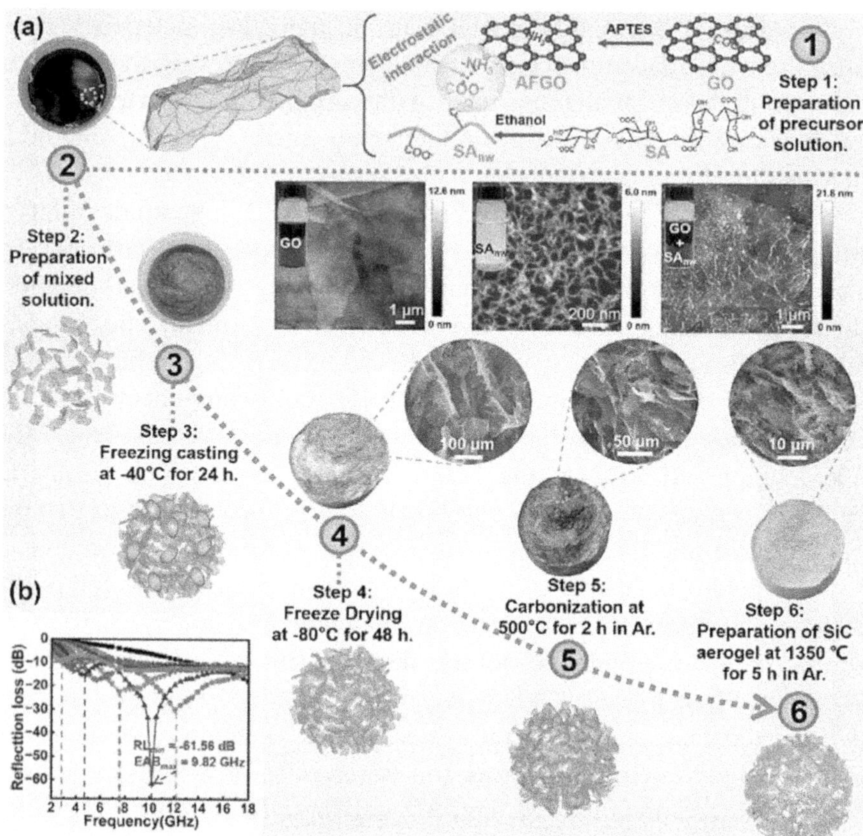

FIG. 3.4   The preparation process of SiC aerogel (a) and *RL* curve of SiC aerogel-2 (b). (Reproduced with permission for Ref.[96] Copyright 2023, Elsevier.)

better improve the loss capacity, the design idea of SiC is mainly to compound it with another component to induce polarization and improve conductive loss.

3.3.3 Two-dimensional materials

Two-dimensional (2D) materials are an emerging class of materials with thicknesses ranging from a single atomic layer to multiple atomic layers. There are numerous kinds, such as graphene, MXene, layered double hydroxide (LDHs), boron nitride (BN), molybdenum disulfide (MoS_2), tungsten disulfide (WS_2), molybdenum diselenide ($MoSe_2$), and tungsten diselenide (WSe_2). Two-dimensional materials have distinctive characteristics, which have spurred numerous research in the past decade for diversified applications. Meanwhile, they can also serve as well-suited building modules for a series of layered structures, membranes, and other composite materials[97]. Two-dimensional materials are extensively used in transparent conductive electrodes, photodetectors, gas detectors, batteries, supercapacitors, EMW absorption, and EM shielding materials. Moreover, 2D nanomaterials can typically adjust their band gap values through optimizing layers, so that they can have broadband absorption, moderate dielectric values and can optimize impedance matching and promote microwave attenuation capabilities. Next, several unique 2D materials such as MXene-based composites, transition metal chalcogenides and LDHs-based composites will be summarized and discussed.

3.3.3.1 MXene-based composites as MAM

2D transition metal carbide/carbonitride material (MXene) has been widely studied because of its special sheet structure and surface chemistry. However, untreated and unengineered MXene can undergo aggregation and stacking, which limits their EM absorption application[17]. MXene is generally prepared by selective extraction of particular atoms from their layered parents $M_{n+1}AX_n$, known as MAX phases. The formula of MXene is $M_{n+1}X_nT_x$, where M stands for transition metal (such as Sc, Zr, Hf, Ti, V, Nb, Ta, Cr, Mo, etc.), A refers to an element in group IIIA or IVA such as Si and Al, X represents C or N, n represents 1–3, and T_x represents surface terminations (OH, O and F groups)[97–100]. MXene allows the construction of multifarious structures and application in various fields, including EMW absorption, EM shielding, energy storage, sensing, photocatalysis, and organic catalysis due to its intriguing properties such as high conductivity, diverse functional groups, abundant surface defects, large specific surface area and dielectric loss characteristics (Figs. 3.5a–c). There are multiple etching methods used to prepare MXene from MAX, such as in-situ hydrofluoric acid (HF) etching, fluorine-free etching, green electrochemical etching, alkali etching and common molten salt etching[101].

It is widely known that the M and A atoms in the MAX precursor are interconnected by metal bonds, which makes it impossible to obtain MXene by simple mechanical stripping of the MAX precursor. For the first time, Gogotsi et al. prepared 2D Ti_3C_2 by etching MAX phase Ti_3AlC_2 with HF in 2011[98]. To date, there are a multitude of methods for the preparation of MXenes including top-down etching-assisted exfoliation and bottom-up

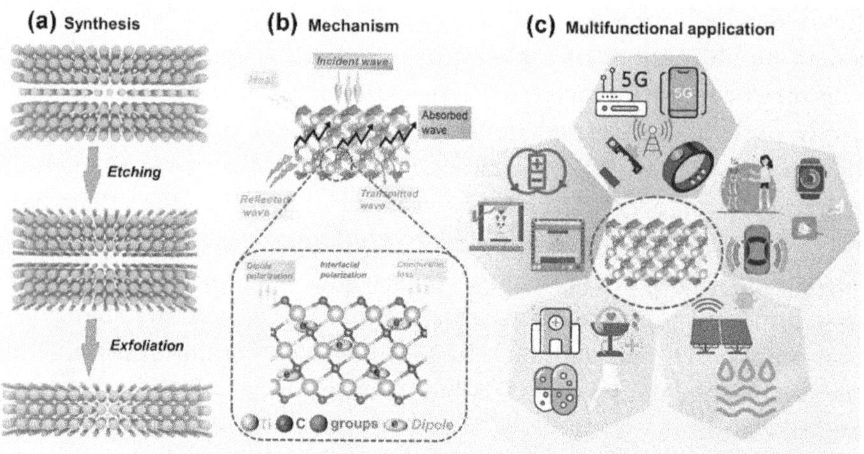

FIG. 3.5   Diagram of (a) the synthesis process, (b) reaction mechanism and (c) multifunctional application of MXene. (Reproduced with permission from Ref.[100] Copyright 2022, John Wiley & Sons.)

synthesis (CVD) and epitaxial growth[98,102,103]. Etching-assisted exfoliation is the primary method for synthesizing monolayer and multilayer MXenes, which is used to synthesize MXenes EM applications. HF etching methods have been triumphantly carried out in the synthesis process of many members with MXene families. Nevertheless, depending on the M-site atoms, A-site atoms, X-site atoms and n-value of the MAX precursor, the etching method requires different HF concentration, etching time, and reaction temperature. Qing et al. first studied the EMW absorption performance of pure Ti_3C_2 nanosheets prepared by immersion of Ti_3AlC_2 powder in HF and subsequent sonication[104]. The Ti_3C_2 nanosheets showed a reflection loss of –11.0 dB (absorption over 92%) in a frequency range of 12.4–18 GHz at a thickness of 1.4 mm.

As a transition metal carbon and nitrogen compound, MXene has a high dielectric constant and ultra-low magnetic permeability, which makes it difficult to have excellent impedance matching ability. Therefore, only by further optimizing the MXene-based composites with dielectric loss materials or EM loss materials can impedance matching be optimized to obtain EMW absorption materials with high absorption capacity. The common disadvantage of MXene is self-stacking and aggregation. Moreover, the interaction between the MXene layers through hydrogen bond hinders the utilization of its surface-active site. Therefore, it is significant to introduce carbon-based materials, conductive polymer-based materials, ceramic-based materials, semiconductor-based materials and other dielectric materials to optimize MA capacity.

3.3.3.1.1 MXene/carbon composites In general, carbon-based materials mainly include CNTs, carbon nanofibers (CNFs), graphene, biomass-derived carbon, and helical carbon nanotubes (HCNTs), and they are widely investigated as novel EMW absorbers because of their excellent stability, electronic conductivity, and low density[105]. CNTs offer multiple merits, such as a high aspect ratio, low seepage threshold, and excellent dielectric losses. Therefore, compounding CNTs with MXene can decrease the aggregation of MXene and enhance the absorption performance. For example, as shown in Fig. 3.6(a), $Ti_3C_2T_x$ MXene was prepared via HCl/LiF etching. Subsequently, a graphite-ring-stacked CNT/MXene with in situ growth of CNTs on $Ti_3C_2T_x$ was firstly synthesized via CVD at different temperatures (450/500/550/600/700 °C)[106]. From Fig. 3.6(b), it can be seen that a uniformly distributed forest of CNTs grows on the surface, and the structure and morphology of MXene are not damaged. The resulting CNT/MXene-500 exhibited an RL_{min} of −47.28 and an EAB of 4.16 GHz at a low thickness of 1.36 mm and 1.16 mm (Fig. 3.6c). As can be seen from Fig. 3.6(c), the appearance of bimodal in composites can be attributed to the ideal quarter-wavelength thickness relationship, the optimal double-band impedance matching, and the maximum polarization loss at the corresponding frequency.

Compared with mechanically mixed CNT materials, CNTs grown in situ on the surface of the material were more evenly distributed with reduced aggregation, which can benefit the multiple reflections and scattering of EMW. Similarly, Liu et al. also fabricated bamboo-shaped CNTs and helical CNFs on $Ti_3C_2T_x$ MXene under different ultra-low temperatures (450/525/600 °C) by CVD method[107]. The CNT/MXene hybrid exhibited excellent EMW absorption performance with an RL_{min} of −52.56 dB in the X-band at a low sample thickness of 2.5 mm and an EAB of 2.16 GHz. Meanwhile, Peng et al. synthesized nacre-like and loofah-like carbonized CNF/MXenes membranes (CCMs) via different drying methods, including an oven and freeze-drying, after carbonization at low temperature (Fig. 3.6d)[108]. The nacre-like and loofah-like CCMs had outstanding absorption performance with an RL of −42.2 dB and −63.8 dB, corresponding to an EAB of 7.12 GHz at 3.6 mm and 7.32 GHz at 2.5 mm, respectively, as shown in

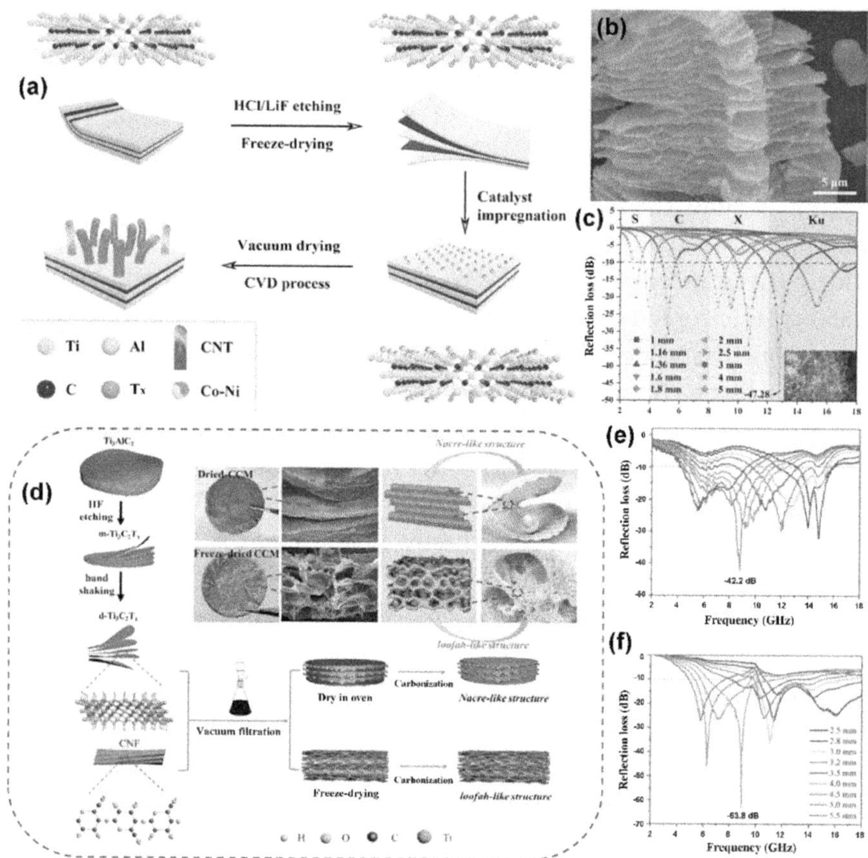

FIG. 3.6   The synthesis process (a), SEM image (b) and *RL* curves with different matching thicknesses (c) of graphite-ring-stacked CNT/MXene. (Reproduced with permission from Ref.[106] Copyright 2023 Elsevier.) The preparation process (d) and *RL* curves (e and f) of the nacre-like and loofah-like CCM. (Reproduced with permission from Ref.[108] Copyright 2022, Elsevier.)

Figs. 3.6(e and f). For nacre-like carbonized CCM, its absorption peak is mainly concentrated in the X-band. Instead, the main absorption peak of loofah-like CCM moves toward the Ku band.

The presence of CNF can adjust EM parameters, improve attenuation capacity, and effectively adjust the absorption band. Therefore, the composite of CNF with MXene self-assembly can better optimize impedance matching, and the biomimetic structures can promote polarization loss, conductivity loss, and multiple reflection and scattering. In 2022, a core-shell MXene/N-doped C heterostructure was constructed from MXene/

PANI by exfoliation, which was synthesized using three processes—HCl and LiF etching, in-situ polymerization, and carbonization[109]. The EAB of MXene/N-doped C hybrid was up to 5.0 GHz at 1.72 mm, corresponding to a low filler loading of 30 wt%.

As a chiral material, HCNTs have low density, high theoretical MA, and a variety of attenuation mechanisms, and they are often considered potential EMW materials[110,111]. A kind of novel HCNTs@$Ti_3C_2T_x$ MXene aerogel microsphere was prepared via a freeze drying-assisted electrostatic spinning technique[112]. The authors explored the absorption properties of HCNTs by adjusting the timing of functionalized HCNTs (from O-HCNTs to F-HCNTs) to form a variety of interlaminar structures. The HCNTs@ $Ti_3C_2T_x$ MXene aerogel possessed an absorption property of −35.5 dB at 10.6 GHz with 2.2 mm, corresponding to an EAB of 2.2 GHz. The successful synthesis of this composite simultaneously expanded the absorption band of HCNT-based materials in the C–Ku and S-bands.

3.3.3.1.2 MXene/polymer composites   For MXene, its self-aggregation issue leads to a reduction in the effective surface area, which needs to be addressed by introducing other materials to provide synergistic effects. However, not all materials are able to provide effective synergies with MXene composites. Instead, introduction of polymers into MXene's intercalation can prevent it from restacking[113]. Organic conductive polymers, such as PANI, PPy and polythiophene (PTh), have excellent conductive controllability, low density, excellent mechanical properties, corrosion resistance and are often used as absorbents. PANI, a common conductive polymer, does not have enough loss mechanisms and struggles to achieve desired EMW performance. However, researchers showed that loading PANI on MXene's nanosheet layer can provide more interfaces and thus can increase interfacial polarization. For instance, Jiang et al. successfully designed a $Ti_3C_2T_x$ MXene/PANI composites with different PANI contents by electrostatic self-assembly process[114]. When the loading of PANI was only 10 wt.%, the composites achieved an RL_{min} of −60.6 dB and an EAB up to 6.0 GHz. The introduction of PANI can balance the impedance matching of MXene and connect single MXene nanosheets, which is conducive to forming multiple heterogeneous interfaces and improving EMW absorption performance.

Unlike PANI, PPy can be designed to form a wide variety of morphologies and can have a low filling rate when functions as a MAM. Meanwhile, PPy has been widely studied for its excellent characteristics such as high

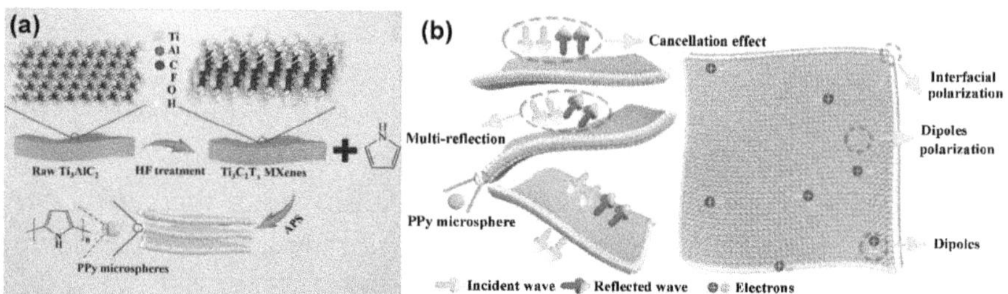

FIG. 3.7   (a) The synthesis process of Ti3C2Tx@PPy composites and (b) their possible absorption mechanism. (Reproduced with permission from Ref.[115] Copyright 2019, Elsevier.)

electrical conductivity, good mechanical properties, lightweight, and good formability[115]. For example, a novel core-shell nanostructured composite was designed by modifying $Ti_3C_2T_x$ MXenes and using ρ-TSA to adjust the conductivity of PPy microspheres, as shown in Figs. 3.7(a and b)[115]. The experimental results display that the $Ti_3C_2T_x$@PPy-2 composite with a filling of 10 wt% had an RL_{min} of −49.5 dB at 7.6 GHz with 3.6 mm and an EAB of 5.14 GHz (6.44–11.58 GHz), covering the whole X band. It is worth noting that conductive polymer composites still have many disadvantages, such as low melting point, limited thermal stability, and inferior high-temperature mechanical properties, which limit their application at high temperatures.

3.3.3.1.3 MXene/ceramics composites   Ceramic-based composites have excellent mechanical properties, corrosion resistance, lightweight flexibility, and heat resistance and have been widely used in high-temperature environments. Ceramic-based materials that have been widely studied include SiC, Si_3N_4, Al_2O_3, SiO_2, SiOC, SiBN, $BaTiO_3$, Ti_3SiC_2, and BCN, etc[116]. The MXenes and ceramic-based composites have also been gradually explored by researchers. For instance, Liu's group designed MXene@SiC$_{NWs}$@Co/C composites using in situ growth, a carbonization process, and an electrostatic self-assembly strategy. Firstly, as shown in Fig. 3.8(a), SiC$_{NWs}$@ZIF-67 (Zeolitic Imidazolate Framework-67) was fabricated, and then SiC$_{NWs}$@Co/C material was synthesized via a carbonization process at 800 °C with a heating rate 5 °C/min for 2 h[117]. Secondly, MXene@SiC$_{NWs}$@Co/C hybrids were freeze-dried by compounding MXene and SiC$_{NWs}$@Co/C. The hybrid displayed a superior impedance matching, mutual dissipation mechanisms, and abundant heterogeneous interface and good absorption performance

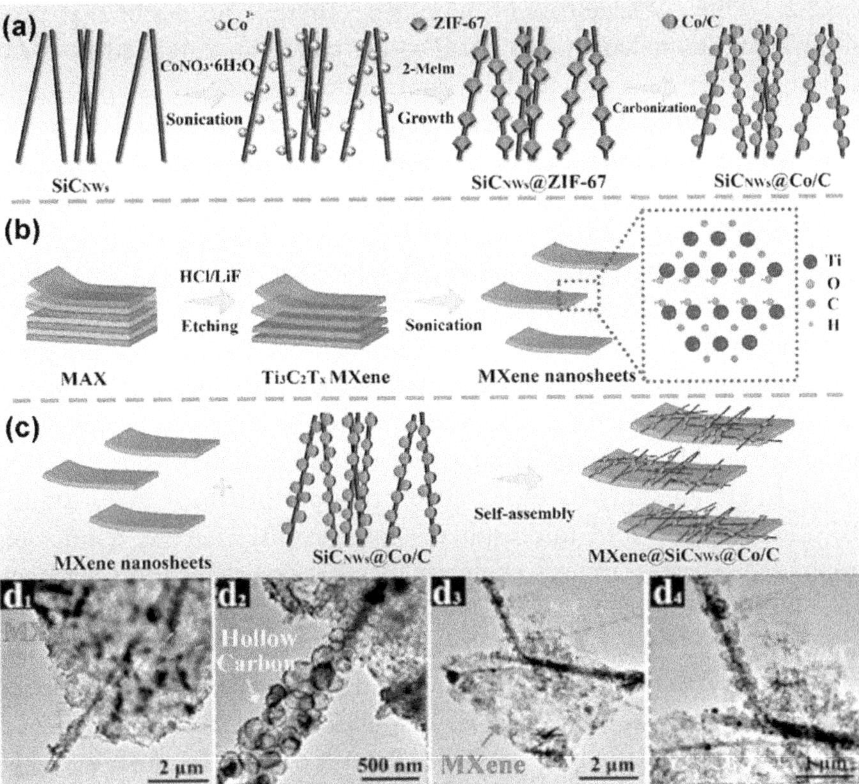

FIG. 3.8　The synthesis process of SiC_{NWs}@Co/C (a), MXene (b) and MXene@ SiC_{NWs}@Co/C (c). (d1–d4) TEM images of MXene@ SiC_{NWs}@Co/C composites material. (Reproduced with permission from Ref.[117] Copyright 2023, American Chemical Society.)

of an RL_{min} of −76.5 dB at 6.36 GHz at 3.9 mm, corresponding to an EAB of 6.2 GHz (11.8–18.0 GHz), covering the whole Ku band, as presented in Figs. 3.10(b and c). From Fig. 3.8(d), it can be seen that the ZIF-67-derived material in the MXene@SiC_{NWs}@Co/C hybrids exhibits a hollow frame structure. The introduction of SiC_{NWs}@Co/C materials can balance the excessive conductivity of MXene, promote the synergistic effect of magnetic loss mechanism and dielectric loss mechanism, promote impedance matching ability, and facilitate EMW absorption. Xiong et al. designed a 3D sandwich-like M-$Ti_3C_2T_x$@SiO_2@C hybrid by inserting SiO_2 hollow spheres into the flower-like $Ti_3C_2T_x$ MXene framework[118]. When the filler loading was 40 wt%, the 3D sandwich-like M-$Ti_3C_2T_x$@SiO_2@C hybrid displayed an RL_{min} of −43.97 dB at 4.5 mm and its EAB was up to 4.12 GHz at 1.5 mm.

3.3.3.1.4 Other MXene-based composites   There is no doubt that transition metal dichalcogenides (TMDs) are widely investigated as EMW absorbing materials due to their semiconductor properties and unique electrical features. TMDs have a layered structure, which gives them a large specific surface area and inherent defects, allowing dielectric loss capacity to be optimized. Their excellent electrical conductivity is mainly attributed to the quantum confinement effect, the relative movement between nanosheets, the reduction of the number of layers, and the influence of carrier transport behavior[119]. The high conductivity of MXene is notoriously detrimental to impedance matching. The rational design of semiconductor-MXene composite materials can enhance EMW absorption by adjusting the dielectric constant and the electrical properties of the material. Many researchers have combined MXene with MoS_2, WS_2, or CeO_2, etc. to study its EMW absorption performance. For instance, 3D MXene/MoS_2 fold microspheres were synthesized via a one-step method[120]. The researchers explored the superior absorption performance by adjusting the mass ratios of MXene and MoS_2. The results show that when the mass ratio of MXene to MoS_2 was 5:1 and the fill rate reached 30 wt%, the 3D MXene/MoS_2 fold microsphere gave a reflection loss of −51.21 dB at 10.4 GHz, corresponding to a matching thickness of 2.5 mm and an EAB up to 4.4 GHz at 1.6 mm. Meanwhile, a 2D/2D MoS_2/$Ti_3C_2T_x$ heterostructure was prepared by microwave-assisted hydrothermal method[121]. When the mass ratio of MoS_2 and $Ti_3C_2T_x$ was 6:1, it obtained an RL_{min} of −41.5 dB at 12.2 GHz and a corresponding EAB of 3.6 GHz.

Hence, through a series of studies, it can be found that when MXene is compounded with other materials, the impedance-matching ability of MXene materials can be optimized. For high dielectric materials such as carbon materials, conductive polymers, and metal NPs, they are capable of causing various effects. Three-dimensional carbon foam not only improves the filler stack, but also extends the multiple reflections and scattering of incident waves, thus facilitating the formation of new interfaces that improve impedance matching. For low-dielectric materials, such as SiC, ZnO, Fe_2O_3, and CeO_2, their composites with MXene also have excellent EMW absorption performance. Based on their low dielectric constant and complex permittivity, the introduction of low-dielectric materials can properly adjust the dielectric characteristics of MXene and optimize impedance matching. Like transition metal oxides with low dielectric constant, they also have semiconductor properties and numerous band gap values. However, large band gap values lead to inferior conductivity and small dielectric values.

Commonly, for the combination of dielectric materials with MXene, EMW absorption capacity can be effectively adjusted through conductivity, heterogeneous interface, and defect engineering. The combination of MXene matrix material and EM loss material can effectively realize the double loss mechanism, provide more magnetic loss mechanism to promote impedance matching, and achieve EM balance. For example, a $Ti_3C_2T_x$ MXene/$Ni_{0.6}Zn_{0.4}Fe_2O_4$ heterogeneous nanocomposite was synthesized by a two-step process with an etching process and in-situ hydrothermal process, which exhibited a double loss mechanism[122]. The heterogeneous $Ti_3C_2T_x$ MXene/$Ni_{0.6}Zn_{0.4}Fe_2O_4$-2 (with 2 mg MXene) showed an excellent EMW performance with an RL_{min} of −66.2 dB at 15.2 GHz and an EAB up to 4.74 GHz. Meanwhile, the construction of MXene and Metal-organic frameworks (MOFs)-derived composite materials can effectively construct the EMW absorption performance, accompanied by dielectric loss and magnetic loss mechanisms. In this composite, MOFs are converted into metals or metal alloys and nanoporous carbon through a high-temperature pyrolysis process, which can be rich in defects. Metals and metal alloys overcome MXene's deficiencies to optimize impedance matching and enhance dipolarization and EM losses. For instance, Wang et al. successfully synthesized an original $Ti_3C_2T_x$/Co-MOF derived TiO_2/Co/C (T/Co/C) composite architecture by *in-situ* solvothermal reaction and pyrolysis. The nanocomposites demonstrated a minimum RL_{min} of −50.45 dB at 7.32 GHz with 3.0 mm thickness, and achieved an EAB of 5.48 GHz at only 1.7 mm[123]. The excellent EMW absorption performance of T/Co/C composites is mainly due to the high-temperature pyrolysis of $Ti_3C_2T_x$ and Co-MOF, and the resulting conductive carbon sheets and graphitic carbon produced respectively, which work together to construct more conductive pathways for enhanced electron migration, resulting in increased conduction loss and EMW energy consumption.

The composite structure and multi-component structure facilitate the formation of internal multiple reflections and scattering when incident EMWs penetrate the absorber, resulting in effective EM dissipation. Meanwhile, dielectric losses related to dipole and interfacial polarization contribute greatly to the conversion of EMWs into thermal energy.

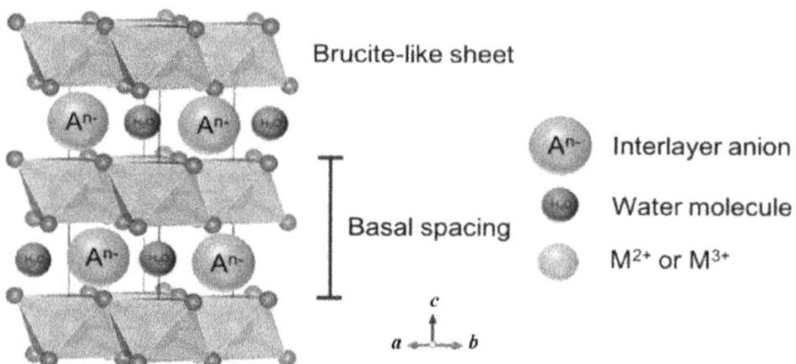

FIG. 3.9 Schematic diagram of LDHs structure. (Open access from Ref.[127] Copyright 2019, MDPI.)

3.3.3.2 LDHs-based composites as MAM

LDHs, composed of positively charged host layers and exchangeable interlayer anions, are structurally similar to brucite $Mg(OH)_2$, in which cations are coordinated with OH^- octahedrons, and are typical anionic layered compounds. As shown in Fig. 3.9, its chemical structure can be expressed as $[M_{1-x}^{2+}M_x^{3+}(OH)_2(A^{n-})_{x/n}\cdot mH_2O]$, where M^{2+} and M^{3+} represent divalent (Cu^{2+}, Ca^{2+}, Mg^{2+}, Zn^{2+}, Ni^{2+}, Co^{2+}) and trivalent (Al^{3+}, Fe^{3+}, Ga^{3+}, Cr^{3+}) metals, respectively and A^{n-} represents interlaminar anions (Cl^-, Br^-, NO_3^-, I^-, OH^-, SO_4^{2-})[124–128]. LDHs are comprehensively researched in the field of EMWs, mainly with various dominating factors, including simple synthesis methods, green and low cost and diverse morphological structures. LDHs can be synthesized through solvothermal methods[129], urea hydrolysis[130], electrodeposition method[131], sol-gel method[130], co-precipitation methods[132], electrostatic interstratification[133] and other methods[128].

LDHs have diverse species, including some binary and ternary (Co, Ni, Fe, Mg, Al, Zn, V, Cu) LDHs and the ion exchange property provide corrosion protection for metals by absorbing Cl^-. Nevertheless, LDHs also have some disadvantages, such as constant morphology, impedance matching imbalance, easy agglomeration, which limit their application in the field of EMWs[134]. Therefore, it is crucial to optimize their EMW absorption performance by rationally designing their composition and microstructure. The ultimate goal of hybridizing LDHs with other components is to alleviate the shortcomings of LDHs such as low conductivity and poor impedance matching.

Confronting the above problems, researchers tried to effectively combine LDHs by carbon networks with large specific surface areas and biomass-derived carbon skeletons, so that LDHs can be well dispersed on the carbon skeleton. For instance, in 2023, Wang et al. skillfully designed a 3D NiAl-LDH/carbon nanofibers (NiAl-LDH/CNFs) structure by an atomic layer deposition (ALD) method as a corrosion-resistant microwave absorber[135]. The size, coating thickness, and content of NiAl-LDH were optimized by varying the number of ALD cycles. The optimized NiAl-LDH/CNFs composite displayed the highest RL_{min} (−55.65 dB) and the widest EAB (4.80 GHz) when the load was only 15 wt%. Furthermore, the optimized NiAl-LDH/CNFs with 50 times cycles number have a lower current density (I_{corr}) and a large polarization resistance (R_p) in neutral solutions, which means that they have strong corrosion resistance. In Che's work, a pine nut shell-derived C@NiCo-LDHs@1D Ni chain aerogel was prepared by solvothermal and freeze-drying method[136], which displayed a RL_{min} as high as −57.4 dB at 13.3 GHz and an EAB up to 6.4 GHz at 2.5 mm.

Additionally, in Wu's work, a heterostructure NiCo-LDHs@ZnO nanorod was reported by a hydrothermal method and then a NiCo@C/ZnO composite derived NiCo-LDHs@ZnO was synthesized via a heat treating method, as shown in Fig. 3.10(a)[137]. In consequence, an eminent absorption performance of both broadband absorption (6.08 GHz) at 2.0 mm and a high-efficiency loss (−60.97 dB) at 2.3 mm were obtained (Figs. 3.10(b and d)). As to the absorption mechanism, the large specific surface area forms a conductive network, which is conducive to the transmission of induced current under the action of the external magnetic field, so that the internal electrons undergo directional migration, which converts EM energy into heat energy and dissipates it. When a large number of incident waves enter the conductive network, the EMWs are reflected and scattered, effectively causing EMW attenuation. Due to the excellent dielectric parameters, the interface between the NiCo alloy and the rod ZnO leads to interface polarization. The presence of NiCo@C composites and O vacancies also leads to depolarization[137].

Wu et al. synthesized three different kinds of LDHs (CoNi-LDHs, FeNi-LDHs and FeCo-LDHs) by a hydrothermal method and then grew them in situ on the surface of short carbon fibers (SCFs), as shown in Fig. 3.10(e)[138]. Then, these LDHs/SCFs were converted into layered double oxides (LDOs) by high-temperature calcination at 500 °C for 4 h in the Ar atmosphere. Through testing and data analysis, it can be obtained that FeCo-LDHs derived composite has excellent absorption performance, accompanied by favorable impedance matching and attenuation ability.

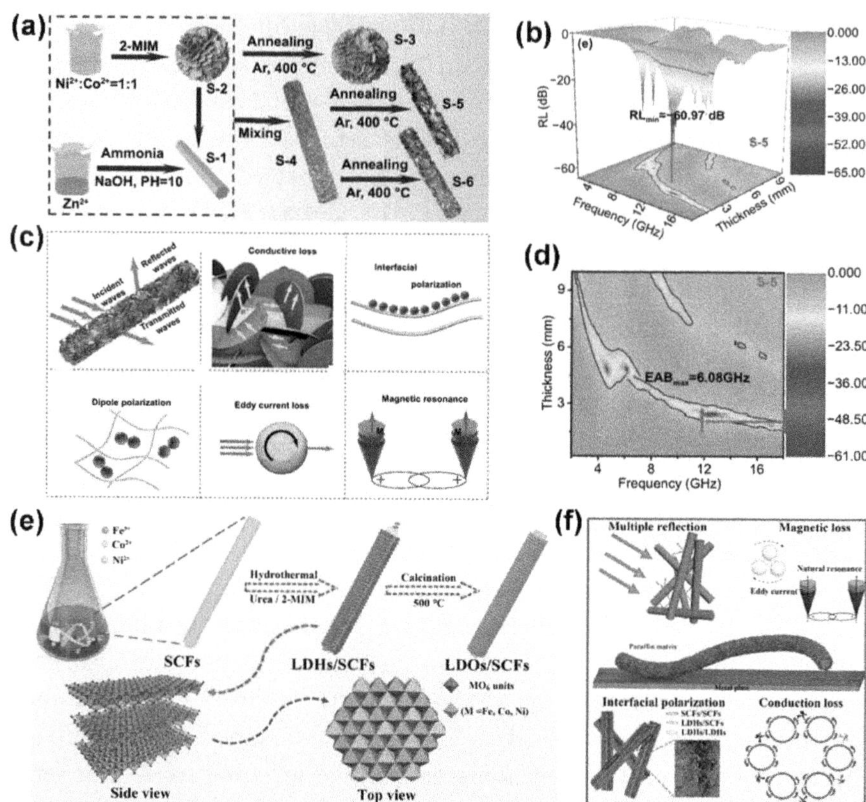

FIG. 3.10 NiCo-LDHs@ZnO nanorod and LDOs/SCF composites: (a and e) Synthesis process diagram; (b and d) 3D *RL* curve and 2D representation of the values of *RL*; (c and f) absorption mechanism. (Reproduced with permission from Ref.[137,138] Copyright 2021, Springer Nature; Copyright 2020, Elsevier.)

The FeCo-LDHs-derived composite displayed a wide EAB of 4.48 GHz at 2 mm, which may be attributed to the interaction between metal oxide shells and SCFs, leading to a moderate complex permittivity. Large-sized SCFs make it more efficient to create multiple reflections and scattering after EMWs enter the surface, resulting in more efficient dissipation. The resonance peak of μ'' is around 4 GHz and the C_0 value is stable at 7–18 GHz, demonstrating the contribution of natural resonance and eddy current losses (Fig. 3.10f). Moreover, the metal oxide coatings on SCFs generate more interfaces, providing intense interface polarization[138].

Carbonyl iron powder (CIP), as a magnetic loss type ultrafine magnetic metal powder absorbent, due to its high magnetic permeability, large saturation magnetization strength, and EM absorption performance in the low-frequency region (S, C band), is widely used in EMWs field. An example is the construction of NiFe-LDH/CIP composite in which two different morphologies of the CIP (spherical or flaky) were used. The authors successfully prepared three kinds of NiFe-LDH/CIP composites by solvothermal method: NiFe-LDH/SCIP, NiFe-LDH/FCIP, and NiFe-LDH/SFCIP[139]. The NiFe-LDH/FCIP displayed an outstanding absorption performance with an optimum RL_{min} of −96.5 dB and an EAB reaching 7.6 GHz at 8.0 mm. Indeed, when the thickness was 3 mm, the RL_{min} value can still reach up to −16.4 dB with an EAB of 7.2 GHz. However, the uniform distribution of nanoflower-like LDHs on the surface of the CIP matrix can effectively improve the dispersion of LDHs, induce stronger EM loss, and form a heterogeneous interface with a heterogeneous structure that can induce the polarization relaxation of the absorber.

In summary, LDH materials have a wide EAB but low reflection loss. Therefore, the research idea of LDHs is mainly focused on how to explore the reaction mechanism by hybridizing LDHs with other functional materials to achieve strong EM absorption capacity. From the above example, it can be concluded that the LDHs hybrid materials can further enhance the reflection loss and have a wide EAB.

3.3.3.3 TMDs as MAM

TMDs, which are MX_2 type semiconductors, where M are transition metal atoms (such as Mo or W) and X are chalcogenide atoms (such as S, Se or Te), exist in one of two structural phases, triangular prism (2H) or octahedral (1T) coordination of metal atoms (Fig. 3.11)[140]. Due to the individual layers formed by different stacking sequences, the 2H phase corresponds to the ABA stacking, where the chalcogenide atoms are located in different atom planes. These atoms occupy the same position A and overlap with each other in directions perpendicular to the layer. In contrast, the 1T phase has the ABC stacking order[140]. Compared with other traditional EMW absorption materials, TMDs have excellent properties, such as the diversity in terms of types and properties, the presence of band gaps in the band structure, the adjustability of the number of layers of the electronic structure, and the polarization state difference caused by the polarization of the energy valley. TMD materials are a 2D semiconductor electronic material with great potential, and is expected to become a key component of a new generation of optoelectronic devices, quantum information devices, and spin devices[141]. In addition, TMD materials have a large specific surface

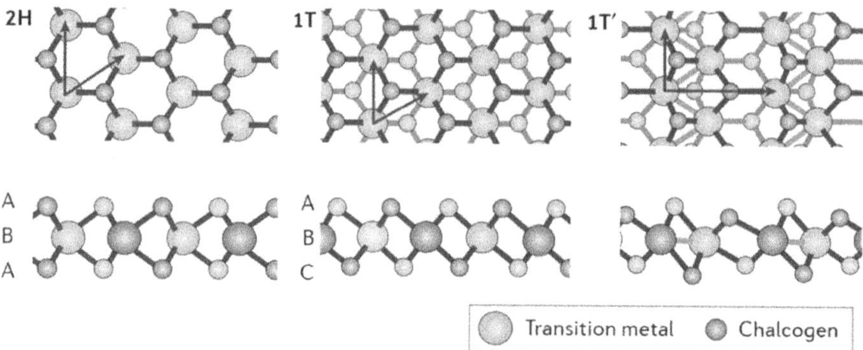

FIG. 3.11 Atomic structure of single layers of TMDCs in their trigonal prismatic (2H), distorted octahedral (1T) and dimerized (1T′) phases. (Reproduced with permission from Ref.[140] Copyright 2017, Springer Nature.)

area and unique chemical and physical properties, which make 2D metal sulfide materials a potential absorption material[142]. There are numerous synthesis methods for metal sulfides, such as solvothermal/hydrothermal approaches, gas-solid reactions, microwave-assisted synthesis and other methods.

3.3.3.4 MoS$_2$-based composites as MAM

MoS$_2$ is one of the most typical 2D transition metal chalcogenides, which consists of S–Mo–S layers stacked by van der Waals interactions[143]. In addition, MoS$_2$ can consume EMWs due to numerous defects caused by Mo and S vacancies and the introduction of multiple interface polarization by high specific surface area[144,145]. Moreover, this allows 2D MoS$_2$ nanosheets to enhance their electrical properties, fast electron transfer, and tunable conductivity. There are two ways to improve the material properties—morphological regulation and the introduction of foreign hybridization. For example, Jin et al. prepared a few-layer MoS$_2$ nanosheet by a top-down exfoliation method, which was derived from bulk MoS$_2$[146]. The data shows that the imaginary dielectric constant of the MoS$_2$ nanosheet is twice that of the bulk MoS$_2$. Moreover, the nanosheet MoS$_2$ exhibited an RL_{min} of −38.42 dB at 2.4 mm, when the filling loading was 60 wt.%, which is four times higher than that of MoS$_2$ bulk. The EAB of nanosheet MoS$_2$ can reach 4.1 GHz. The high absorption performance of the MoS$_2$ nanosheet is attributed to the vacancies between Mo–S and the excellent specific surface area. However, the high density, low charge transfer rate, and few exposed active sites of MoS$_2$ make it more challenging to prepare

EMW-absorbing materials with high performance, low thickness, and strong absorption[147,148].

The 2H-phase MoS_2 has stable semiconductor properties, but small layer spacing due to van der Waals force limitations. Therefore, the active site of unsaturated sulfur atoms exists only on the 2D layer edge. To increase the active sites of MoS_2 nanosheets, the researchers employed different strategies such as defect engineering[149], preferential exposure of active facets[150], and amorphization[151]. For instance, Fe-doped 1T/2H-MoS_2 few-layer nanosheets were fabricated by microwave-assisted hydrothermal treatment and it was found that Fe-doping of the pristine 2H-MoS_2 lattice structure can cause a gradual increment in atomic defects[152]. Fe-doped pristine 2H-MoS_2 composite has a high EMW absorption performance, i.e., RL_{min} of −60.03 dB at 5.65 mm and an EAB of up to 5.33 GHz. The excellent energy absorption of Fe-doped 1T/2H-MoS_2 nanosheets is attributed to the doping effect and phase transition, which enhances the dielectric constant and magnetic permeability, optimizes impedance matching, and improves attenuation characteristics. Wang et al. synthesized a flower-like 1T/2H-MoS_2@α-Fe_2O_3 with different amounts of α-Fe_2O_3 (40, 60, 80 mg) via hydrothermal reaction[153]. The result suggests that the flower-like 1T/2H-MoS_2@α-Fe_2O_3 has excellent EMW capabilities, leading to a perfect impedance matching, resulting in a RL_{min} value of −61.18 dB at 6.52 GHz and an EAB of 4.68 GHz at 1.83 mm.

However, it is still difficult to prepare high-performance absorbers with desirable impedance matching and attenuation characteristics under small thicknesses. Dong et al. rationally designed a MoS_2/multi-walled carbon nanotube (MWCNT) composite that can improve the accumulation of MoS_2 nanosheets and have more active sites, by adjusting the concentration of L-cysteine acid, unlocking the base surface of MoS_2, and expanding the layer spacing from 0.62 nm to 0.99 nm[154]. The obtained MoS_2/MWCNT-2 with a L-cysteine quality of 0.280 g exhibited that the optimal RL_{min} value achieved was −49.38 dB and the EAB can reach up to 4.64 GHz at 1.7 mm.

For 2D EMW absorbing materials, although a large number of results have been achieved, they are only present in graphene, MXene, transition metal sulfides, and LDHs. Therefore, further research on the development of low-cost, easy-to-synthesize other 2D materials and their composite materials is still needed in the future, and finally, EMW absorbing materials with excellent impedance matching and strong absorption capacity will be realized.

3.3.4 Three-dimensional materials

Three-dimensional nanostructures with multiple dimensional combinations (such as 0D, 1D, 2D or multiple structural units) are diverse. The 3D hierarchical structures were widely investigated due to their large surface area, good impedance matching, and diverse morphologies[155]. For instance, Chen et al. successfully designed 3D hierarchical structures with $SiO_2@$ NCSiNi@NCNT (NCNT stands for nitrogen-doped CNT) via a three-step process, as shown in Figs. 3.12(a and b), which includes an oil bath-based heat treatment, a hydrothermal reaction, and high-temperature pyrolysis[156]. The RL_{min} value reached −59.90 dB at 1.4 mm (Fig. 3.12c). The 3D hierarchical structure consists of SiO_2 spheres as the yolk material, Ni NPs, and N-doped carbon layers of amorphous $NiSiO_3$ as shell materials. The NCNT array that encapsulates the Ni NPs branches out from the yolk-shell sphere to form a 3D hierarchical structure. This 3D SiO_2@NCSiNi@NCNT structure has excellent absorption properties, which can be attributed to the conductivity loss provided by the Ni NPs and NCNTs, the improved impedance matching of SiO_2, the rich interfaces, and defects, as shown in Fig. 3.12(d). Moreover, the formation of the core-shell structure and tubular characteristics also reduce the filling ratio.

Similarly, Cheng et al. fabricated a 1D CNT-decorated 3D crucifix carbon framework embedded with $Co_7Fe_3/Co_{5.47}N$ NPs by in situ pyrolysis of the precursor ZIFZnFeCo[158]. The authors regulated the structure of the 1D CNT on the surface of the 3D crucifix carbon framework with different pyrolysis temperatures of 800, 900, and 1000 °C, and achieved superior EM absorption characteristics, with an RL_{min} of −54.0 dB at 2.25 mm under a filler loading of 30 wt%, and an EAB of 5.4 GHz under a thickness of 1.65 mm. The excellent MA performances are mainly attributed to the conductivity loss formed by the surface of the CNT and the 3D crucifix carbon framework, and the interfacial polarization and magnetic loss induced by the $Co_7Fe_3/Co_{5.47}N$ NPs. Similarly, a 3D hierarchical CNTs/VO_2 composite microsphere was prepared via an ultrasonic atomization process[159]. The 3D CNTs/VO_2 microsphere achieved an RL_{min} of −58.2 dB and an EAB of 3.94 GHz in the X-band. The great attenuation ability originates from the conductive loss caused by the highly conductive CNTs and the polarization loss caused by the heterogeneous interface formed between the VO_2 and the CNTs. The results show that this 3D hierarchical structure improves structural stability and enhances the multiple reflection and scattering paths of EMWs, promoting the energy consumption of EMWs.

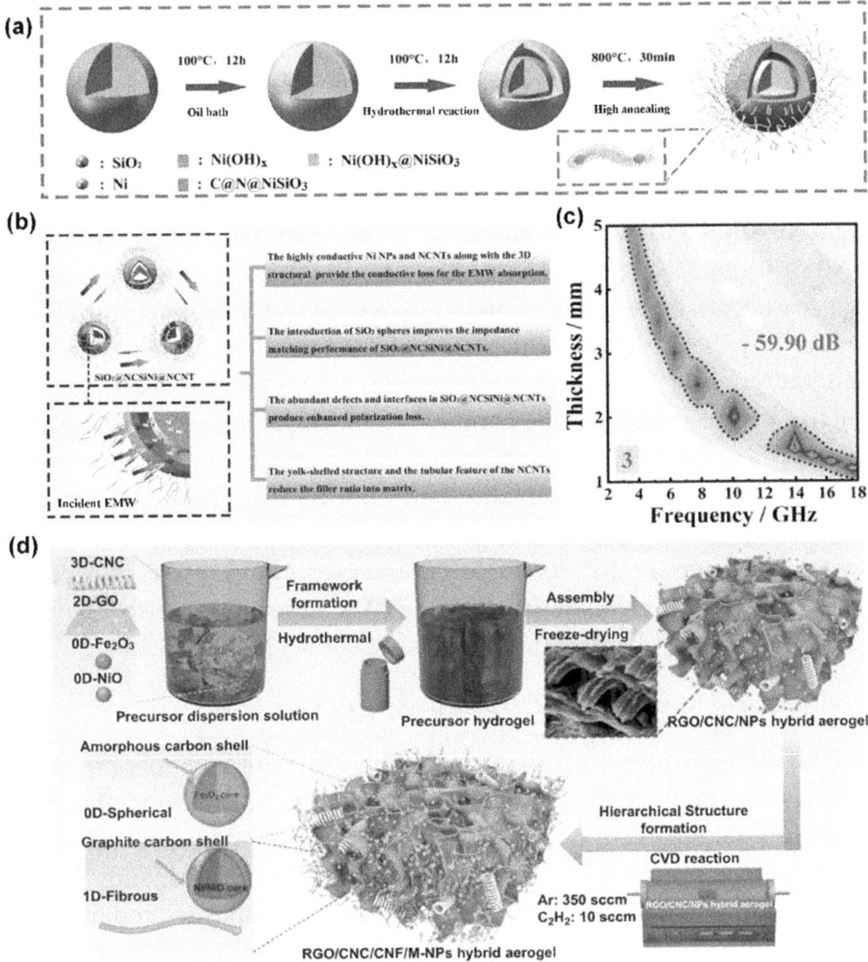

FIG. 3.12 The synthesis process (a), electromagnetic wave reaction mechanism (b) and 2D *RL* projection plots (c) of SiO₂@NCSiNi@NCNT. (Reproduced with permission from Ref.[156] Copyright 2021, Elsevier.) (d) The synthesis procedure of the rGO/CNC/CNF/M-NPs hierarchical aerogel. (Reproduced with permission from Ref.[157] Copyright 2021, Springer Nature.)

Recently, Zhao et al. designed a 3D helix (carbon nanocoils (CNCs))–2D sheet (graphene oxide (GO))–1D fiber (CNTs)–0D (Fe₂O₃/NiO) dot hierarchical aerogels with CNCs/graphene/CNFs/core-shell particles via a hydrothermal process and an in-situ CVD method[157]. In detail, high-purity CNCs were prepared via a CVD method, which used porous Fe₂O₃/SnO₂ NPs as catalysts. Subsequently, Fe₂O₃ and NiO NPs were synthesized with the same catalyst by a hydrothermal method. The rGO/CNC/M-NPs aerogels were fabricated via a hydrothermal method and a freeze-drying process. Finally, rGO/CNC/CNF/M-NPs aerogels were constructed by a CVD reaction. Surprisingly, the GCA-$M_{0.3}$-20 had a splendid EMW absorption performance with an RL_{min} of −71.5 dB at 9.5 GHz under a thickness of 2.95 mm, and an EAB of up to 4.5 GHz, covering the entire X-band. Additionally, GCA–$M_{0.2}$–10 had an RL_{min} of −55.1 dB at 13.8 GHz and an EAB of up to 5 GHz at only 1.9 mm. The EMW absorption mechanism was mainly attributed to the following aspects: the cross-polarization loss induced by the 3D CNCs with a chiral structure; the multiple reflections and scattering of EMWs promoted by the 2D rGO nanosheets; the conductive loss promoted by the conductive network built via the 1D CNFs; the interfacial polarization and dielectric loss enhanced by the heterogeneous interfaces. In addition, the Ni@CNF and the Fe_3O_4@C can also facilitate magnetic loss to optimize impedance matching. Moreover, the synergy between these different component structures results in excellent EMW absorption characteristics. The EMW absorption performance of various 2D EMW absorption materials are shown in Table 3.1.

3.4 POROUS EMW ABSORPTION MATERIALS

For EMW absorbing materials, the pores can induce dipole relaxation polarization to improve dielectric loss capacity. In addition, the pores in the crystal structure can reconstruct chemical bonds, creating crystal defects at the pore edges[160]. At present, porous materials mainly involve two types, namely MOFs-derived porous carbon composites and biomass-derived porous carbon materials. Below, we will analyze and discuss these two composite materials in EM absorption fields.

3.4.1 MOFs-derived porous composites

MOFs, as a new 3D porous material, are considered one of the most ideal precursor candidates for MAM due to their tunable structure, high porosity, uniform pore size, and large specific surface area. However, the pristine MOFs have weak conductivity and inferior stability, which cannot provide desirable dielectric loss, limiting its application in EMWs. The design of MOFs materials in EMW applications is mainly through the following aspects: (1) Single metal MOFs as precursors; (2) Bimetallic MOFs as precursors; (3) MOFs-derived carbon composites; (4) MOFs

TABLE 3.1 The electromagnetic absorption properties of the various 2D absorbers.

MAM type	Matrix and filler weight (wt.%)	Synthesis process	EAB (GHz)	RL_{min} (dB)	Refs.
MXenes-based materials	Ti_3C_2	HF etching	12.4–18 GHz	−11.0 dB@1.4 mm	104
	graphite-ring-stacked CNT/ MXene (39 wt%)	HCl/LiF/CVD	4.16 GHz@1.36/ 1.16 mm	−47.28 dB@1.36 mm	106
	CNT/MXene hybrid-450(40 wt%)	HCl/LiF/CVD	2.16 GHz	−52.56 dB@2.5 mm	107
	nacre-like CCM (15 wt%)	HF/vacuum-assisted filtration.	7.12 GHz@3.6 mm	−42.2 dB@8.82 GHz	108
	loofah-like CCM (15 wt%)		7.32 GHz@2.5 mm	−63.8 dB@8.91 GHz	
	core-shell MXene/N-doped C (30 wt%)	HCl/LiF/in-situ polymerization/and carbonization	5.0 GHz@1.72mm (13.0–18.0 GHz)	−13.6 dB@1.62 mm	109
	F-HCNTs@$Ti_3C_2T_x$ MXene-3.0h	a freeze drying-assisted electrostatic spinning technique	2.2 GHz (9.7–11.9 GHz)	−35.5 dB@10.6 GHz	112
	$Ti_3C_2T_x$ MXene/PANI (10 wt%)	electrostatic self-assembly process	6.0 GHz (8.0–14.0 GHz)	−60.6 dB@2.6 mm	114
	$Ti_3C_2T_x$ @PPy-2 (10 wt%)	ice-bath	5.14 GHz (6.44–11.58 GHz)	−49.5 dB@3.6 mm	115
	MXene@SiC$_{NWs}$@Co/C hybrids	in situ growth/ a carbonization process/ an electrostatic self-assembly strategy	6.2 GHz (11.8–18.0 GHz)	−76.5 dB@3.9 mm	117
	3D sandwich-like M-$Ti_3C_2T_x$@ SiO_2@C hybrids (40 wt%)	ion intercalation technology	4.12 GHz @1.5 mm	−43.97 dB@4.5 mm	118
	3D MXene/MoS_2 fold microspheres	a one-step method	4.4 GHz@1.6 mm	−51.21 dB@2.5 mm	120
	2D/2D MoS_2/$Ti_3C_2T_x$ heterostructure	microwave-assisted hydrothermal method	3.6 GHz	−41.5 dB@12.2 GHz	121
	$Ti_3C_2T_x$ MXene/ $Ni_{0.6}Zn_{0.4}Fe_2O_4$-2	etching process @ in-situ hydrothermal process	4.74 GHz@ 1.6 mm	-66.2 dB@15.2 GHz	122
	TiO_2/Co/C (T/Co/C) composite	in-situ solvothermal reaction and pyrolysis	5.48 GHz @1.7 mm	−50.45 dB@3.0 mm	123

(Continued)

TABLE 3.1 (Continued)

MAM type	Matrix and filler weight (wt.%)	Synthesis process	EAB (GHz)	RL_{min} (dB)	Refs.
LDHs-based composites	50 NiAl-LDH/CNFs@15 wt%	ALD method	4.80GHz@1.6mm	−55.65 dB@14.80 GHz	135
	C@NiCo-LDHs@1D Ni chain@30 wt%	a simple solvothermal/freeze-drying method	6.40 GHz@2.5 mm (10.6–17.0 GHz)	−57.40 dB@2.5 mm	136
	NiCo@C@ZnO@33.3wt%	hydrothermal method/annealing process	6.08 GHz@2.0 mm	−60.97 dB@2.3 mm	137
	LDOs/SCFs@30wt%	hydrothermal method/calcination treatment	4.48 GHz@2.0 mm	−29.9 dB@2 mm	138
	NiFe-LDH/FCIP@30wt%	hydrothermal method	7.6 GHz@8.0 mm	−96.5 dB@5 mm	139
MoS$_2$-based composites	a few-layer MoS$_2$ nanosheet (60 wt%)	a top-down exfoliation method	4.10 GHz (9.6–13.76 GHz)	−38.42 dB@2.4 mm	146
	Fe-doped 1T/2H-MoS$_2$ few-layer nanosheets-2(30 wt%)	microwave-assisted hydrothermal	/	−60.03 dB@5.65 mm	152
	flower-like 1T/2H-MoS$_2$@α-Fe$_2$O$_3$(60 wt%)	hydrothermal reaction	4.68 GHz@1.82mm (13.31−18GHz)	−61.18 dB@3.94 mm	153
	MoS$_2$/MWCNT composite(20wt%)	solvothermal reaction/freeze-dried	4.64 GHz@1.7 mm	−49.38 dB@14.80 mm	154
Carbon-based composites	SiO$_2$@NCSiNi@NCNT	include oil bath/hydrothermal reaction/ high temperature pyrolysis	/	−59.90 dB@1.4 mm	156
	GCA-M$_{0.3}$-20	CVD reaction	4.5 GHz (covering the entire X-band)	−71.5 dB@2.95 mm	157
	1D CNT-decorated 3D crucifix carbon framework with Co$_7$Fe$_3$/Co$_{5.47}$N NPs	in situ pyrolysis (30 wt%)	5.4 GHz @1.65 mm	−54.0 dB@2.25 mm	158
	CNTs/VO$_2$ composite microsphere	an ultrasonic atomization process	3.94 GHz (X-band)	−58.2 dB	159

composite materials for constructing core-shell structure, hollow structure, and yolk-shell structure.

3.4.1.1 Single metal MOFs as precursors

Typical single metal MOFs include zeolitic imidazolate framework (ZIF-67, ZIF-8)[161], MOF-74[162], UiO-66[163] and MILs (MIL-101, MIL-88, etc.)[164,165]. It is well known that the presence of magnetic metal elements in MOFs is crucial for MA systems. MOFs with magnetic metal elements have conductive and EM properties that optimize impedance matching and microwave attenuation. Iron (Fe), cobalt (Co), nickel (Ni), and their alloys demonstrate good magnetic properties, and MOFs with these elements have been studied extensively in the EMW absorption field. For instance, monometallic MOFs have been used as precursors to prepared EMW absorption materials mainly through pyrolysis at different temperatures, and the resulting composites can induce and/or enhance magnetic loss and dielectric loss.

Recently, a special hexagonal biconical Fe/Fe_3C embedded in the N-doped porous carbon layer ($Fe/Fe_3C@NC$) was constructed using a Fe-MOF as the precursor, via high-temperature pyrolysis, as shown in Fig. 3.13(a)[166]. The $Fe/Fe_3C@NC$-700 displayed the highest RL_{min} of −70.8 dB at a thickness of 2.5 mm, and an EAB of 5.15 GHz at a matching thickness of 1.7 mm, as shown in Fig. 3.13(b). The authors first adjusted the synthesis temperature of the Fe-MOF to achieve stable MOFs and vary their morphology. Then, the graphitization degree of the MOF derivatives was adjusted by using different pyrolysis temperatures to effectively control the dielectric properties. MA capacity was attributed to the synergistic effect of multiple mechanisms, including multiple scattering, dielectric loss, magnetic loss, and impedance matching (Fig. 3.13c).

In 2021, a Ni/CNTs composite was synthesized by in-situ pyrolysis of a Ni-MOF, as shown in Fig. 3.13(d). The Ni/CN-700 composite synthesized under a 700 ºC calcination displayed an RL_{min} of −65 dB with an EAB of about 4.6 GHz at 1.9 mm (Fig. 3.13e). When Ni^{2+} in the Ni-MOF is converted to metal Ni, it can promote the formation of CNTs during heat treatment. The excellent MA capacity of the Ni/CNT composites was mainly attributed to the synergistic effect of CNTs and Ni NPs, as illustrated in Fig. 3.13(f)[167]. Yu et al. prepared a Fe/C-900 composite from a Fe-MOF via an in-situ pyrolysis process, as shown in Fig. 3.13(g). The composite exhibited an RL_{min} of −37.6 dB with a matching thickness of 1.5 mm and an EAB of 4.4 GHz[168]. The superior EMW absorption performance of the

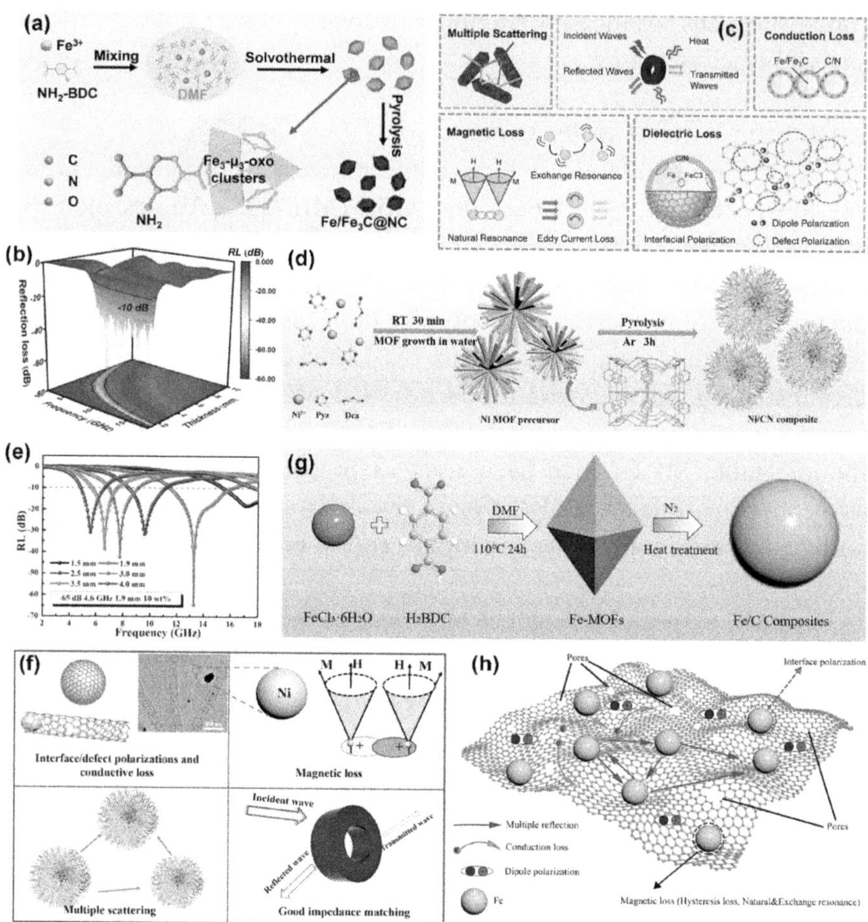

FIG. 3.13   The schematic synthesis (a), *RL* plots (b) and electromagnetic wave absorption mechanism (c) of Fe/Fe$_3$C@NC composites. (Reproduced with permission from Ref.[166] Copyright 2023, Elsevier.) The synthesis process (d), *RL*$_{min}$ values (e) and electromagnetic absorption mechanism (f) of Ni/CN composites. (Reproduced with permission from Ref.[167] Copyright 2021, Elsevier.) The synthesis process (g) and electromagnetic absorption mechanism (h) of Fe/C composites. (Reproduced with permission from Ref.[168] Copyright 2022, Elsevier.)

Fe/C-900 composite was attributed to the synergistic effect between the dielectric loss of the porous carbon matrix and the magnetic loss of the metal iron, which optimize impedance matching and thus improve the MA performance, as shown in Fig. 3.13(h).

In the field of EMW absorption, monometallic MOFs are usually converted to metal/metal carbide-based porous carbon composites by high-temperature pyrolysis based on the fact that MOFs are 3D porous structures composed of metal ions and organic ligands. Because of this, pyrolyzed monometallic MOFs can facilitate dielectric loss and magnetic loss.

3.4.1.2 Single nonmagnetic-metal-based MOFs as precursors

There are also non-magnetic monometallic MOFs such as ZIF-8 (Zn-MOF)[161], UiO-66 (Zr-MOF)[163] and Ti-MOF. These non-magnetic single-metal MOFs (such as Zn/Ti/Zr/Ce-MOFs) are hindered in EMW absorption applications because they do not have magnetic elements. In the EMW application of single-metal non-magnetic MOFs, the selection of precursors and the design of experiments are primarily based on the adjustment of dielectric constant and microstructure due to the absence of a magnetic loss medium. Therefore, introducing magnetic components into the MOFs having non-magnetic metals and magnetic metals as mixing centers to form polymetallic MOFs as precursors have attracted enormous attention in the field of EMW absorption[169]. Zinc-based MOFs have a relatively low melting point (420 °C) and boiling point (908 °C), and the Zn element in Zn-based MOFs can be converted to ZnO when calcined above 550 °C[170]. Wang et al. synthesized N-doped highly porous carbon composites from a Zn-based MOF using a thermal treatment with three different temperatures (NPC-700/800/900°C). The NPC-800 displayed an RL_{min} of −39.7 dB at 4 mm with an EAB covering the whole X band (8–12 GHz)[170]. The authors concluded that the porous structure, graphitization degree, and graphite-N level are significant factors that can improve dielectric loss capacity and MA performance.

Since Zn-MOF-based materials do not involve an EM loss mechanism, they are greatly limited in EMW applications. However, the introduction of magnetic NPs can enhance the performance of non-magnetic metal-based MOFs materials. For instance, Liu et al.[163] prepared cobalt-decorated porous ZrO_2/C hybrid octahedrons by pyrolysis of $Co(NO_3)_2$ impregnated NH_2-UIO-66 and the $Co/ZrO_2/C$ nanocomposite, as shown in Fig. 3.14(a). As can be clearly observed in Fig. 3.14(b), the Co NPs embedded in the ZrO_2/C can provide magnetic loss to improve the impendence matching and attenuation capacity. The composite exhibited the strongest absorption

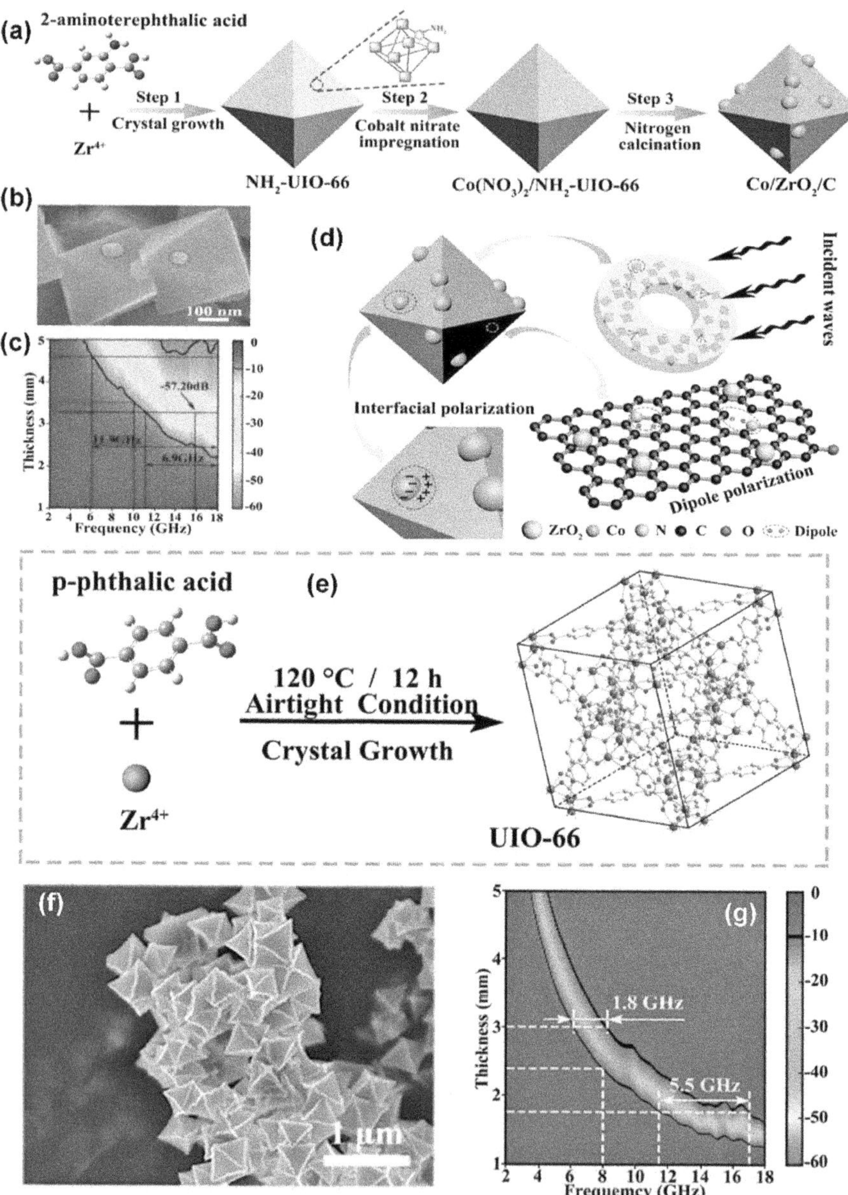

FIG. 3.14 The schematic synthetic route (a), SEM image (b), 2D absorption platforms (c) of Co/ZrO$_2$/C nanocomposites and their corresponding EMW absorption mechanism (d). (Reproduced with permission from Ref.[163] Copyright 2019, American Chemical Society.) The synthetic process of UIO-66 (e), the SEM image (f) and the corresponding 2D absorption platforms (g) of ZrO$_2$/C-800. (Reproduced with permission from Ref.[171] Copyright 2020, Royal Society of Chemistry.)

of −57.2 dB at 15.8 GHz with a matching thickness of 3.3 mm. Meanwhile, the EAB reached 11.9 GHz (6.1−18 GHz), covering 74.4% of the whole measured bandwidth (Fig. 3.14c). The excellent performance of the Co/ZrO$_2$/C can be attributed to strong interface polarization and suitable impedance matching, resulting from the synergistic effects of different components, as shown in Fig. 3.14(d). In Figs. 3.14(e and f), MOF-derived ZrO$_2$/C octahedra were prepared from UIO-66 via high-temperature pyrolysis (700−900 °C)[171]. The MOF-derived ZrO$_2$/C-800 octahedra displayed an *RL* value of −58.7 dB at 16.8 GHz with a thickness of 1.5 mm and EAB covering 91.3% of the measured frequency (3.4−18.0 GHz) within the thickness range of 1.0−5.0 mm, as shown in Fig. 3.14(g). The befitting carbonization temperature help enhance the EMW absorption performance of the ZrO$_2$/C-800, which is the main factor contributing to strong attenuation ability and improved impedance matching.

3.4.1.3 Mixed-metal MOFs as precursors

Herein, mixed-metal MOFs include bimetallic and polymetallic MOFs. For single metal-based MOFs, the type and number of metal ions are relatively fixed, so only the permeability and permittivity can be adjusted by changing the pyrolysis temperature. For bimetallic MOFs, EM parameters can be efficaciously adjusted by changing the proportion of different metals, thereby improving the EMW absorption performance. For instance, by controlling the Ni/Co ratio, flower-like MOFs-derived N-doped CNTs encapsulated magnetic NiCo composites (NCNT/NiCo/C) were synthesized through a hydrothermal reaction followed by thermal CVD with melamine-assisted catalyst, as shown in Fig. 3.15(a) [172]. As shown in Fig. 3.15(c), the NCNT/Ni$_1$Co$_1$/C composite exhibits an excellent EMW absorption performance with an RL_{min} of −66.1 dB and an EAB of 4.64 GHz with a thickness of 1.5 mm. Importantly, the derived N-doped 3D nanoflower CNTs firmly connected contiguous nanosheets to build a specific 3D conductive network, which can effectively expedite charge transfer and improve conduction loss (Fig. 3.15b). Likewise, Tu et al. prepared 3D flower-shape Co/Cu bimetallic nanocomposites by thermolysis of MOFs precursors and tuning the ratio of Co/Cu (Fig. 3.15d)[173]. When the molar ratio of Co:Cu salts was 4:1, the CoCu-CNF-10 displayed a minimum *RL* of −51.7 dB at 12.6 GHz with a thickness of 2.7 mm. Meanwhile, as shown in Fig. 3.15(e), the multi-component and structure of the nanoflower CoCu-CNF-X composites promote the entry of EMWs into the material, which

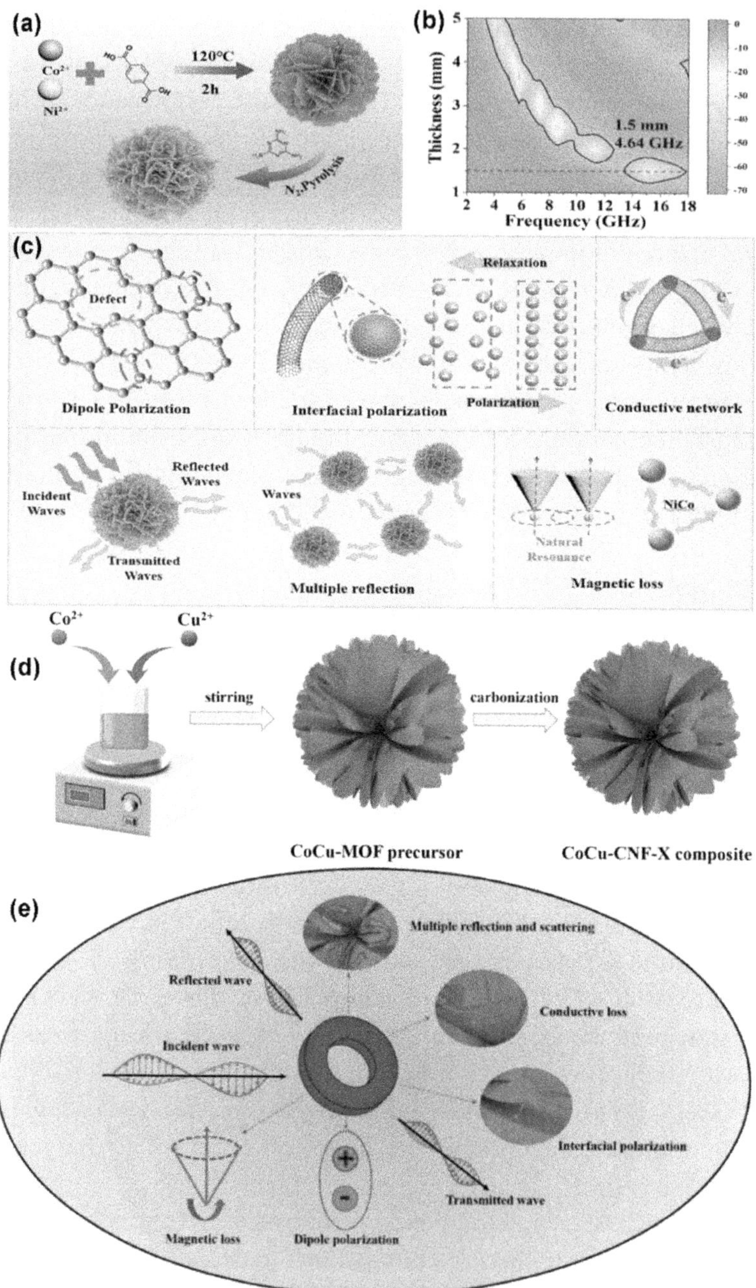

FIG. 3.15 The schematic diagram of the preparation process (a), 2D *RL* projection plots (b) and EMW absorption mechanism diagram (c) of NCNT/NiCo/C composite. (Reproduced with permission from Ref.[172] Copyright 2023, Elsevier.) The schematic preparation process (d) and EMW absorption mechanism diagram (e) of CoCu-CNF-X composite. (Reproduced with permission from Ref.[173] Copyright 2023, Elsevier.)

is conducive to the impedance matching properties. The 3D structure and large BET-specific surface area are conducive to enhancing the multiple reflection and scattering of EMWs.

The growth of MOFs of different shapes is affected by metal sources, organic ligands (bond length, bond angle, chirality, etc.), synthesis conditions (such as pH, temperature, solvent, metal source/organic ligand molar ratio), and template agents. It is due to the fact that MOFs are constructed by the principle of coordination chemistry, and some chemical and physical stimuli will produce different effects in the crystal structure, thereby synthesizing MOFs with different morphologies. However, the major shortcoming of MOFs derivatives is inferior conductivity and weak EM properties. For example, in 2022, a series of MOFs-derived metal/carbon-based materials with different morphologies were constructed by adding stencils, changing the metal source and then pyrolyzing under a high-temperature, as shown in Fig. 3.16(a)[174].

These morphologies show the transition from six-pointed star, flower-shape, cubic-shape to random stone-shape, corresponding to the absence of metal sources and different proportions, respectively. It is noteworthy that CNTs grew on the sample surface under the catalysis of the metal source Co to construct a 3D conductive network and enhance the dielectric loss in the high-temperature pyrolysis process. Importantly, the Co@ZnO@NC-a, Co@ZnO@NC-b and Co@NC composites had an RL_{min} of −61.9, −32.6, and −53.9 dB, respectively, and the EAB was 5.5, 5.2, and 5.2 GHz, respectively. The EMW absorption mechanism of Co@ZnO@NC-a mainly includes two aspects, one is that in the interior of the material, CNTs act as a bridge to make Co@ZnO@NC-a monomers interconnected, which can not only generate self-polarization to enhance dielectric loss, but also promote the construction of 3D conductive networks and enhance the EMW reflection and scattering path. Secondly, the nitrogen-doped carbon (NC) layer acts as a connector, and the Co, ZnO NPs act as nodes, which are conducive to the mutual transport of electrons and increases conductive loss[174].

He et al. successfully prepared hierarchical nest-like CoFe@C composites, by first synthesizing the CoFe-MOF-74 *via* adjusting the molar ratio of Co/Fe (1:0, 3:1 and 2:2) and then pyrolysis at 800 °C in Ar atmosphere (Fig. 3.16b)[175]. When the molar ratio of Co:Fe was 3:1, the hierarchical nest-like CoFe-MOF-74 nanorods were constructed. The CoFe@C composites displayed an RL_{min} of −61.8 dB at 12.7 GHz with a thickness

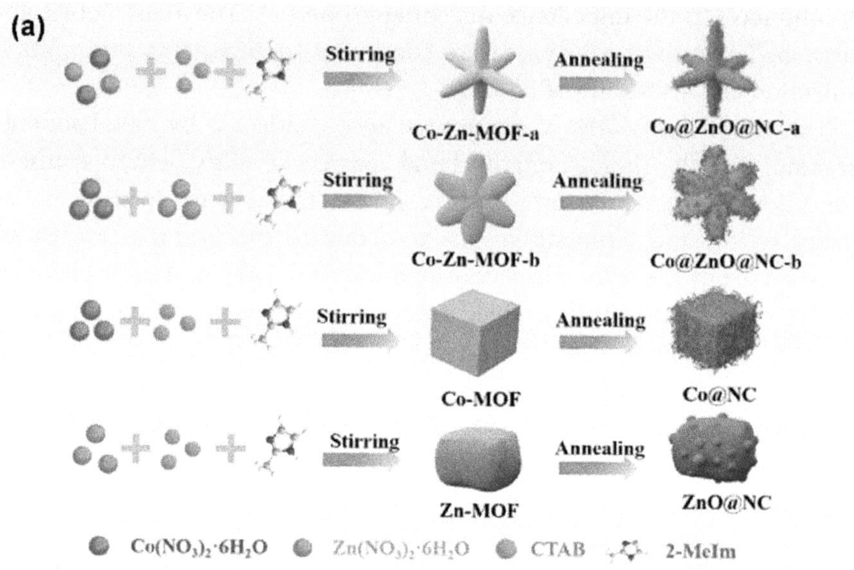

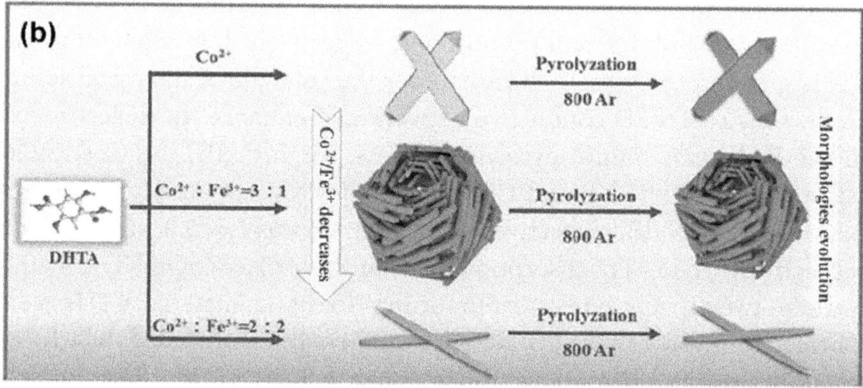

FIG. 3.16 Synthesis and formation mechanism of Co@ZnO@NC-a, Co@ZnO@NC-b, Co@NC and ZnO@NC (a). (Reproduced with permission from Ref.[174] Copyright 2022, Elsevier.) (b) Schematic illustration of the preparation process of hierarchical nest-like structures of CoFe@C composites. (Reproduced with permission from Ref.[175] Copyright 2020, Elsevier.)

of 2.8 mm and an EAB of 9.2 GHz (8.8–18.0 GHz). CoFe@C composites' unique layered nest structure improved impedance matching, the formation of multi-scale pore structure, and multiple reflection and scattering, all of which help improve their MA performance.

The inferior impedance matching of MOF-derived MAM is caused by sufficient permittivity and insufficient permeability. Liu et al.

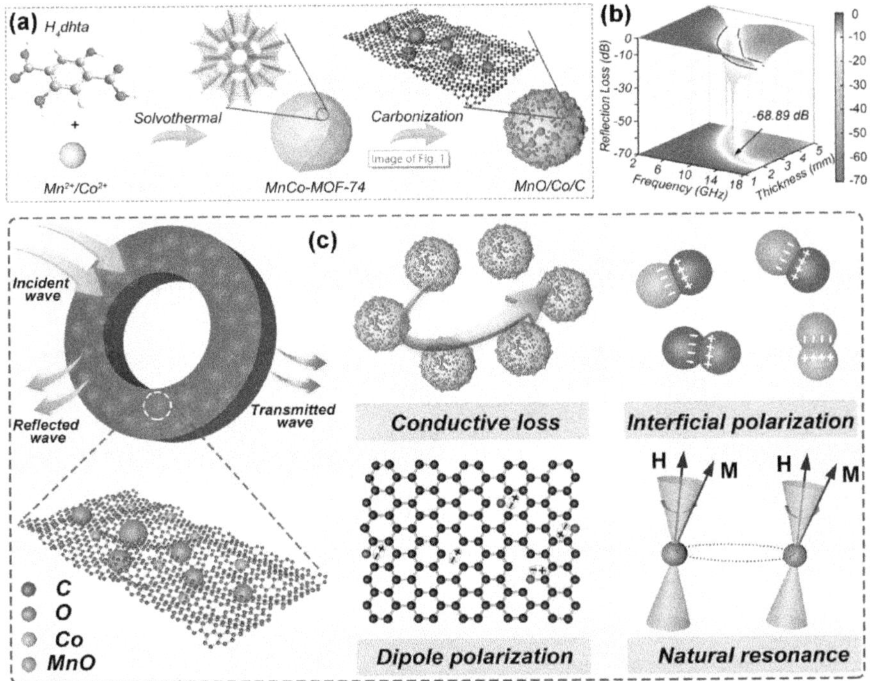

FIG. 3.17 Schematic diagram of the synthesis process (a), 3D *RL* plots (b) and the diagram of the reaction mechanism (c) of MnO/Co/C materials. (Reproduced with permission from Ref.[176] Copyright 2022, Elsevier.)

successfully synthesized MOF-74-derived nanocomposites (MnO/Co/C) via hydrothermal and carbonization processes (Fig. 3.17a). The MnO/Co/C nanocomposites displayed an RL_{min} of −68.89 dB at the thickness of 2.64 mm and an EAB of 5.3 GHz at the thickness of 2.3 mm (Fig. 3.17b), which was attributed to the synergistic effect of dielectric MnO NPs, magnetic Co NPs, carbon matrix, and the porous structure derived from the MOFs, as shown in Fig. 3.17(c)[176]. In a similar work conducted by Liu et al., 3D porous MOF-derived Fe/C/carbon foam was successfully fabricated using carbon foam as a template and Prussian Blue (PB) as a precursor. The Fe/C/carbon foam exhibited an RL_{min} of −66.7 dB and an EAB of 6.34 GHz. Through mechanism analysis, PB-derived Fe/C nano-cubes can improve impedance matching ability, and the 3D network structure of Fe/C/CF composites is conducive to the multiple scattering of microwaves, which promote the absorption performance[177].

Liu et al. successfully fabricated a bamboo-like CNT connected carbon nanorods (FeNi@CNT/CNRs) by pyrolyzing melamine and

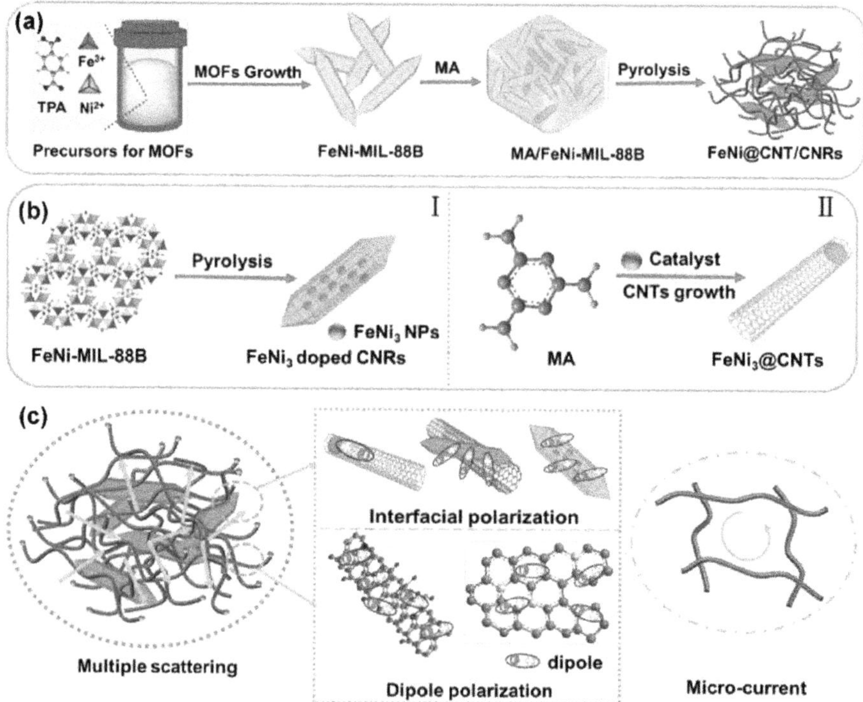

FIG. 3.18 The schematic illustration of the synthesis process of FeNi@CNT/CNRs (a), FeNi₃/CNTs (b) and their electromagnetic absorption mechanisms (c). (Reproduced with permission from Ref.[178] Copyright 2019, American Chemical Society.)

FeNiMIL-88B in one step. The resulting product demonstrated a strong absorption of −47.0 dB at 2.3 mm and an EAB of 4.5 GHz at 1.6 mm in thickness, as illustrated in Fig. 3.18(a)[178]. As shown in Fig. 3.18(b), the FeNi₃ NPs obtained by pyrolysis of FeNi MIL-88B can act as catalysts to promote the conversion of melamine to CNTs. Accordingly, the excellent absorption performance of FeNi@CNT/CNRs was mainly attributed to the synergistic effect of 3D interconnected CNT networks, MOFs-derived M/C on multiple transmission paths, impedance matching, dielectric losses, multi-polarization, and microcurrent, as shown in Fig. 3.18(c). In 2022, the authors used 1,3,5-trimethylbenzene (TMB) and dopamine (DA) to regulate the morphology of bimetallic MIL-53 such that MIL-53 changed from rod to spindle[179]. Subsequently, the spindle-like MIL-53 was modulated and pyrolyzed to successfully synthesize FeCo/C composites with splendid absorption properties. The FeCo/C composites demonstrated an RL_{min} of −78.6 dB with an EAB of 5.1 GHz at 2.6 mm thickness[179]. The carbonization of spindle-like MIL-53 caused carbon component defects in the functional groups of polydopamine and TMB, further enhancing the dipole polarization of FeCo/C composites. Similarly, the high specific surface area increased the interfacial polarization strength of FeCo/C composites.

For mixed-metal MOFs, there are more diverse design ideas. For example, researchers can design the ratio of different metal salts to obtain MOFs with various morphologies, and then calcinate MOFs with different morphologies at high temperatures. Regardless of whether it is a mono-metal or multi-metal MOF, there are a variety of design ideas, such as in situ etching, N, P and S-heteroatom doping, introducing various carbon materials such as CNT, MXene, graphene and LDHs, and their combination. Therefore, rational design of metal-organic framework materials is necessary to improve the absorption performance of EMWs.

3.4.2 Biomass-derived porous carbon materials

Biomass materials have attracted attention in MAMs applications due to their low cost and high dielectric loss, as well as ultra-light, green, and pollution-free characteristics. Numerous biomass materials such as walnut shell[180], spinach stem[181], wood[182], rapeseed flower[183], bamboo[184], peanut shell[185] and apricot shells[186] have been directly used for MA. The application of biomass materials in the field of EMW absorption is mainly achieved by converting them into porous carbon frame materials through high-temperature pyrolysis under inert gas. It is worth noting that the intrinsic structures of biomass sources are different, and the composition and structure after carbonization are also different[187]. It was found that biomass sources contain abundant heteroatomic elements (N, P, S, etc.). After carbonization, heteroatom-doped biomass porous carbon materials can be formed. These heteroatom doping can effectively change the electronic structure or dipole relaxation strength, thereby affecting the absorption performance of EMW.

The preparation process of biomass-derived carbon-based composite absorbing materials is relatively simple, which usually involves pyrolysis of biomass precursors at high temperatures under inert gas protection, an oxygen-free environment or a reducing atmosphere such as H_2. In detail, high-temperature pyrolysis can be divided into three stages. The first stage involves mainly the elimination of water and the decomposition of groups such as COH, COC, CH, etc. Mass decreases in this stage; In the second stage, the separation of hydrocarbons from carbide products leads to the formation of a porous structure, resulting in a constant shape and a decrease in volume. The third stage involves a heat preservation process, during which the biomass precursor is fully converted into biomass carbon material[188].

3.4.2.1 Activation of biomass-derived carbon

The biomass carbon obtained by the simple and direct method (i.e., high temperature calcination and solvothermal carbonization) has poor porosity and limited pore volume, which may cause weak dipole relaxation loss behavior. Therefore, in order to better improve porosity and pore volume, activation of biomass-derived materials is worth studying. The activation methods of biomass derived materials include physical activation, chemical activation, and other methods.

Physical activation is achieved by the skeletal etching of biomass derived materials under oxidizing gases (such as air, CO_2, steam and a mixture of these gases). CO_2 gas is most commonly used for the physical activation, because it is easy to handle and can effectively etch the biomass derived materials. The physical activation is generally conducted at a high temperature of 600–1200 °C. For example, date petioles, a source of biomass lignocellulose, were activated by pyrolysis at a high temperatures of 1000 °C under the flow of N_2 and then treated at different temperatures (750, 850 and 900 °C) under the flowing gas of CO_2. The specific surface area of the activated sample increased from 225 m^2 g^{-1} to 546 m^2 g^{-1}, and the micropore volume increased from 0.09 cm^3 g^{-1} to 0.23 cm^3 g^{-1} [189]. Note, it typically takes a longer time to achieve a highly porous structure by physical activation because the relatively large dimension of the CO_2 molecule restricts its entry into microspores[190].

Chemical activation, one of the most commonly used activation methods, involves mixing of chemical activator with matrix carbon, then etching the matrix carbon in a specific atmosphere by oxidation and dehydrogenation reactions, and then cleaning with acid or water to obtain biomass-derived porous carbon. In particular, it involves mixing the biomass material with the activator in a certain proportion, through grinding or impregnation, and pyrolysis in a protective gas at high temperature.

Commonly used activators are usually divided into three types, which are acids (HCl, H_3PO_4, H_2SO_4, HNO_3), alkalis (KOH, NaOH), and salts (NaSnO$_3$, ZnCl$_2$, K$_2$CO$_3$ and Na$_2$CO$_3$)[191]. Alkali is the most commonly used activator, which typically requires a lower activation temperature[192]. When KOH is mixed with biomass carbon, the biomass carbon reacts with KOH to form K_2CO_3 and K, and then K continues to react with the remaining KOH to form K_2O during the activation. When the pyrolysis temperature exceeds 700 °C, K_2CO_3 and K_2O further react with carbon to generate K and CO. CO plays an important role in generating porosity[19]. What's more, the ratio of activators to biomass carbon is critical. For instance, Zhao et al. fabricated 3D lamellar skeletal network porous carbons by controlling the ratio of KOH to the hull of water chestnut[193]. When the carbonization temperature was 600 °C, the 3D lamellar skeletal network porous carbon had an RL_{min} of −60.76 dB at 2.97 mm and an EAB of 6.0 GHz at 1.9 mm with a filling ratio of 35 wt%, covering the whole Ku band. With the increase of KOH content, the specific surface area of 3D lamellar skeletal network porous carbon increased from 123.46 m^2 g^{-1} to 926.15 m^2 g^{-1}, which was higher than other composites without added KOH content (123.46 m^2 g^{-1}). The data show that the KOH amount has an important effect on the specific surface area and porosity of the resulting biomass-derived porous carbon materials.

3.4.2.2 Biomass-derived carbon as MAM

In order to improve the MA performance of biomass-derived porous carbon materials, researchers usually consider two strategies, one is to introduce heteroatom, and the other is to introduce magnetic metal materials, metal oxides, and/or conductive polymers. The functions of different doped heteroatoms on carbon materials are different, such as the doping of B atoms can change the electron distribution of the carbon materials, the doping of N atoms can change the inherent electrical properties and hydrophilicity, and the doping of O atoms can stabilize the oxygen functional groups of the carbon materials. Moreover, the doped heteroatoms can also be introduced from external sources, such as through the introduction of auxiliary agents containing heteroatoms in the pyrolysis process[194]. For example, Yu et al. fabricated eggshell membrane-derived Co–Co$_x$S$_y$–Ni/ N, S-codoped carbon composites by a thermal treatment in an atmosphere of N_2. The synthesis process is shown in Fig. 3.19(a)[195]. The multiple

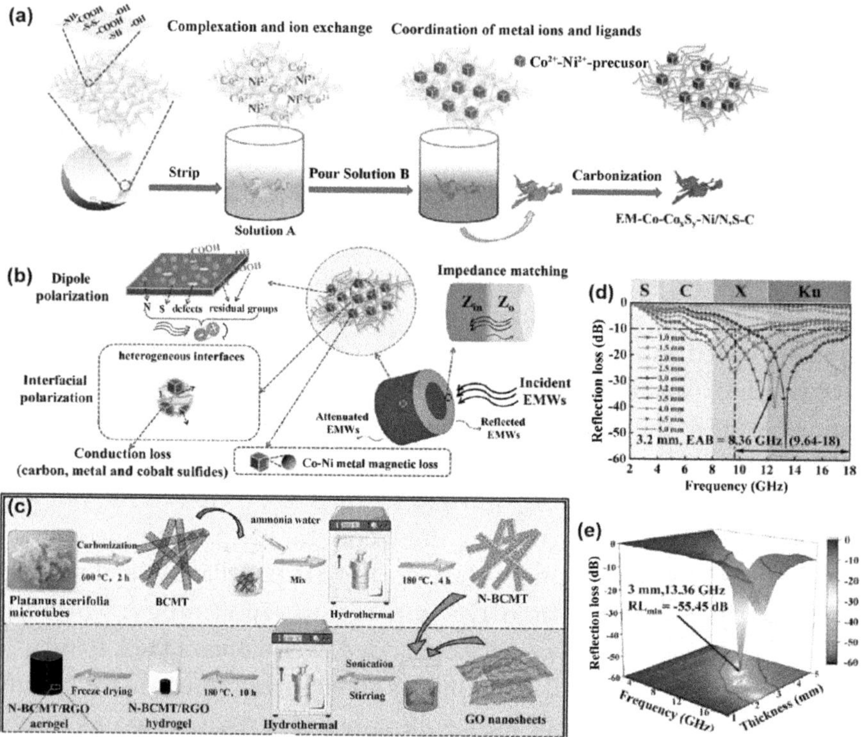

FIG. 3.19 The synthesis process (a) and schematic diagram of electromagnetic wave mechanism (b) of EM-Co-Co$_x$S$_y$-Ni/N, S-C composites. (Reproduced with permission from Ref.[195] Copyright 2022, Elsevier.) The preparation process (c) and RL_{min} curves (d, e) of N-BCMT/rGO aerogel. (Reproduced with permission from Ref.[196] Copyright 2023, Elsevier.)

loss mechanism and good impedance matching achieved by adjusting the molar ratio between the metal salts and ligands are the main reasons for the excellent EMW absorption performance (Fig. 3.19b). Carbon, nitrogen and sulfides were obtained from the eggshell membrane itself, rather than from external sources. The EM-Co-Co$_x$S$_y$-Ni/N, S-C composites exhibited an excellent absorption performance with an RL_{min} of −48.5 dB at 10.1 GHz with 3.5 mm and an EAB of 7.4 GHz (10.3–17.7 GHz) at 3 mm[195]. Recently, Kong et al. synthesized an ultralight N-doped Platanus acerifolia biomass carbon microtubes (BCMTs)/rGO composite aerogel via carbonation and freeze drying, as in Fig. 3.19(c)[196]. When the low filler loading was 2 wt%, the N-BCMT/rGO aerogel displayed an RL_{min} of −55.45 at 13.36 GHz with 3.0 mm and an EAB reaching 8.36 GHz (9.64–18 GHz), as shown in Figs. 3.19(d and e).

In summary, although biomass-derived porous carbon composites can be prepared by chemical activation to produce composites with large specific surface area and high porosity, there are still some issues in the activation process, such as the complex washing process to remove various impurities and the difficulty in adjusting pore parameters. Many studies have shown that porous structure and porosity can enhance EMW absorption. Therefore, it is essential to optimize the porous structure. More importantly, heteroatom doping can enhance defects and cause defect polarization. Enhancing the multiple loss mechanism is still the main strategy for the design of biomass-derived porous carbon composites. In the field of biomass carbon materials, it is significant to investigate the relationship between porous structure and absorption performance or the relationship between multi-component synergy. The EMW absorption performance of various porous carbon materials are shown in Table 3.2.

3.5 HOLLOW STRUCTURAL EMW ABSORPTION MATERIALS

3.5.1 MOFs-derived hollow materials

A hollow structure, defined as a solid structure, consists of a void and one or more different shells[19]. The relatively low density of hollow nanomaterials is conducive to the preparation of lightweight EMW absorbing materials with excellent impedance matching and absorption capacity. For example, Chu et al. synthesized solid and hollow carbon fibers with different diameters via various spinning techniques and investigated the influence of diameter and hollow structure on EM parameters[197]. The results show that the hollow structure can reduce the increasing rate of the imaginary part, which is beneficial to impedance matching. As is known, dielectric materials themselves often have high conductivity (such as graphene, carbon fiber, and MXene, etc.), making it challenging for them to function effectively as electromagnetic wave absorbers. According to the theory of free electrons ($\varepsilon'' = \sigma/2\pi\varepsilon_0 f$), the value of ε'' is positively correlated with conductivity (σ). Therefore, the introduction of hollow structures can create rich free space inside nanomaterials and reduce conductivity, which provides more possibilities for EMW absorbers. Similarly, hollow structures can optimize impedance matching by modifying the complex permittivity and permittivity[198,199]. In addition, hollow structures with large specific surface areas can produce the Maxwell–Wagner effect by heterogeneous interfaces while enhancing interface polarization[200].

TABLE 3.2 Porous EMW absorption materials.

Types	Precursor	Synthesis process	EAB (GHz)	RL_{min} (dB)	Ref.
Single metal MOFs	Fe/Fe$_3$C@NC(Fe-MOF)	a solvothermal reaction/high-temperature pyrolysis	5.1 5GHz@1.7 mm	−70.8 dB@2.5 mm	166
	Ni/CN(Ni-MOF)	in-situ pyrolysis	4.6 GHz@1.9 mm	−65 dB	167
	Fe/C-900(Fe-MOF)	solvothermal reactions/in situ pyrolysis processes	4.4 GHz	−37.63 dB@1.5 mm	168
Single nonmagnetic-metal-based MOFs	NPC-800 (Zn-MOF) 50 wt%	in situ pyrolysis processes	4.3 GHz	−39.7 dB@4.0 mm	170
	Co/ZrO$_2$/C (Zr-MOF)	hydrothermal method/high temperature pyrolysis	11.9 GHz	−57.2 dB@3.3 mm	163
	ZrO$_2$/C (UiO-66)	hydrothermal method/ high temperature pyrolysis	16.8 GHz@1.5 mm	−58.7 dB	171
Mixed-metal MOFs	NCNT/NiCo/C(15 wt%)	hydrothermal method/thermal CVD	4.64 GHz	−66.1 dB@1.5 mm	172
	CoCu-CNF-10(30 wt%)	carbonization processes	6.4 GHz	−51.7 dB@2.7 mm	173
	Co/ZnO/NC-a(30 wt%)	room temperature synthesis and carbonization processes	5.5 GHz@2.3 mm (11.6–17.1 GHz)	−61.9 dB@8.7 GHz	174
	Nest-like CoFe@C (CoFe-MOF) (10 wt%)	room temperature synthesis/high temperature pyrolysis	9.2 GHz (8.8–18.0GHz)	−61.8 dB@2.8 mm	175
	MnO/Co/C (MnCo-MOF-74) (50 wt%)	hydrothermal process/subsequent carbonization	5.3 GHz@2.3 mm	−68.89 dB @2.64 mm	176
	Fe/C/carbon foam (PB)	carbonization processes	6.34 GHz@4.08 mm (8.20–14.54) GHz	−66.7 dB@4.18 mm	177
	FeNi@CNT/CN Rs (FeNiMIL-88B) (20 wt %)	one-step pyrolyzing	4.5 GHz@1.6 mm	−47.0 dB@2.3 mm	178
	FeCo/C (MIL-53) (20 wt%)	hydrothermal process/ carbonization processes	5.1 GHz@2.6 mm	−78.6 dB	179

Pure biomass derived carbon materials	3D lamellar skeletal network porous carbon (35%)	carbonization process	6.0 GHz@1.9 mm	−60.76 dB@2.97 mm	193
	the hull of water chestnut 600 °C (35 wt%)	a one-step carbonization process	6.0 GHz@1.9 mm (12–18 GHz)	−60.76 dB@2.97 mm	193
Heteroatom doping biomass-derived porous carbon materials	EM-Co-Co$_x$S$_y$-Ni/N,S-C composites 700 °C (16.7 wt%)	co-precipitation reaction/heating	7.4 GHz@3 mm (10.3–17.7 GHz)	−48.5 dB@3.5 mm	195
	N-BCMT/rGO aerogel 600 °C (2wt%)	carbonization process/ hydrothermal/freeze drying	8.36 GHz@3.2 mm (9.64–18 GHz)	−55.45 dB@3.0 mm	195

There are many methods available for the construction of hollow structures, including template-guided methods (hard formwork, soft formwork and self-template), interface induction method (gas–liquid interface, liquid–liquid interface, and solid–liquid interface), and local etching method (competitive coordinated etching of inner core etching and selective phase transition)[197,201]. The self-forming strategy is to directly transform solid structures into hollow structures, and it has several formation principles, namely the Kirkendall effect, galvanic replacement, and Ostwald ripening, etc.[19]. The Kirkendall effect, which originally refers to the formation of defects in two metals with different diffusion rates, has become a facile preparation method for hollow NPs. Galvanic displacement, a self-template method, involves the synthesis of an anode metal that reacts with cations in the cathode[202,203]. Ostwald ripening refers to the dissolution of small crystals or sol particles and the re-deposition of dissolved substances on the surface of larger crystals or sol particles, resulting in hollow structures[204,205]. This can solve the issue of template removal for the preparation of hollow structures by traditional methods. Below, we will elaborate and analyze several different formation mechanisms.

Firstly, based on the Kendall effect, the void is formed by the uneven flow of matter caused by the diffusion of metal atoms from the core outward rather than inward, which is often applied to the conversion of metal NPs to oxides, Sulfides, Selenides and Phosphides[206,207]. For instance, Liu et al. prepared a hollow CoS/diatoms co-doped carbon aerogel via physical crosslinking, one-direction freezing, Kirkebdall effect, and heteroatomic doping[208]. The composites displayed an RL_{min} of −51.96 dB and an EAB of 6.4 GHz. Wang et al. synthesized S-doped $Cu_{2-x}Se$ micro-cubic boxes derived from Cu_2O microcubes by template-directed in situ selenization/sulfidation reaction, which combined Kirkendall effect and Ostwald ripening. The S-doped $Cu_{-x}Se$-1 had splendid absorption capacities with an RL_{min} of −44.5 dB at 1.9 mm and an EAB of 4.3 GHz[209]. Zhang et al. fabricated hollow flower-like CuS hierarchical microspheres by Galvanic replacement[210]. When the filling loading was 50 wt%, the hollow flower-like CuS exhibited an RL_{min} of −17.5 dB and an EAB of 3.0 GHz at 1.1 mm.

3.5.2 MOF on MOF

For MOF-on-MOF building structures, it is done by combining two or more MOFs into one composite material. There are many synthetic strategies for MOF on MOF, such as epitaxial growth, surfactant assistant growth, heteroepitaxial growth, ligand/metal ion exchange, and nucleation kinetic guided growth[211]. MOF on MOF can build different structures, including core-shell, yolk-shell, core-satellite, hollow multi-shell, asymmetric structure, and film-on-film[212]. In addition, MOF-on-MOF materials not only have the intrinsically unique characteristics of each MOF, but also exhibit excellent synergy in a single system, so they are used in various fields. Thanks to the selective combination of functional components, absorbers with multiple MOF-on-MOF structures can easily adjust impedance matching and supplement the EM loss mechanism, thus exhibiting desired MA characteristics[213–215]. In 2021, two binary hybrid MOF heterostructures were constructed through a simple carbonization process. In the carbonization process, the heterostructures of two binary hybrid MOFs were converted into magnetic porous carbon materials, as illustrated in Fig. 3.20(a)[216]. These heterogeneous components allow the absorber to have a layered pore structure, first-rank impedance matching, and eminent EM attenuation. The special hybrid structure of hexagonal pyramid columns and octahedra was mainly obtained by the selective growth of guest MOF on the specific crystal planes of the host MOF. Furthermore, the DM-700 exhibited an RL_{min} of − 65.2 dB with an EAB of 4.8 GHz at a thickness of only 2.0 mm.

In addition, Zhang et al. first successfully constructed binary components DUT-52@MIL-88B (DM) and DUT-52@MIL-88 (DMC) by combining two unique anisotropic epitaxial growth strategies[217]. Then, the ternary component DUT-52@MIL-88B@MIL-88C (DMM) was also synthesized at 110 °C for 10 min. Subsequently, the DMM was converted into magnetic porous carbon-based absorbent (DMM-700) by a high-temperature carbonization process. The DM-700 displayed an RL_{min} of −67.5 dB at 3.6 mm with an EAB of 2.0 GHz at a filler loading of 55%. Compared with DM-700, it is worth noting that the optimized DMM-700 not only exhibited a high RL_{min} value of −56.4 dB, but also reduced the matching thickness to 2.4 mm, and displayed an EAB of 4.0 GHz, which is twice that of the DM-700 (Fig. 3.20b). The large BET and aperture of the ternary MOF-derived composite structure can improve impedance matching and produce multiple scattering and scattering effects on microwaves. Holes and defects were introduced into the DMM-700 during carbonization, while the metal source in DMM was converted to metal NPs to induce dipolar polarizations (Fig. 3.20c).

The formation of heterogeneous interfaces such as metal oxides/C, metallic carbon/C, and absorbers/paraffins provides strong interfacial

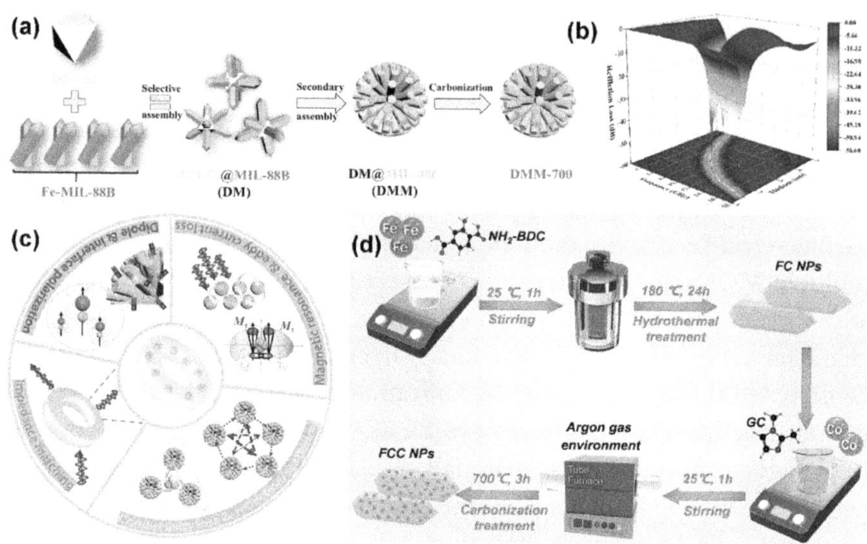

FIG. 3.20 The synthesis process (a), 3D *RL* plots (b) and schematic diagram of MA mechanisms (c) of DMM-700. (Reproduced with permission from Ref.[217] Copyright 2022, Elsevier.) (d) The synthesis process of Fe/Co/C (FCC) multiple composite absorbents. (Reproduced with permission from Ref.[218] Copyright 2023, Elsevier.)

polarization (Fig. 3.20c). Therefore, the multi-component composite structure can optimize impedance matching, and can equip the absorber with a strong micro-absorption capacity[217]. For instance, in 2023, a MOF-on-MOF design idea was proposed to synthesize Fe/Co/C (FCC) composite. Specifically, Fe-MOF (NH$_2$-MIL-88B (Fe)) was synthesized *via* a solvothermal method, and Co-MOF was then attached to the Fe-MOF surface by *in-situ* growth method, and finally FCC multiple composite absorbent was synthesized by high-temperature carbonization (Fig. 3.20d)[218]. Under the high temperature pyrolysis process, the Co-MOF and Fe-MOF were converted into magnetic metal NPs Co and Ni, respectively, and the ligands were converted into carbon layers. When the magnetic NPs Fe and Co were coated in the carbon layer, a unique spindle-shaped structure was formed. The FCC's unique porous structure, rich heterogeneous interface, and N-element doping are the primary factors that heighten the properties of dielectric and magnetism, which provides various advantages for EMW conduction and loss. What's more, the composite absorbent can cover the absorption band of 2–18 GHz with an adjustable thickness.

3.5.3 Core (yolk)-shell

For the core (yolk)-shell structure, it is widely used in the field of EMW absorption due to its multi-interface characteristics. The core-shell structure designed sequentially by dielectric and magnetic materials facilitates the optimization of the absorber and free space impedance matching[219,220]. Moreover, the core-shell structure can provide multiple reflection paths, broaden the effective absorption frequency band, and promote EM loss. The shell thickness, coating sequence, and high-temperature materials of the core-shell structure are all important factors in the design of high-performance EMW absorbing materials[219,221]. MOFs-derived porous carbon materials are significant for MA, since they can provide a uniformly dispersed matrix for metal NPs, thereby forming a large heterogeneous interface, promoting impedance matching and enhancing the interface polarization loss. Simultaneously, core-shell and hollow structures are important ways to form multi-stage heterogeneous interface-induced polarization loss, which is often used in MOFs-derived carbon composites due to the large specific surface area, high porosity, numerous active sites, and uniform pore size distribution of MOFs.

For traditional MOFs materials, the construction of core-shell nanocomposites is mainly through high-temperature pyrolysis of the pristine MOFs, and the synthesis of MOFs-derived metal carbon and metal carbon-nitrogen doping (introduction of heteroatom S, N, P). For example, in 2022, core-shell structured Cu/NC@Co/NC composites were fabricated by a high-temperature pyrogenation of Cu-MOF@Co-MOF precursor. The Cu/NC@Co/NC-3.75 using 3.75 mmol of Co $(NO_3)_2 \cdot 6H_2O$ displayed excellent EMW adsorption abilities, which showed an RL_{min} of -54.13 dB at a 3 mm thickness, and an EAB of 5.19 GHz (10.18–15.37 GHz) at 2.5 mm[222]. Guo et al.[223] synthesized core-shell $Co_9S_8@MoS_2$ nanocubes by self-assembly and pyrolysis. The core-shell $Co_9S_8@MoS_2$ nano cubes were coincidentally anchored to rGO nanosheets (A-G/$Co_9S_8@MoS_2$) to form a layered heterostructure. The A-G/$Co_9S_8@MoS_2$ demonstrated an RL_{min} of -55.13 dB and an EAB of 6.61 GHz with only 8 wt% filler loading. In another instance, Wang et al.[224] constructed micro-flower like core-shell structured ZnCo@C@1T-2H-MoS_2 composites by MOF self-template method, which can achieve an RL_{min} of -35.83 dB at 5.83 GHz with a 5.0 mm thickness and an EAB of up to 4.56 GHz at 2.0 mm. Urchin-like multiple core-shelled Co/CoS_2@C@MoS_2 (CCM) composites were fabricated through a pyrogenation process and a solvothermal reaction[225]. The CCM composites gave excellent EMW performance with an RL_{min} of −43.9 dB and an EAB of 4.4 GHz at 1.5 mm thickness. The core-shell structure with a large number of interfaces affects polarization relaxation loss, defects of MoS_2 and NC substrates lead to dipole orientation polarization loss.

For MOFs of MIL-68 (In), increasing the pyrolysis temperature can enhance conductivity, but as metal indium (In) precipitates from the nanorods, the structure will collapse, thereby affecting the EMW performance. Therefore, constructing a core-shell structure in which the MIL-68 (In) nanorods core layers are coated with other NPs can solve these issues[226]. In 2023, a core-shell ZnO/C@In/C heterostructure was fabricated *via* coating the MIL-68 (In) with ZIF-8 (Zn) dodecahedral NPs by a microwave-assisting hydrothermal reaction[226]. The ZnO/C@In/C-0.136 displayed an RL_{min} of −56.75 dB at 16.34 GHz and a wide EAB of 6.76 GHz at 2.36 mm, which is owing to the impedance matching and high attenuation ability coming from massive interfaces.

In the MA analysis of core-shell structures synthesized using MOFs as precursors, thermal decomposition temperature is critical for the EMW absorption performance. For instance, CoFe alloy@N-doped carbon shell composites were fabricated by a thermal decomposition process of CoFe PBA with different temperatures (550/600/650 °C)[227]. When the pyrolysis temperature was 600°C, the CoFe@NC-600 composites showed an RL_{min} value of −50.77 dB with an EAB of 7.2 GHz at 2.1 mm, which is superior to CoFe@NC-500/550 °C. The dielectric loss provided by NC as the shell and the magnetic loss and conduction loss provided by CoFe alloy as the core are the primary elements for the outstanding EMW performance of CoFe alloy@N-doped carbon shell composites. In another instance, Wu et al. prepared a multi-scale core-shell Ni-MOF@N-doped carbon composites (Ni-MOF@N–C) via a solvothermal reaction and a high temperature heat treatment using dicyanamide (DCDA) as a nitrogen source[228]. The Ni-MOF@N-C-500 displayed an RL_{min} of −69.6 dB at a thickness of 3 mm and an EAB up to 6.8 GHz.

In addition, a core-shell Ag@C sphere derived from Ag-MOFs was synthesized through adjustable phase inversion[229]. The adjustable phase inversion is the method for ligand exchange, which is to replace the organic ligand terephthalic acid (H_2BDC) with 2-methylimidazole (Hmim), so that Ag-MOF-5 gradually changes to ZIF-L series, thereby changing the morphological structure of the Ag@C. Furthermore, with the transformation of the phase, the conductivity gradually diminishes, resulting in a gradual reduction in conductivity loss. When the mole ratio of Ag-MOF-5 to Hmim is 1:2, the core-shell Ag@C exhibited an RL_{min} of −50.14 dB at 3 mm and an EAB of 4 GHz at 3.0 mm. In summary, the construction of a core-shell structure has the synergistic effects of multiple reflection and scattering, impedance matching, and heterogeneous structure, which is conducive to improving the absorption performance of EMWs.

A core-shell structure with a special core@void@shell has attracted numerous attentions in high-performance MAM designing and application, which not only has the characteristics of a core-shell structure but also a hollow structure[230]. In addition, the yolk-shell structure is a new multi-interface structure derived from the core-shell structure. In these microstructures, the yolk-shell structure consists of a movable core as the yolk and a hollow shell as the shell[231]. The yolk-shell structure was constructed in early designs mainly from different inorganic compounds, with Fe_3O_4 as the core, such as $Fe_3O_4@ZrO_2$[232], $Fe_3O_4@SnO_2$[233], $Fe_3O_4@TiO_2$[234], $Fe_3O_4@C$[235], $ZnFe_2O_4@rGO@TiO_2$[236], and $Fe_3O_4@NC@2H/1T-MoS_2$[237]. The yolk-shell structure with low density, large surface area and core-shell synergistic effect can facilitate multiple reflection and refraction of incident waves, which is conducive to the absorption performance of EMWs[238,239]. The optimization of the yolk-shell structural material involves adjusting its yolk and shell size, composition, and morphology, which also provides new ideas for the preparation of high-performance absorbent materials. In recent years, by pyrolysis of MOFs materials, a series of nanocomposites with yolk-shell structures have been derived, such as $ZnCo/NC@TiO_2$[240], $Co@ZnO/Ni@NC$[241], HF-Ni-MOF[242], NiO/Ni/GN@Air@NiO/Ni/GN[243], CoFe@C[244], and ZnO-Ni@CNT[245]. These MOFs-derived diversified multi-component structures provide favorable approaches for preparing EMW absorption materials and open up opportunities.

For instance, Zhuang et al.[246] successfully constructed yolk-shell $Fe_3O_4@C@Co/N$-doped C microspheres (FCCNC-x) using layered construction strategies and quantitative construction techniques, as shown in Fig. 3.21(a). The FCCNC-2 absorber demonstrated an RL_{min} of −66.39 dB at a thickness of 1.9 mm with a surpassing EAB of 6.49 GHz at 1.6 mm, suggesting that the synergistic effects of electrical and magnetic components are beneficial for enhanced absorption ability (Fig. 3.21b). Concretely, the surface conductive networks acquired by quantitative construction of carbonized ZIF-67, combined with the Fe_3O_4 magnetic core and dielectric carbon layer linked by the cavitat, results in the cooperative enhancement of impedance matching optimization and

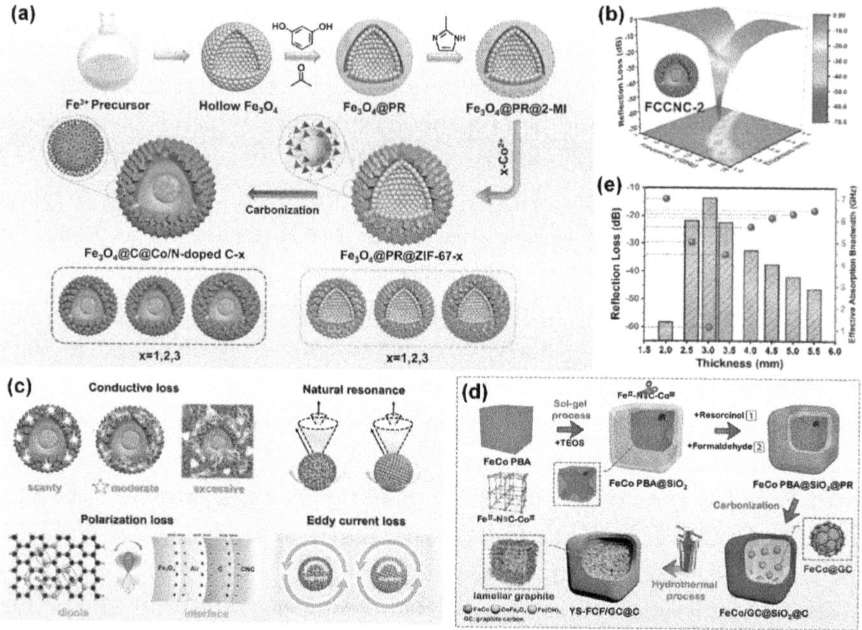

FIG. 3.21 Schematic diagram of the synthesis process (a), 3D *RL* plot (b) and diagram of the reaction mechanism (c) of FCCNC-2 absorbers (yolk-shell Fe_3O_4@C@Co/N-doped C). (Reproduced with permission from Ref.[246] Copyright 2023, Wiley.) The synthesis process (d) and absorption performances with different thicknesses of YS-FCF/GC@C (e). (Reproduced with permission from Ref.[247] Copyright 2022, Royal Society of Chemistry.)

synergistic attenuation in the FCCNC-doped absorber. The introduction of heteroatoms into CNC elements produces dipole polarization, and the yolk-shell structure promotes interfacial polarization, which can both be classified as polarization modes. Moreover, the numerous magnetic components ferric oxide and cobalt particles in both shells can significantly boost the eddy current losses and natural resonance loss (Fig. 3.21c). Zhuang et al. skillfully designed yolk-shell structured YS-FeCo/CoFe$_2$O$_4$/Fe(OH)$_3$/GC@C (YSFCF/GC@C) nanoboxes composites via a "coating-coating-carbonization-etching" route, as illustrated in Fig. 3.21(d)[247]. The FeCo-PBA@SiO$_2$@PR was first constructed via a twice-coating craft process with the classic Stöber method and polycondensation. Afterwards, the FeCo/GC@SiO$_2$@C was obtained by carbonization under N$_2$ atmosphere at 650 °C. The YS-FCF/GC@C nanoboxes were then successfully prepared with a hydrothermal process. The YSFCF/GC@C nanoboxes exhibited preeminent EMW performance with an RL_{min} of as high as −60 dB and an EAB of 7.04 GHz (9.76–16.80 GHz) at 3.02 mm, which can be attributed to various loss mechanism, reflection loss effects, and affluent interface polarization stimulated by a yolk-shell structure, as shown in Fig. 3.21(e)[247].

In conclusion, the synthesis of MOFs still relies primarily on solvothermal or hydrothermal methods. However, these methods are not conductive to large-scale production or effective practical applications. The search for low-cost, mass-production synthetic methods is essential. At present, the EMW absorption mechanism is still ambiguous, such as dielectric loss, resistance loss, and magnetic loss mechanism. It was found that the construction of a hollow structure can effectively improve the absorption performance of EMWs, which may be because the constructed hollow structure can optimize impedance matching. The huge cavity of the hollow structure promotes multiple reflections and scattering, thereby extending the propagation path of the incident wave and dissipating it as much as possible. Regardless of whether it is a hollow structure core-shell, yolk-shell or multi-shell structure, the Maxwell–Wager effect, induced by the multi-component heterogeneous interface resulting from their large specific surface area, is beneficial to enhance interface polarization. In future research, efforts will be made to reduce costs, improve efficiency, and promote practical application.

3.6 SKELETON STRUCTURAL EMW ABSORPTION MATERIALS

Researchers have investigated the application of skeleton structures in EMW absorption. Skeleton structures composed of nanocrystals can be divided into gel-like structures, hierarchical and fractal microstructures, and crosslinking networks. The pore size of the skeleton structure varies from tens to hundreds of microns, which is much larger than the pore size of many ordinary porous materials[248]. The skeleton structure is considered to enhance conductive loss in the field of EMW absorption. In addition, the porosity of the skeleton structure results in its relatively low permittivity, which is conducive to impedance matching[249]. Below, we will introduce the EMW application of skeleton materials from two aspects, namely aerogel-based materials and foam-based materials.

3.6.1 Aerogel-based composites

Aerogels mainly include graphene-based, CNT-based, biomass derived carbon aerogels, MXene-based, polymer-based, silica aerogels, and aerogel composites[250]. Traditional powder aerogels generally involve the application of powder materials to the matrix, which benefits practical applications. Compared with traditional powder materials, the construction of carbon aerogels can create a variety of morphological structures such as microspheres, fibers, and films, etc. As the most widely studied class of unique carbon-based materials, carbon aerogels have low density, light weight, high conductivity, and ultra-high specific surface area[251–253]. Regardless of the type of aerogel, it possesses a 3D network that interconnects with each other. These 3D networks exhibit a high specific surface area and a large number of pores. Moreover, they are widely used in the field of EMW absorption, as shown in Fig. 3.22. Aerogels can optimize the matching thickness and broaden the EAB, offering significant advantages in EMW absorption[254].

For example, a 3D porous graphene aerogel was prepared via self-assembly and annealing processes, which had an ultralow density and high compressibility[255]. When the filling ratio was 30 wt%, the 3D porous graphene aerogel displayed an outstanding absorption performance with an RL_{min} of −61.09 dB at 6.30 GHz under a thickness of 4.81 mm. However, the narrow EAB is the main bottleneck of carbon-based aerogels. Yu et al. prepared carbon aerogels modified by carbon microspheres with different morphologies by an in-situ polymerization and carbonization process[256]. The experimental data manifest that the unique skeleton and morphology of carbon aerogels can boost the absorption performance of EMWs. When the filling rate was 7.7 wt%, it had an RL_{min} of −24.17 dB and an effective absorption broadband of 8.47 GHz at 3.0 mm. The excellent absorption performance of carbon aerogels is mainly attributed to the presence of carbon aerogels in the skeleton framework, which accelerates the transport of free electrons on the substrate, promotes conductivity loss, and optimizes impedance matching of porous structures. In addition, heteroatom N doping and the presence of oxygen vacancies can induce defect polarization, further enhancing EMW absorption[256].

For graphene-based aerogel, it has a high degree of graphitization and high conductivity, which increases the complex permittivity, achieves good conduction loss, and promotes dielectric loss, thereby improving the EMW absorption performance[5]. Compared with traditional graphene, graphene

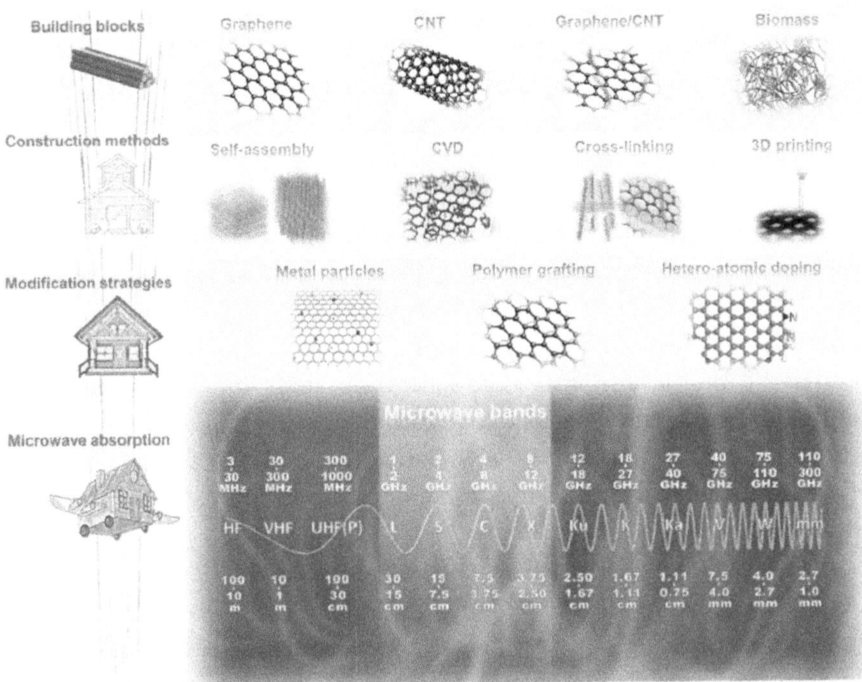

FIG. 3.22 The construction and modification of carbon aerogels towards superior EMW absorption performance. (Reproduced with permission from Ref.[250] Copyright 2023, Royal Society of Chemistry.)

aerogels have excellent absorption properties. However, their impedance matching ability is poor, which may be due to its relatively high complex dielectric constant. Although single-component graphene aerogels can be successfully obtained by direct pyrolysis, and they have improved absorption performance and effective broadband, it is still difficult for them to achieve conditions that offer a wide absorption broadband and strong absorption with lower thickness. Therefore, in order to better achieve high-performance EMW absorbers, composite aerogels have been widely studied.

For example, Ji et al. synthesized a cellulose skeleton with a 3D structure with PANI conductive polymer on its surface[257]. The cellulose-chitosan framework with a 3D skeleton exhibited an RL_{min} of −54.76 dB at 2.1 mm and an EAB of 6.04 GHz at 1.9 mm. The excellent absorption performance and infrared stealth performance of the cellulose-chitosan/PANI aerogel were attributed to the backbone provided by the cellulose aerogels, the dielectric properties provided by the loaded PANI, and the infrared stealth properties. This 3D structure enhances conductivity loss. The porous structure promotes charge polarization and promotes the absorption of EMWs. Composite aerogels include not only conductive polymer-based aerogels, but also magnetic metals, ferrites, ceramics, and multicomponents composites aerogels. Zhao et al. synthesized ultralight CoNi/rGO aerogels via an in-situ solvothermal and carbonization method and the CoNi NPs were uniformly distributed on the surface of the rGO aerogel skeleton[258]. As a result, the CoNi/rGO aerogel displayed an RL_{min} of –53.3 dB at an ultrathin thickness of 0.8 mm and an EAB of up to 3.5 GHz. The composition of the multilayer porous rGO sheet decorated with CoNi nanospheres resulted in moderate dielectric loss and magnetic loss for the CoNi/rGO aerogels. More importantly, the introduction of CoNi nanospheres can effectively adjust the impedance matching, resulting in stronger MA and weaker reflection.

Although researchers have achieved considerable advancements in aerogel composites, there are still some challenges to overcome. For example, further study and confirmation are needed to determine whether the loss mechanism of aerogel composites is consistent with or different from that of ordinary porous materials.

3.6.2 Foam-based composites

Foam-based materials have low density, high porosity, and chemical stability, and an interconnected network-like structure. Foam-based materials mainly include metal-based foams, polymer-based foams, and carbon-based foams. It is generally unsuitable to directly use metal-based foams as EMW absorbers because their high relative permittivity value results in poor impedance matching. One efficient way to address this issue is to compound metal-based foams with materials that have a low dielectric constant and/or are magnetic.

For example, Wu et al. designed a lightweight Ni foam with $NiO/NiFe_2O_4$ via a leaven dough method and an in-*situ* generation approach, which demonstrated excellent EMW absorption, mainly due to the synergistic effect of the foam structure and the $NiO/NiFe_2O_4/Ni$ components[259]. The $NiO/NiFe_2O_4/Ni$ foam had a wide absorption bandwidth of 14.24 GHz at 0.6 mm and a specific EAB of 19444.4 $GHz \cdot g^{-1} \cdot cm^{-2}$. In another example, a $Ni@MnO_2$ nanosheet (NS) foam was fabricated by a hydrothermal process. It can be clearly observed in Figs. 3.23 (a and b) that numerous MnO_2 nanosheets were loaded on the smooth surface of Ni foam[249]. The

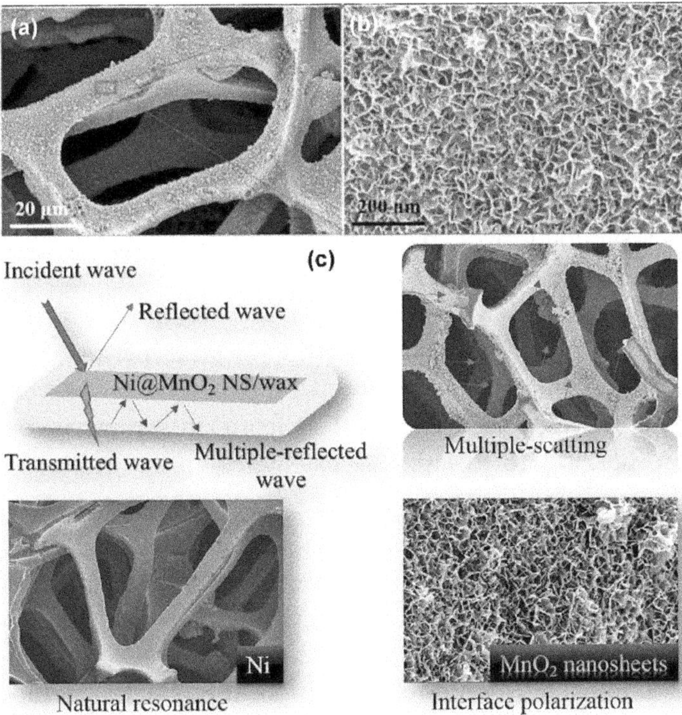

FIG. 3.23 The SEM of Ni@MnO₂ NS foam (a and b) and electromagnetic wave attenuation mechanism (c). (Reproduced with permission from Ref.[249] Copyright 2019, Elsevier.)

experimental data show that Ni@MnO$_2$ NS foam has a minimum reflection loss of −37.55 dB at a thickness of 3.6 mm, and an ultra-wide effective absorption broadband of 11.2 GHz. The excellent EMW absorption performance was attributed to the EM loss of Ni foam's natural resonance and dielectric loss with interfacial polarization and dipole polarization, as revealed in Fig. 3.23c. Therefore, when the metal foam is compounded with a metal oxide with low dielectric characteristics or a magnetic material, the impedance matching can be optimized, thereby improving the EMW absorption performance.

For carbonaceous foams, 3D graphene and reduced graphene oxide have been widely studied in the field of EMW absorption because of their low density, interlinked conductive network, and compressibility[260]. However, rGO itself shows a relatively weak antioxidant capability, which limits its applications. In order to further enhance the EMW attenuation ability of foam-based composites, it is necessary to properly introduce dielectric or EM components into carbonaceous foam materials. Researchers have investigated the incorporation of SiC NWs into rGO foam to improve the absorption performance and thermal stability. For example, a hierarchical rGO/SiC NWs foam was synthesized via freeze-drying and heating processes, which exhibited effective absorption in the whole X-band (8.2–12.4 GHz) at 3.0 mm[260]. Similarly, Chen et al. designed a 3D $NiCo_2O_4$/rGO composite foam by a solvothermal treatment followed by a calcination process and freeze drying. The foam exhibited an RL_{min} of –52.2 dB and an EAB of 7.04 GHz (10.96–18 GHz), covering the entire Ku band, at 2.6 mm[261]. The excellent and efficient absorption performance may be attributed to the reflection and scattering of $NiCo_2O_4$ and rGO foam, and the synergy between $NiCo_2O_4$ and rGO foam. In summary, foam-based EMW absorbers are worth exploring extensively in the future.

CONCLUSION

In this chapter, traditional EMW absorbing materials are analyzed and discussed from different perspectives, including phase composition, heterostructure design, different dimensions (0D, 1D, 2D and 3D), as well as porous and hollow structures. In the past few decades, many researchers have devoted themselves to studying and exploring the absorption mechanisms and practical applications of traditional EMW-absorbing materials. These materials have demonstrated excellent performance in absorbing and dissipating electromagnetic energy, thereby reducing the reflection and transmission of EMWs. They also have played a crucial role in various applications such as stealth technology, radar absorption, and electromagnetic interference (EMI) shielding. The development of traditional absorbing materials has significantly contributed to the advancements in communication, defense and electronic industries.

To satisfy the requirements of intelligence and versatility of EM absorbing materials, traditional materials are currently facing various challenges in their applications. Firstly, traditional EMW absorbing materials typically exhibit good absorption performance only in certain frequency bands and may not perform well over a wide range of frequencies. Secondly, some of them require a large volume or mass-filling ratio to achieve better absorbing performance, which is not suitable for some special applications, such as those that require a lightweight and compact design. Thirdly, the preparation of some traditional EMW absorbing materials needs complex and expensive processes, which limits their commercialization and large-scale application. Lastly, traditional EMW absorbing materials show limited stability and durability, such as performance degradation or damage during prolonged use or in harsh environments.

However, benefitting from the advanced characterization and theoretical simulation methods, such as density functional theory calculation (DFT), in-situ Raman, in-situ X-ray diffraction (XRD), electron holography and synchrotron radiation, one can analyze the electromagnetic absorption mechanism more clearly and optimize the design of material structure more efficiently. In this sense, the future of traditional EMW absorbing materials is promising, with ongoing research efforts focusing on enhancing performance, expanding applications, and addressing emerging challenges in the ever-evolving landscape of electromagnetic wave technology.

REFERENCES

1. Harris, V.G., Geiler, A., Chen, Y., et al. (2009). Recent advances in processing and applications of microwave ferrites. *Journal of Magnetism and Magnetic Materials 321*: 2035–47.

2. Pullar, R.C. (2012). Hexagonal ferrites: A review of the synthesis, properties and applications of hexaferrite ceramics. *Progress in Materials Science 57*: 1191–334.

3. Wang, G.Z., Gao, Z., Tang, S.W., et al. (2012). Microwave absorption properties of carbon nanocoils coated with highly controlled magnetic materials by atomic layer deposition. *ACS Nano 6*: 11009–17.

4. Wang, R., He, M., Zhou, Y., et al. (2020). Metal–organic frameworks self-templated cubic hollow Co/N/C@MnO$_2$ composites for electromagnetic wave absorption. *Carbon 156*: 378–88.

5. Meng, F., Wang, H., Huang, F., et al. (2018). Graphene-based microwave absorbing composites: A review and prospective. *Composites Part B: Engineering 137*: 260–77.

6. Zhang, Y., Ruan, K., Shi, X., et al. (2021). Ti$_3$C$_2$T$_x$/rGO porous composite films with superior electromagnetic interference shielding performances. *Carbon 175*: 271–80.

7. Putri, L.K., Ong, W.J., Chang, W.S., et al. (2015). Heteroatom doped graphene in photocatalysis: A review. *Applied Surface Science 358*: 2–14.

8. He, S.M., Huang, C.C., Liou, J.W., et al. (2019). Spectroscopic and electrical characterizations of low-damage phosphorous-doped graphene via ion implantation. *ACS Applied Materials & Interfaces 11*: 47289–98.

9. Liu, Y., Feng, Q., Xu, Q., et al. (2013). Synthesis and photoluminescence of F and N co-doped reduced graphene oxide. *Carbon 61*: 436–40.

10. Liu, P., Zhang, Y., Yan, J., et al. (2019). Synthesis of lightweight N-doped graphene foams with open reticular structure for high-efficiency electromagnetic wave absorption. *Chemical Engineering Journal 368*: 285–98.

11. Zhou, J., Chen, Y., Li, H., et al. (2017). Facile synthesis of three-dimensional lightweight nitrogen-doped graphene aerogel with excellent electromagnetic wave absorption properties. *Journal of Materials Science 53*: 4067–77.

12. Yang, H., Cao, M., Li, Y., et al. (2014). Enhanced dielectric properties and excellent microwave absorption of SiC powders driven with NiO nanorings. *Advanced Optical Materials 2*: 214–19.

13. Yang, H.J., Cao, W.Q., Zhang, D.Q., et al. (2015). NiO hierarchical nanorings on SiC: enhancing relaxation to tune microwave absorption at elevated temperature. *ACS Applied Materials & Interfaces 7*: 7073–77.

14. Wang, L., Xing, H., Gao, S., et al. (2017). Porous flower-like NiO@graphene composites with superior microwave absorption properties. *Journal of Materials Chemistry C 5*: 2005–14.

15. Yang, H., Ye, T., Lin, Y. (2015). Microwave absorbing properties based on polyaniline/magnetic nanocomposite powders. *RSC Advances 5*: 103488–93.

16. Wang, Y., Wu, X., Zhang, W., et al. (2017). Fabrication and enhanced electromagnetic wave absorption properties of sandwich-like graphene@ NiO@PANI decorated with Ag particles. *Synthetic Metals 229*: 82–88.

17. Yin, Y., Zeng, M., Liu, J., et al. (2016). Enhanced high-frequency absorption of anisotropic Fe_3O_4/graphene nanocomposites. *Scientific Reports 6*: 25075.

18. Chen, C., Bao, S., Zhang, B., et al. (2019). Development of sulfide-doped Graphene/Fe_3O_4 absorber with wide band electromagnetic absorption performance. *Journal of Alloys and Compounds 770*: 90–97.

19. Lv H.L., Yang Z.H., Pan H.G., et al. (2022). Electromagnetic absorption materials: Current progress and new frontiers. *Progress in Materials Science 127*: 100946.

20. Wu F., Liu Z.H., Wang J.Q., et al. (2021). Template-free self-assembly of MXene and CoNi-bimetal MOF into intertwined one-dimensional heterostructure and its microwave absorbing properties. *Chemical Engineering Journal 422*: 130591.

21. Jiang, X.W., Niu, H.H., Li, J.L., et al. (2023). Construction of core-shell structured SiO_2@MoS_2 nanospheres for broadband electromagnetic wave absorption. *Applied Surface Science 628*: 157355.

22. Zhi, D.D., Li, T., Qi, Z.H., et al. (2022). Core-shell heterogeneous graphene-based aerogel microspheres for high-performance broadband microwave absorption via resonance loss and sequential attenuation. *Chemical Engineering Journal 433*: 134496.

23. Yan, C., Wang, J., Wang, X., et al. (2014). An intrinsically stretchable nanowire photodetector with a fully embedded structure. *Advanced Materials 26*: 943–50.

24. Wang, N., Wang, Y., Lu, Z., et al. (2023). Hierarchical core-shell FeS_2/Fe_7S_8@C microspheres embedded into interconnected graphene framework for high-efficiency microwave attenuation. *Carbon 202*: 254–64.

25. Yao, C., Wu, Z., Guo, X., et al. (2023). In-situ construction of $Cu_{7.2}S_4$ microsphere embedded into biomass-derived porous carbon for electromagnetic wave absorption. *Materials Today Sustainability 23*: 100421.

26. Lu, X.K., Li, X., Zhu, W.J., et al. (2022). Construction of embedded heterostructures in biomass-derived carbon frameworks for enhancing electromagnetic wave absorption. *Carbon 191*: 600–09.

27. Wang, F.Y., Cui, L.R., Zhao, H.H., et al. (2021). High-efficient electromagnetic absorption and composites of carbon microspheres. *Journal of Applied Physics 130*: 230902.

28. Chen, X., Wu, Y., Gu, W., et al. (2022). Research progress on nanostructure design and composition regulation of carbon spheres for the microwave absorption. *Carbon 189*: 617–33.

29. Wang, N., Wu, F., Xie, A.M., et al. (2015). One-pot synthesis of biomass-derived carbonaceous spheres for excellent microwave absorption at the Ku band. *RSC Advances 5*: 40531–35.

30. Li N., Liu L.L., Duan Y., et al. (2023). Exploration of magnetic media modulation engineering on heterogeneous carbon spheres for optimized electromagnetic wave absorption. *Journal of Alloys and Compounds 943*: 169109.

31. Song, Y., Yin, F.X., Zhang, C.W., et al. (2021). Three-dimensional ordered mesoporous carbon spheres modified with ultrafine zinc oxide nanoparticles for enhanced microwave absorption properties. *Nano-Micro Letters 13*: 76.

32. Zhou, W., Hu, X., Bai, X., et al. (2011). Synthesis and electromagnetic, microwave absorbing properties of core–shell Fe_3O_4–Poly(3,4-ethylenedioxythiophene) microspheres. *ACS Applied Materials & Interfaces 3*: 3839–45.

33. Wu, G., Cheng, Y., Yang, Z., et al. (2018). Design of carbon sphere/magnetic quantum dots with tunable phase compositions and boost dielectric loss behavior. *Chemical Engineering Journal 333*: 519–28.

34. Liao, Y., E, Y.F., Zhou, X., et al. (2022). Quantum dots with Mott-Schottky effect embedded in crystal-amorphous carbon for broadband electromagnetic wave absorption. *Journal of Alloys and Compounds 929*: 167246.

35. Hu, Q., Fang, Y., Du, Z., et al. (2021). Controllable synthesis and enhanced microwave absorption properties of novel lightweight graphene quantum dots/hexagonal boron nitride composites. *Carbon 182*: 134–43.

36. Gao, Z.G., Xu, B.H., Ma, M.L., et al. (2019). Electrostatic self-assembly synthesis of $ZnFe_2O_4$ quantum dots (ZnFe2O4@C) and electromagnetic microwave absorption. *Composites Part B: Engineering 179*: 107417.

37. Wei, B., Wei, X.F., Wang, M.Q., et al. (2023). Ultra-broadband microwave absorption of honeycomb-like three-dimensional carbon foams embedded with zero-dimensional magnetic quantum dots. *Journal of Alloys and Compounds 939*: 168781.

38. Mo, Z., Yang, R., Lu, D., et al. (2019). Lightweight, three-dimensional carbon Nanotube@TiO2 sponge with enhanced microwave absorption performance. *Carbon 144*: 433–39.

39. Chen, X., Liu, H., Hu, D., et al. (2021). Recent advances in carbon nanotubes-based microwave absorbing composites. *Ceramics International 47*: 23749–61.

40. Ren, Q.Q., Yao, L.W., Sun, Y.S., et al. (2023). Enhanced electromagnetic wave absorption properties of ultrathin SnO nanosheets/carbon nanotube hybrids. *Journal of Alloys and Compounds 960*: 170694.

41. Liang, H.S., Xing, H., Qin, M., et al. (2020). Bamboo-like short carbon fibers@ Fe_3O_4@phenolic resin and honeycomb-like short carbon fibers@Fe_3O_4@FeO composites as high-performance electromagnetic wave absorbing materials. *Composites Part A: Applied Science and Manufacturing 135*: 105959.

42. Bi, Z.S., Yao, L.W., Wang, X.Z., et al. (2022). Experimental and theoretical study on broadband electromagnetic wave absorption of algae-like NiO/carbon nanotubes absorbers. *Journal of Alloys and Compounds 926*: 166821.

43. Zhang, H., Zhao, B., Dai, F.Z., et al. (2021). $(Cr_{0.2}Mn_{0.2}Fe_{0.2}Co_{0.2}Mo_{0.2})$ B: A novel high-entropy monoboride with good electromagnetic interference shielding performance in K-band. *Journal of Materials Science & Technology 77*: 58–65.

44. Zhao, J., Zhang, J.L., Wang, L., et al. (2020). Fabrication and investigation on ternary heterogeneous MWCNT@TiO_2-C fillers and their silicone rubber wave-absorbing composites. *Composites Part A: Applied Science and Manufacturing 129*: 105714.

45. Che, R.C., Peng, L.M., Duan, X.F., et al. (2004). Microwave absorption enhancement and complex permittivity and permeability of Fe encapsulated within carbon nanotubes. *Advanced Materials 16*: 401–05.

46. Zhao, D.L., Zhang, J.M., Li, X., et al. (2010). Electromagnetic and microwave absorbing properties of Co-filled carbon nanotubes. *Journal of Alloys and Compounds 505*: 712–16.

47. Zou, T., Li, H., Zhao, N., et al. (2010). Electromagnetic and microwave absorbing properties of multi-walled carbon nanotubes filled with Ni nanowire. *Journal of Alloys and Compounds 496*: L22–L24.

48. Zhao, D.L., Li, X., Shen, Z.M. (2008). Electromagnetic and microwave absorbing properties of multi-walled carbon nanotubes filled with Ag nanowires. *Materials Science and Engineering: B 150*: 105–10.

49. Singh, B.P., Saket, D.K., Singh, A.P., et al. (2015). Microwave shielding properties of Co/Ni attached to single walled carbon nanotubes. *Journal of Materials Chemistry A 3*: 13203–09.

50. Hu, Y., Jiang, R., Zhang, J., et al. (2018). Synthesis and properties of magnetic multi-walled carbon nanotubes loaded with Fe 4 N nanoparticles. *Journal of Materials Science & Technology 34*: 886–90.

51. Zeng, X.J., Sang, Y.X., Xia, G.H., et al. (2022). 3-D hierarchical urchin-like Fe_3O_4/CNTs architectures enable efficient electromagnetic microwave absorption. *Materials Science and Engineering: B 281*: 115721.

52. Jia, T., Qi, X., Wang, L., et al. (2023). Constructing mixed-dimensional light-weight flexible carbon foam/carbon nanotubes-based heterostructures: An effective strategy to achieve tunable and boosted microwave absorption. *Carbon 206*: 364–74.

53. Wu, N., Lv, H., Liu, J., et al. (2016). Improved electromagnetic wave absorption of Co nanoparticles decorated carbon nanotubes derived from synergistic magnetic and dielectric losses. *Phys Chem Chem Phys 18*: 31542–50.

54. Yu, R.H., Wen, X., Liu, J., et al. (2021). A green and high-yield route to recycle waste masks into CNTs/Ni hybrids via catalytic carbonization and their application for superior microwave absorption. *Applied Catalysis B: Environmental 298*: 120544.

55. Lu, M.M., Cao, W.Q., Shi, H.L., et al. (2014). Multi-wall carbon nanotubes decorated with ZnO nanocrystals: mild solution-process synthesis and highly efficient microwave absorption properties at elevated temperature. *Journal of Materials Chemistry A 2*: 10540.

56. Ning, M., Li, J., Kuang, B., et al. (2018). One-step fabrication of N-doped CNTs encapsulating M nanoparticles (M = Fe, Co, Ni) for efficient microwave absorption. *Applied Surface Science 447*: 244–53.

57. Wang, S.P., Zhang, H., Liu, Q.C., et al. (2022). Magnetic carbon nanotubes-based microwave absorbents: Review and perspective. *Synthetic Metals 291*: 117198.

58. Kuang, D.T., Wang, S.L., Hou, L.Z., et al. (2020). A comparative study on the dielectric response and microwave absorption performance of FeNi-capped carbon nanotubes and FeNi-cored carbon nanoparticles. *Nanotechnology 32*: 105701.

59. Kuang, D., Hou, L., Wang, S., et al. (2019). Large-scale synthesis and outstanding microwave absorption properties of carbon nanotubes coated by extremely small FeCo-C core-shell nanoparticles. *Carbon 153*: 52–61.

60. Li, N., Huang, G.W., Li, Y.Q., et al. (2017). Enhanced microwave absorption performance of coated carbon nanotubes by optimizing the Fe_3O_4 nanocoating structure. *ACS Appl Mater Interfaces 9*: 2973–83.

61. Hosseini, H., Mahdavi, H. (2018). Nanocomposite based on epoxy and MWCNTs modified with $NiFe_2O_4$ nanoparticles as efficient microwave absorbing material. *Applied Organometallic Chemistry 32*: e4294.

62. Sabet, M., Jahangiri, H., Ghashghaei, E. (2018). Synthesis of carbon nanotube, graphene, $CoFe_2O_4$, and $NiFe_2O_4$ polypyrrole nanocomposites and study their microwave absorption. *Journal of Materials Science: Materials in Electronics 29*: 10853–63.

63. Pang, H., Sahu, R.P., Duan, Y., et al. (2019). $MnFe_2O_4$-coated carbon nanotubes with enhanced microwave absorption: Effect of CNT content and hydrothermal reaction time. *Diamond and Related Materials 96*: 31–43.

64. Che, R.C., Zhi, C.Y., Liang, C.Y., et al. (2006). Fabrication and microwave absorption of carbon nanotubes/$CoFe_2O_4$ spinel nanocomposite. *Applied Physics Letters 88*: 033105.

65. Sarkar, D., Bhattacharya, A., Nandy, P., et al. (2014). Enhanced broadband microwave reflection loss of carbon nanotube ensheathed Ni–Zn–Co-ferrite magnetic nanoparticles. *Materials Letters 120*: 259–62.

66. Yuan, Y., Wei, S.C., Liang, Y., et al. (2021). Solvothermal assisted synthesis of $CoFe_2O_4$/CNTs nanocomposite and their enhanced microwave absorbing properties. *Journal of Alloys and Compounds 867*: 159040.

67. Ali, A., Rahimian Koloor, S.S., Alshehri, A.H., et al. (2023). Carbon nanotube characteristics and enhancement effects on the mechanical features of polymer-based materials and structures–A review. *Journal of Materials Research and Technology 24*: 6495–521.

68. Chen, B., Li, S., Imai, H., et al. (2015). An approach for homogeneous carbon nanotube dispersion in Al matrix composites. *Materials & Design 72*: 1–8.

69. Simões, S., Viana, F., Reis, M.A.L., et al. (2015). Influence of dispersion/mixture time on mechanical properties of Al–CNTs nanocomposites. *Composite Structures 126*: 114–22.

70. Ren, F.J., Yu, H.J., Wang, L., et al. (2014). Current progress on the modification of carbon nanotubes and their application in electromagnetic wave absorption. *RSC Advances 4*: 14419.

71. Yin, Z.B., Yuan, J.T., Xu, W.W., et al. (2019). Improvement in microstructure and mechanical properties of Ti(C, N) cermet prepared by two-step spark plasma sintering. *Ceramics International 45*: 752–58.

72. Wang, H., Meng, F., Huang, F., et al. (2019). Interface modulating CNTs@PANi hybrids by controlled unzipping of the walls of CNTs to achieve tunable high-performance microwave absorption. *ACS Appl Mater Interfaces 11*: 12142–53.

73. Wu, R., Zhou, K., Yue, C.Y., et al. (2015). Recent progress in synthesis, properties and potential applications of SiC nanomaterials. *Progress in Materials Science 72*: 1–60.

74. Chen, S., Li, W., Li, X., et al. (2019). One-dimensional SiC nanostructures: Designed growth, properties, and applications. *Progress in Materials Science 104*: 138–214.

75. Sun, Y., Cui, H., Yang, G.Z., et al. (2010). The synthesis and mechanism investigations of morphology controllable 1-D SiCnanostructuresvia a novel approach. *CrystEngComm 12*: 1134–38.

76. Chen, S.L., Ying, P.Z., Wang, L., et al. (2014). Controlled growth of SiC flexible field emitters with clear and sharp tips. *RSC Advances 4*: 8376–82.

77. Sun, X.H., Li, C.P., Wong, W.K., et al. (2002). Formation of silicon carbide nanotubes and nanowires via reaction of silicon (from disproportionation of silicon monoxide) with carbon nanotubes. *Journal of the American Chemical Society 124*: 14464–71.

78. Meng, A., Zhang, M., Zhang, J.L., et al. (2012). Synthesis and field emission properties of silicon carbide nanobelts with a median ridge. *CrystEngComm 14*: 6755–60.

79. Yang, T., Chang, X., Chen, J., et al. (2015). B-doped 3C-SiC nanowires with a finned microstructure for efficient visible light-driven photocatalytic hydrogen production. *Nanoscale 7*: 8955–61.

80. Hao, Y.J., Wagner, J.B., Su, D.S., et al. (2006). Beaded silicon carbide nanochains via carbothermal reduction of carbonaceous silica xerogel. *Nanotechnology 17*: 2870–74.

81. Shen, G., Bando, Y., Ye, C., et al. (2006). Synthesis, characterization and field-emission properties of bamboo-like β-SiC nanowires. *Nanotechnology 17*: 3468–72.

82. Wu, R.B., Pan, Y., Yang, G.Y., et al. (2007). Twinned SiC zigzag nanoneedles. *Journal of Physical Chemistry C 111*: 6233–37.

83. Wu, C., Liao, X., Chen, J. (2010). The formation of symmetric SiC bi-nanowires with a Y-shaped junction. *Nanotechnology 21*: 405303.

84. Zhang, D.Q., Alkhateeb, A., Han, H.M., et al. (2003). Silicon carbide nanosprings. *Nano letters 3*: 983–87.

85. Liu, B.D., Bndo, Y., Tang, C.C. et al. (2008). Mn–Si-Catalyzed synthesis and tip-end-induced room temperature ferromagnetism of SiC/SiO$_2$ core–shell heterostructures. *The Journal of Physical Chemistry C* *112*: 18911–15.

86. Liu, H.T., Huang, Z.H., Huang, J.T., et al. (2014). Thermal evaporation synthesis of SiC/SiO$_x$ nanochain heterojunctions and their photoluminescence properties. *Journal of Materials Chemistry C 2*: 7761–67.

87. Qi, X., Zhai, G., Liang, J., et al. (2014). Preparation and characterization of SiC@CNT coaxial nanocables using CNTs as a template. *CrystEngComm 16*: 9697–703.

88. Wang, L., Li, C., Yang, Y., et al. (2014). Large-scale growth of well-aligned SiC tower-like nanowire arrays and their field emission properties. *ACS Applied Materials & Interfaces 7*: 526–33.

89. Han, T., Luo, R., Cui, G., et al. (2019). Effect of SiC nanowires on the high-temperature microwave absorption properties of SiCf/SiC composites. *Journal of the European Ceramic Society 39*: 1743–56.

90. Wu, R., Yang, Z., Fu, M., et al. (2016). In-situ growth of SiC nanowire arrays on carbon fibers and their microwave absorption properties. *Journal of Alloys and Compounds 687*: 833–38.

91. Kumar, A., Singh, S., and Singh, D. (2019). Effect of heat treatment on morphology and microwave absorption behavior of milled SiC. *Journal of Alloys and Compounds 772*: 1017–23.

92. Kuang, J., Jiang, P., Hou, X., et al. (2019). Dielectric permittivity and microwave absorption properties of SiC nanowires with different lengths. *Solid State Sciences 91*: 73–76.

93. Zhang, B., Zhong, Z., Tang, J., et al. (2023). Ultra-light 3D bamboo-like SiC nanowires felt for efficient microwave absorption in the low-frequency region. *Ceramics International 49*: 6368–77.

94. Xiang, Z.N., Xu, B.K., He, Q.C., et al. (2023). Synergistic magnetic/dielectric loss of Fe$_3$Si/SiC composites for efficient electromagnetic wave absorption. *Chemical Engineering Journal 457*: 141198.

95. Cai, Z., Su, L., Wang, H., et al. (2020). Hydrophobic SiC@C nanowire foam with broad-band and mechanically controlled electromagnetic wave absorption. *ACS Applied Materials & Interfaces 12*: 8555–62.

96. Song, L.M., Wu, C.W., Zhi, Q., et al. (2023). Multifunctional SiC aerogel reinforced with nanofibers and nanowires for high-efficiency electromagnetic wave absorption. *Chemical Engineering Journal 467*: 143518.

97. Anasori, B., Lukatskaya, M.R., Gogotsi, Y. (2017). 2D metal carbides and nitrides (MXenes) for energy storage. *Nature Reviews Materials 2*: 16098.

98. Naguib, M., Kurtoglu, M., Presser, V., et al. (2011). Two-dimensional nanocrystals produced by exfoliation of Ti_3AlC_2. *Advanced Materials* 23: 4248–53.

99. Huang, K., Li, Z.J., Lin, J., et al. (2018). Two-dimensional transition metal carbides and nitrides (MXenes) for biomedical applications. *Chemical Society Reviews 47*: 5109–24.

100. Guan, X.M., Yang, Z.H., Zhou, M., et al. (2022). 2D MXene nanomaterials: Synthesis, mechanism, and multifunctional applications in microwave absorption. *Small Structures 3*: 2200102.

101. Huang, P.F., Han, W.Q. (2023). Recent advances and perspectives of Lewis acidic etching route: An emerging preparation strategy for MXenes. *Nano-Micro Letters 15*: 2311.

102. Xu, C., Wang, L.B., Liu, Z.B., et al. (2015). Large-area high-quality 2D ultrathin Mo_2C superconducting crystals. *Nature Materials 14*: 1135–41.

103. Jeon, J., Park, Y., Choi, S., et al. (2018). Epitaxial synthesis of molybdenum carbide and formation of a Mo_2C/MoS_2 hybrid structure via chemical conversion of molybdenum disulfide. *ACS Nano 12*: 338–46.

104. Qing, Y., Zhou, W.C., Luo, F., et al. (2016). Titanium carbide (MXene) nanosheets as promising microwave absorbers. *Ceramics International 42*: 16412–16.

105. Wang, G.Z., Gao, Z., Tang, S.W. et al. (2012). Microwave absorption properties of carbon nanocoils coated with highly controlled magnetic materials by atomic layer deposition. *ACS Nano 6*: 11009–17.

106. Yue, Y., Wang, Y.X., Xu, X.D., et al. (2023). Graphite-ring-stacked carbon nanotubes synthesized during the rescue of $Ti_3C_2T_x$ MXene for dual-peak electromagnetic wave absorption. *Journal of Alloys and Compounds 945*: 169342.

107. Yue, Y., Wang, Y.X., Xu, X.D., et al. (2022). In-situ growth of bamboo-shaped carbon nanotubes and helical carbon nanofibers on $Ti_3C_2T_x$ MXene at ultra-low temperature for enhanced electromagnetic wave absorption properties. *Ceramics International 48*: 6338–46.

108. Peng, H., He, M., Zhou, Y.M., et al. (2022). Low-temperature carbonized biomimetic cellulose nanofiber/MXene composite membrane with excellent microwave absorption performance and tunable absorption bands. *Chemical Engineering Journal 433*: 133269.

109. Xu, X.D., Wang, Y.X., Yue, Y., et al. (2022). Core-shell MXene/nitrogen-doped C heterostructure for wide-band electromagnetic wave absorption at thin thickness. *Ceramics International 48*: 30317–24.

110. Wang, D.C., Lei, Y., Jiao, W., et al. (2020). A review of helical carbon materials structure, synthesis and applications. *Rare Metals 40*: 3–19.

111. Wu, N.N., Hu, Q., Wei, R.B., et al. (2021). Review on the electromagnetic interference shielding properties of carbon based materials and their novel composites: Recent progress, challenges and prospects. *Carbon* *176*: 88–105.

112. Li, Y., Guo, S.R., Li, Y.D., et al. (2023). Electrostatic-spinning construction of HCNTs@$Ti_3C_2T_x$ MXenes hybrid aerogel microspheres for tunable microwave absorption. *Reviews on Advanced Materials Science* *62*: 20230339.

113. Abdah, M., Awan, H., Mehar, M., et al. (2023). Advancements in MXene-polymer composites for high-performance supercapacitor applications. *Journal of Energy Storage 63*: 106942.

114. Yu, M., Huang, Y., Liu, X., et al. (2023). In situ modification of MXene nanosheets with polyaniline nanorods for lightweight and broadband electromagnetic wave absorption. *Carbon 208*: 311–21.

115. Liu, T.S., Liu, N., An, Q.D., et al. (2019). Designed construction of $Ti_3C_2T_x$@PPY composites with enhanced microwave absorption performance. *Journal of Alloys and Compounds 802*: 445–57.

116. Li, W., Yu, Z., Wen, Q., et al. (2022). Ceramic-based electromagnetic wave absorbing materials and concepts towards lightweight, flexibility and thermal resistance. *International Materials Reviews 68*: 487–520.

117. Wang, H., Zhao, J., Wang, Z., et al. (2023). Bird-nest-like multi-interfacial MXene@SiCNWs@Co/C hybrids with enhanced electromagnetic wave absorption. *ACS Applied Materials & Interfaces 15*: 4580–90.

118. Wang, S.J., Zhang, Z.Y., Fan, X.X., et al. (2023). Embedment of hollow SiO_2 spheres into flower-like $Ti_3C_2T_x$ MXene framework with decoration of carbon for efficient microwave absorption. *Journal of Alloys and Compounds 960*: 170724.

119. Zhang, S., Li, J., Jin, X., et al. (2023). Current advances of transition metal dichalcogenides in electromagnetic wave absorption: A brief review. *International Journal of Minerals, Metallurgy and Materials 30*: 428–45.

120. Liu, Z.H., Cui, Y.H., Li, Q., et al. (2022). Fabrication of folded MXene/MoS_2 composite microspheres with optimal composition and their microwave absorbing properties. *Journal of Colloid and Interface Science 607*: 633–44.

121. Fan, B.B., Ansar, M.T., Chen, Q.Q., et al. (2022). Microwave-assisted hydrothermal synthesis of 2D/2D MoS_2/$Ti_3C_2T_x$ heterostructure for enhanced microwave absorbing performance. *Journal of Alloys and Compounds 923*: 166253.

122. Guo, S.Y., Guan, H.L., Li, Y., et al. (2021). Dual-loss $Ti_3C_2T_x$ MXene/$N_{i0.6}Zn_{0.4}Fe_2O_4$ heterogeneous nanocomposites for highly efficient electromagnetic wave absorption. *Journal of Alloys and Compounds 887*: 161298.

123. Lv, Y.H., Ye, X.Y., Chen, S., et al. (2023). $Ti_3C_2T_x$/Co-MOF derived TiO_2/ Co/C composites for broadband and high absorption of electromagnetic wave. *Applied Surface Science 622*: 156935.

124. Browne, M.P., Sofer, Z., Pumera, M. (2019). Layered and two dimensional metal oxides for electrochemical energy conversion. *Energy & Environmental Science 12*: 41–58.

125. Qin, M., Liang, H.S., Zhao, X.R., et al. (2020). Glycine-assisted solution combustion synthesis of $NiCo_2O_4$ electromagnetic wave absorber with wide absorption bandwidth. *Ceramics International 46*: 22313–20.

126. Deng, X.H., Kang, X.M., Li, M., et al. (2020). Coupling efficient biomass upgrading with H_2 production via bifunctional Cu_xS@NiCo-LDH core–shell nanoarray electrocatalysts. *Journal of Materials Chemistry A 8*: 1138–46.

127. Wijitwongwan, R., Intasa-ard, S., Ogawa, M. (2019). Preparation of layered double hydroxides toward precisely designed hierarchical organization. *ChemEngineering 3*: 68.

128. Li, X., Ren, J., Sridhar, D., et al. (2023). Progress of layered double hydroxide-based materials for supercapacitors. *Materials Chemistry Frontiers 7*: 1520–61.

129. Wan, L., Wang, Y.M., Du, C., et al. (2022). NiAlP@Cobalt substituted nickel carbonate hydroxide hetero structure engineered for enhanced supercapacitor performance. *Journal of Colloid and Interface Science 609*: 1–11.

130. Kameliya, J., Verma, A., Dutta, P., et al. (2023). Layered double hydroxide materials: A review on their preparation, characterization, and applications. *Inorganics 11*: 121.

131. Islam, S., Mia, M.M., Shah, S.S., et al. (2022). Recent advancements in electrochemical deposition of metal-based electrode materials for electrochemical supercapacitors. *The Chemical Record 22*: e202200013.

132. Li, B., Guo, M., Chen, X., et al. (2022). Hydrothermally synthesized N and S co-doped mesoporous carbon microspheres from poplar powder for supercapacitors with enhanced performance. *Advanced Composites and Hybrid Materials 5*: 2306–16.

133. Ban, S., Xie, J., Wang, Y.J., et al. (2015). Insight into the nanoscale mechanism of rapid H_2O transport within a graphene oxide membrane: Impact of oxygen functional group clustering. *ACS Applied Materials & Interfaces 8*: 321–32.

134. Wen, B., Yang, H.B., Wang, L., et al. (2020). Hierarchical Co_xAl_y layered double hydroxide@carbon composites derived from metal–organic frameworks with efficient broadband electromagnetic wave absorption. *Journal of Materials Chemistry C 8*: 16418–26.

135. Zhang, Y., Li, L., Du, C.L., et al. (2023). Controllable coating NiAl-layered double hydroxides on carbon nanofibers as anticorrosive microwave absorbers. *Journal of Materials Science & Technology* 151: 109–18.

136. Wang, Y., Di, X C., Chen, J., et al. (2022). Multi-dimensional C@NiCo-LDHs@Ni aerogel: Structural and componential engineering towards efficient microwave absorption, anti-corrosion and thermal-insulation. *Carbon 191*: 625–35.

137. Wang, J.W., Jia, Z.R., Liu, X.H., et al. (2021). Construction of 1D heterostructure NiCo@C/ZnO nanorod with enhanced microwave absorption. *Nano-Micro Letters 13*: 175.

138. Zhao, Z.H., Kou, K.C., Zhang, L.M., et al. (2020). High efficiency electromagnetic wave absorber derived from transition metal layered double hydroxides. *Journal of Colloid and Interface Science 579*: 733–40.

139. Dai, X.J., Zhou, Q., Dong, L.C., et al. (2023). Nickel iron layered double hydroxide nanostructures composited with carbonyl iron powder for microwave absorption. *ACS Applied Nano Materials 6*: 3472–83.

140. Manzeli, S., Ovchinnikov, D., Pasquier, D., et al. (2017). 2D transition metal dichalcogenides. *Nature Reviews Materials 2*: 17033.

141. Wang, P.J., Yang, D.R., Pi, X.D. (2021). Toward wafer-scale production of 2D transition metal chalcogenides. *Advanced Electronic Materials 7*: 2100278.

142. Li, B., Wang, F., Wang, K., et al. (2022). Metal sulfides based composites as promising efficient microwave absorption materials: A review. *Journal of Materials Science & Technology 104*: 244–68.

143. Hu, Y.X., Dai, M.J., Feng, W., et al. (2021). Ultralow power optical synapses based on MoS_2 layers by indium-induced surface charge doping for bio-mimetic eyes. *Advanced Materials 33*: 2104960.

144. Wang, J.Q., Liu, L., Jiao, S.L., et al. (2020). Hierarchical carbon fiber@MXene@MoS_2 core-sheath synergistic microstructure for tunable and efficient microwave absorption. *Advanced Functional Materials 30*: 2002595.

145. Ning, M.Q., Jiang, P.H., Ding, W., et al. (2021). Phase manipulating toward molybdenum disulfide for optimizing electromagnetic wave absorbing in gigahertz. *Advanced Functional Materials 31*: 2011229.

146. Ning, M.Q., Lu, M.M., Li, J.B., et al. (2015). Two-dimensional nanosheets of MoS_2: a promising material with high dielectric properties and microwave absorption performance. *Nanoscale 7*: 15734–40.

147. Li, S.W., Liu, Y.C., Zhao, X.D., et al. (2021). Molecular engineering on MoS_2 enables large interlayers and unlocked basal planes for high-performance aqueous Zn-Ion storage. *Angewandte Chemie International Edition 60*: 20286–93.

148. Wang, J.C., Zhang, L.Y., Sun, K, et al. (2019). Improving ionic/electronic conductivity of MoS_2 Li-ion anode via manganese doping and structural optimization. *Chemical Engineering Journal 372*: 665–72.

149. Ye, G.L., Gong, Y.J., Lin, J.H., et al. (2016). Defects engineered monolayer MoS_2 for improved hydrogen evolution reaction. *Nano letters 16*: 1097–103.

150. Kong, D.S., Wang, H.T., Cha, J.J., et al. (2013). Synthesis of MoS_2 and $MoSe_2$ films with vertically aligned layers. *Nano letters 13*: 1341–47.

151. Zhu, P., Chen, Y., Zhou, Y., et al. (2018). A metallic MoS_2 nanosheet array on graphene-protected Ni foam as a highly efficient electrocatalytic hydrogen evolution cathode. *Journal of Materials Chemistry A 6*: 16458–64.

152. Zhang, J.F., Zhao, M., Zhao, Y.B., et al. (2023). Microwave-assisted hydrothermal synthesis of Fe-doped $1T/2H$-MoS_2 few-layer nanosheets for efficient electromagnetic wave absorbing. *Journal of Alloys and Compounds 947*: 169544.

153. Wu, M., Wang, H.C., Liang, X.H., et al. (2023). Flower-like $1T/2H$-$MoS_2@\alpha$-Fe_2O_3 with enhanced electromagnetic wave absorption capabilities in the low frequency range. *Journal of Materials Chemistry C 11*: 2897–910.

154. Geng, H.R., Guo, Y., Zhang, X., et al. (2023). Combination strategy of large interlayer spacing and active basal planes for regulating the microwave absorption performance of MoS2/MWCNT composites at thin absorber level. *Journal of Colloid and Interface Science 648*: 12–24.

155. Xing, L.S., Wu, Z.C., Wang, L., et al. (2020). Polarization-enhanced three-dimensional $Co_3O_4/MoO_2/C$ flowers as efficient microwave absorbers. *Journal of Materials Chemistry C 8*: 10248–56.

156. Cao, F.H., Xu, J., Zhao, Z.B., et al. (2021). Hierarchically three-dimensional structure assembled with yolk-shelled spheres-supported nitrogen-doped carbon nanotubes for electromagnetic wave absorption. *Carbon 185*: 177–85.

157. Zhao, Y.P., Zuo, X.Q., Guo, Y., et al. (2021). Structural engineering of hierarchical aerogels comprised of multi-dimensional gradient carbon nanoarchitectures for highly efficient microwave absorption. *Nano-Micro Letters 13*: 2311.

158. Zhao, H.Q., Jin, C.Q., Yang, X., et al. (2023). Synthesis of a one-dimensional carbon nanotube-decorated three-dimensional crucifix carbon architecture embedded with $Co_7Fe_3/Co_{5.47}N$ nanoparticles for high-performance microwave absorption. *Journal of Colloid and Interface Science 645*: 22–32.

159. Mao, F.Z., Fan, X.K., Long, L., et al. (2023). Constructing 3D hierarchical CNTs/VO_2 composite microspheres with superior electromagnetic absorption performance. *Ceramics International 49*: 16924–31.

160. Dong, Y., Zhu, X., Pan, F., et al. (2021). Fire-retardant and thermal insulating honeycomb-like NiS_2/SnS_2 nanosheets@3D porous carbon hybrids for high-efficiency electromagnetic wave absorption. *Chemical Engineering Journal 426*: 131272.

161. Gu, W., Lv, J., Quan, B., et al. (2019). Achieving MOF-derived one-dimensional porous ZnO/C nanofiber with lightweight and enhanced microwave response by an electrospinning method. *Journal of Alloys and Compounds 806*: 983–91.

162. Wang, K., Chen, Y., Tian, R., et al. (2018). Porous Co–C core–shell nanocomposites derived from Co-MOF-74 with enhanced electromagnetic wave absorption performance. *ACS Applied Materials & Interfaces 10*: 11333–42.

163. Zhang, X., Qiao, J., Zhao, J., et al. (2019). High-efficiency electromagnetic wave absorption of cobalt-decorated NH_2-UIO-66-derived porous ZrO_2/C. *ACS Applied Materials & Interfaces 11*: 35959–68.

164. Miao, P., Zhou, R., Chen, K., et al. (2020). Tunable electromagnetic wave absorption of supramolecular Isomer-derived nanocomposites with different morphology. *Advanced Materials Interfaces 7*: 1901820.

165. Xiang, Z., Song, Y., Xiong, J., et al. (2019). Enhanced electromagnetic wave absorption of nanoporous Fe_3O_4@carbon composites derived from metal-organic frameworks. *Carbon 142*: 20–31.

166. Li, Z., Zhou, J., Wei, B., et al. (2023). Fe/Fe_3C@N-doped carbon composite materials derived from MOF with improved framework stability for strong microwave absorption. *Synthetic Metals 293*: 117272.

167. Qiu, Y., Yang, H., Wen, B., et al. (2021). Facile synthesis of nickel/carbon nanotubes hybrid derived from metal organic framework as a lightweight, strong and efficient microwave absorber. *Journal of Colloid and Interface Science 590*: 561–70.

168. Liang, X., Wang, C., Yao, Z., et al. (2022). A facile synthesis of Fe/C composite derived from Fe-metal organic frameworks: Electromagnetic wave absorption with thin thickness. *Journal of Alloys and Compounds 922*: 166299.

169. Ren, S., Yu, H., Wang, L., et al. (2022). State of the art and prospects in metal-organic framework-derived microwave absorption materials. *Nano-Micro Letters 14*: 68.

170. Wu, Q., Jin, H., Chen, W., et al. (2018). Graphitized nitrogen-doped porous carbon composites derived from ZIF-8 as efficient microwave absorption materials. *Materials Research Express 5*: 065602.

171. Zhang, X., Qiao, J., Liu, C., et al. (2020). A MOF-derived ZrO_2/C nanocomposite for efficient electromagnetic wave absorption. *Inorganic Chemistry Frontiers 7*: 385–93.

172. Hou, W., Peng, K., Li, S., et al. (2023). Designing flower-like MOFs-derived N-doped carbon nanotubes encapsulated magnetic NiCo composites with multi-heterointerfaces for efficient electromagnetic wave absorption. *Journal of Colloid and Interface Science* 646: 265–74.

173. Lu, C., Geng, H., Zhu, Y., et al. (2023). 3D flower-shape Co/Cu bimetallic nanocomposites with excellent wideband electromagnetic microwave absorption. *Applied Surface Science* 615.

174. Yang, K., Cui, Y., Wan, L., et al. (2022). MOF-derived magnetic-dielectric balanced Co@ZnO@N-doped carbon composite materials for strong microwave absorption. *Carbon* 190: 366–75.

175. Wang, L., Wen, B., Yang, H., et al. (2020). Hierarchical nest-like structure of Co/Fe MOF derived CoFe@C composite as wide-bandwidth microwave absorber. *Composites Part A: Applied Science and Manufacturing* 135: 105958.

176. Zhang, X., Tian, X.L., Liu, C., et al. (2022). MnCo-MOF-74 derived porous MnO/Co/C heterogeneous nanocomposites for high-efficiency electromagnetic wave absorption. *Carbon* 194: 257–66.

177. Jiang, C., Wang, Y., Zhao, Z., et al. (2023). Multifunctional three-dimensional porous MOFs derived Fe/C/carbon foam for microwave absorption, thermal insulation and infrared stealth. *Ceramics International* 49: 18861–69.

178. Xu, X., Ran, F., Lai, H., et al. (2019). In situ confined bimetallic metal–organic framework derived nanostructure within 3D interconnected bamboo-like carbon nanotube networks for boosting electromagnetic wave absorbing performances. *ACS Applied Materials & Interfaces* 11: 35999–6009.

179. Jin, L., Lin, Y., Wen, B., et al. (2022). Modification of bimetallic MIL-53 derived FeCo/C composite for high electromagnetic wave absorber. *Journal of Alloys and Compounds* 906: 164330.

180. Gao, S., An, Q., Xiao, Z., et al. (2018). Significant promotion of porous architecture and magnetic Fe_3O_4 NPs inside honeycomb-like carbonaceous composites for enhanced microwave absorption. *RSC Advances* 8: 19011–23.

181. Wu, Z., Tian, K., Huang, T., et al. (2018). Hierarchically porous carbons derived from biomasses with excellent microwave absorption performance. *ACS Applied Materials & Interfaces* 10: 11108–15.

182. Xi, J., Zhou, E., Liu, Y., et al. (2017). Wood-based straightway channel structure for high performance microwave absorption. *Carbon* 124: 492–98.

183. Rong, J., Qiu, F., Zhang, T., et al. (2017). A facile strategy toward 3D hydrophobic composite resin network decorated with biological ellipsoidal

structure rapeseed flower carbon for enhanced oils and organic solvents selective absorption. *Chemical Engineering Journal 322*: 397–407.

184. Gong, Y., Li, D., Luo, C., et al. (2017). Highly porous graphitic biomass carbon as advanced electrode materials for supercapacitors. *Green Chemistry 19*: 4132–40.

185. Lv, W., Wen, F., Xiang, J., et al. (2015). Peanut shell derived hard carbon as ultralong cycling anodes for lithium and sodium batteries. *Electrochimica Acta 176*: 533–41.

186. Zhu, Y., Chen, M., Li, Q., et al. (2018). A porous biomass-derived anode for high-performance sodium-ion batteries. *Carbon 129*: 695–701.

187. Gaddam, R.R., Yang, D., Narayan, R., et al. (2016). Biomass derived carbon nanoparticle as anodes for high performance sodium and lithium ion batteries. *Nano Energy 26*: 346–52.

188. Cheng, Y., Zhao, H., Lv, H., et al. (2019). Lightweight and flexible cotton aerogel composites for electromagnetic absorption and shielding applications. *Advanced Electronic Materials 6*: 1900796.

189. Rezma, S., Birot, M., Hafiane, A., et al. (2017). Physically activated microporous carbon from a new biomass source: Date palm petioles. *Comptes Rendus Chimie 20*: 881–87.

190. Lu, H., Zhao, X.S. (2017). Biomass-derived carbon electrode materials for supercapacitors. *Sustainable Energy & Fuels 1*: 1265–81.

191. Tang, D., Luo, Y., Lei, W., et al. (2018). Hierarchical porous carbon materials derived from waste lentinus edodes by a hybrid hydrothermal and molten salt process for supercapacitor applications. *Applied Surface Science 462*: 862–71.

192. Wang, H., Zhang, Y., Wang, Q., et al. (2019). Biomass carbon derived from pine nut shells decorated with NiO nanoflakes for enhanced microwave absorption properties. *RSC Advances 9*: 9126–35.

193. Shi, Q., Zhao, Y., Li, M., et al. (2023). 3D lamellar skeletal network of porous carbon derived from hull of water chestnut with excellent microwave absorption properties. *Journal of Colloid and Interface Science 641*: 449–58.

194. Liu, C., Shi, G., Wang, G., et al. (2019). Preparation and electrochemical studies of electrospun phosphorus doped porous carbon nanofibers. *RSC Advances 9*: 6898–906.

195. Chang, Q., Li, C., Sui, J., et al. (2022). Cage-like eggshell membrane-derived Co-Co$_x$S$_y$-Ni/N,S-codoped carbon composites for electromagnetic wave absorption. *Chemical Engineering Journal 430*: 132650.

196. Wu, P., Feng, Y., Xu, J., et al. (2023). Ultralight N-doped platanus acerifolia biomass carbon microtubes/RGO composite aerogel with enhanced

mechanical properties and high-performance microwave absorption. *Carbon 202*: 194–203.

197. Chu, Z.Y., Cheng, H.F., Xie, W., et al. (2012). Effects of diameter and hollow structure on the microwave absorption properties of short carbon fibers. *Ceramics International 38*: 4867–73.

198. Cheng, Y., Li, Z., Li, Y., et al. (2018). Rationally regulating complex dielectric parameters of mesoporous carbon hollow spheres to carry out efficient microwave absorption. *Carbon 127*: 643–52.

199. Wen, G., Zhao, X., Liu, Y., et al. (2018). Simple, controllable fabrication and electromagnetic wave absorption properties of hollow Ni nanosphere. *Journal of Materials Science: Materials in Electronics 30*: 2166–76.

200. Zhang, S., Cheng, B., Jia, Z., et al. (2022). The art of framework construction: hollow-structured materials toward high-efficiency electromagnetic wave absorption. *Advanced Composites and Hybrid Materials 5*: 1658–98.

201. Guo, Z.Y., Li, C.X., Gao, M., et al. (2021). Mn-O covalency governs the intrinsic activity of Co-Mn spinel oxides for boosted peroxymonosulfate activation. *Angewandte Chemie International Edition 60*: 274–80.

202. Zhang, Q., Wang, W., Goebl, J., et al. (2009). Self-templated synthesis of hollow nanostructures. *Nano Today 4*: 494–507.

203. Zeng, H.C. (2006). Synthetic architecture of interior space for inorganic nanostructures. *Journal of Materials Chemistry 16*: 649–62.

204. Zeng, H.C. (2007). Ostwald ripening: A synthetic approach for hollow nanomaterials. *Current Nanoscience 3*: 177–81.

205. Weng, W., Lin, J., Du, Y., et al. (2018). Template-free synthesis of metal oxide hollow micro-/nanospheres via Ostwald ripening for lithium-ion batteries. *Journal of Materials Chemistry A 6*: 10168–75.

206. Anderson, B.D., Tracy, J.B. (2014). Nanoparticle conversion chemistry: Kirkendall effect, galvanic exchange, and anion exchange. *Nanoscale 6*: 12195–216.

207. El Mel, A.A., Buffière, M., Tessier, P.Y., et al. (2013). Highly ordered hollow oxide nanostructures: The Kirkendall effect at the nanoscale. *Small 9*: 2838–43.

208. Wang, J., Han, M., Liu, Y., et al. (2023). Multifunctional microwave absorption materials of multiscale cobalt sulfide/diatoms co-doped carbon aerogel. *Journal of Colloid and Interface Science 646*: 970–79.

209. Huang, W., Wang, X., Wang, Y., et al. (2023). Construction of hollow copper selenide boxes and S-doping for enhanced electromagnetic wave absorption. *Journal of Alloys and Compounds 962*: 171168.

210. Zhao, B., Guo, X., Zhou, Y.Y., et al. (2017). Constructing hierarchical hollow CuS microspheres via a galvanic replacement reaction and their use as wide-band microwave absorbers. *CrystEngComm 19*: 2178–86.

211. Gu, Y., Wu, Y.N., Li, L.C., et al. (2017). Controllable modular growth of hierarchical MOF-on-MOF architectures. *Angewandte Chemie International Edition 56*: 15658–62.

212. Liu, C., Wang, J., Wan, J., et al. (2021). MOF-on-MOF hybrids: Synthesis and applications. *Coordination Chemistry Reviews 432*: 213743.

213. Wu, M.X., Wang, Y., Zhou, G., et al. (2020). Core-shell MOFs@ MOFs: Diverse designability and enhanced selectivity. *ACS Applied Materials & Interfaces 12*: 54285–305.

214. Hong, D.H., Shim, H.S., Ha, J., et al. (2021). MOF-on-MOF architectures: applications in separation, catalysis and sensing. *Bulletin of the Korean Chemical Society 42*: 956–69.

215. Zhang, Z., Cai, Z., Wang, Z., et al. (2021). A review on metal–organic framework-derived porous carbon-based novel microwave absorption materials. *Nano-Micro Letters 13*: 56.

216. Wu, F., Li, Q., Liu, Z., et al. (2021). Fabrication of binary MOF-derived hybrid nanoflowers via selective assembly and their microwave absorbing properties. *Carbon 182*: 484–96.

217. Wu, F., Wan, L., Li, Q., et al. (2022). Ternary assembled MOF-derived composite: Anisotropic epitaxial growth and microwave absorption. *Composites Part B: Engineering 236*: 109839.

218. Sun, M., Li, Z., Wei, B., et al. (2023). MOFs derived Fe/Co/C heterogeneous composite absorbers for efficient microwave absorption. *Synthetic Metals 292*: 117229.

219. Tian, C., Du, Y., Xu, P., et al. (2015). Constructing uniform core–shell PPy@PANI composites with tunable shell thickness toward enhancement in microwave absorption. *ACS Applied Materials & Interfaces 7*: 20090–99.

220. Wu, H., Wu, G., Ren, Y., et al. (2015). Co^{2+}/Co^{3+} ratio dependence of electromagnetic wave absorption in hierarchical $NiCo_2O_4$–$CoNiO_2$ hybrids. *Journal of Materials Chemistry C 3*: 7677–90.

221. Cheng, J., Zhang, H., Xiong, Y., et al. (2021). Construction of multiple interfaces and dielectric/magnetic heterostructures in electromagnetic wave absorbers with enhanced absorption performance: A review. *Journal of Materiomics 7*: 1233–63.

222. Zhu, H., Jiao, Q., Fu, R., et al. (2022). Cu/NC@Co/NC composites derived from core-shell Cu-MOF@Co-MOF and their electromagnetic wave absorption properties. *Journal of Colloid and Interface Science 613*: 182–93.

223. Li, F., Li, C., Xie, Q., et al. (2023). MOF-derived core-shell Co_9S_8 @MoS_2 nanocubes anchored on RGO to construct heterostructure for high-efficiency microwave attenuation. *Journal of Alloys and Compounds 935*: 168106.

224. Yu, R., Xia, Y., Pei, X., et al. (2022). Micro-flower like core-shell structured ZnCo@C@1T-2H-MoS$_2$ composites for broadband electromagnetic wave absorption and photothermal performance. *Journal of Colloid and Interface Science 622*: 261–71.

225. Sun, X., Li, X., Chen, P., et al. (2023). Construction of urchin-like multiple core-shelled Co/CoS$_2$@NC@MoS$_2$ composites for effective microwave absorption. *Journal of Alloys and Compounds 936*: 168243.

226. Zhang, R., Guo, D., Liu, Q., et al. (2023). Engineering core–shell heterocomposites of integrated 0D ZIF-8 (Zn) attached on 1D MIL-68 (In) nanorods for high performance electromagnetic wave absorption. *Applied Surface Science 621*: 156898.

227. Wei, S., Shi, Z., Li, X., et al. (2022). Bimetallic MOF-derived CoFe@ nitrogen-doped carbon composites for wide bandwidth and excellent microwave absorption. *Journal of Alloys and Compounds 910*: 164861.

228. Jiang, R., Wang, Y., Wang, J., et al. (2023). Controlled formation of multiple core-shell structures in metal-organic frame materials for efficient microwave absorption. *Journal of Colloid and Interface Science 648*: 25–36.

229. Song, Y., Liu, X., Gao, Z., et al. (2022). Core-shell Ag@C spheres derived from Ag-MOFs with tunable ligand exchanging phase inversion for electromagnetic wave absorption. *Journal of Colloid and Interface Science 620*: 263–72.

230. Liu, J., Cheng, J., Che, R., et al. (2013). Synthesis and microwave absorption properties of yolk–shell microspheres with magnetic iron oxide cores and hierarchical copper silicate shells. *ACS Applied Materials & Interfaces 5*: 2503–09.

231. Miao, P., Chen, J., Chen, J., et al. (2023). Review and perspective of tailorable metal-organic framework for enhancing wicrowave absorption. *Chinese Journal of Chemistry 41*: 1080–98.

232. Yu, M., Liang, C., Liu, M., et al. (2014). Yolk–shell Fe$_3$O$_4$@ZrO$_2$ prepared by a tunable polymer surfactant assisted sol–gel method for high temperature stable microwave absorption. *Journal of Materials Chemistry C 2*: 7275–83.

233. Liu, J., Cheng, J., Che, R., et al. (2012). Double-shelled yolk–shell microspheres with Fe$_3$O$_4$ cores and SnO$_2$ double shells as high-performance microwave absorbers. *The Journal of Physical Chemistry C 117*: 489–95.

234. Liu, J., Xu, J., Che, R., et al. (2013). Hierarchical Fe$_3$O$_4$@TiO$_2$ yolk–shell microspheres with enhanced microwave-absorption properties. *Chemistry – A European Journal 19*: 6746–52.

235. Wang, K., Wan, G., Wang, G., et al. (2018). The construction of carbon-coated Fe$_3$O$_4$ yolk-shell nanocomposites based on volume shrinkage from

the release of oxygen anions for wide-band electromagnetic wave absorption. *Journal of Colloid and Interface Science 511*: 307–17.

236. Feng, J., Wang, Y., Hou, Y., et al. (2017). Tunable design of yolk–shell $ZnFe_2O_4$@RGO@TiO_2 microspheres for enhanced high-frequency microwave absorption. *Inorganic Chemistry Frontiers 4*: 935–45.

237. Ning, M., Lei, Z., Tan, G., et al. (2021). Dumbbell-like Fe_3O_4@N-doped carbon@2H/1T-MoS_2 with tailored magnetic and dielectric loss for efficient microwave absorbing. *ACS Applied Materials & Interfaces 13*: 47061–71.

238. Priebe, M., Fromm, K.M. (2014). Nanorattles or yolk–shell nanoparticles— What are they, how are they made, and what are they good for? *Chemistry – A European Journal 21*: 3854–74.

239. Liu, Q., Cao, Q., Bi, H., et al. (2015). CoNi@SiO_2@TiO_2 and CoNi@ Air@TiO_2 microspheres with strong wideband microwave absorption. *Advanced Materials 28*: 486–90.

240. Yang, K., Cui, Y., Li, Q., et al. (2021). Bimetallic MOFs-derived yolk-shell structure ZnCo/NC@TiO_2 and its microwave absorbing properties. *Applied Surface Science 556*: 149715.

241. Cui, Y., Liu, Z., Li, X., et al. (2021). MOF-derived yolk-shell Co@ZnO/ Ni@NC nanocage: Structure control and electromagnetic wave absorption performance. *Journal of Colloid and Interface Science 600*: 99–110.

242. Qiu, Y., Yang, H., Cheng, Y., et al. (2022). MOFs derived flower-like nickel and carbon composites with controllable structure toward efficient microwave absorption. *Composites Part A: Applied Science and Manufacturing 154*: 106772.

243. Liang, X., Quan, B., Sun, Y., et al. (2017). Multiple interfaces structure derived from Metal-Organic Frameworks for excellent electromagnetic wave absorption. *Particle & Particle Systems Characterization 34*: 1700006.

244. Qiu, Y., Yang, H., Cheng, Y., et al. (2022). Structure design of Prussian blue analogue derived CoFe@C composite with tunable microwave absorption performance. *Applied Surface Science 571*.

245. Zhang, Y., Gao, S., Xing, H., et al. (2019). In situ carbon nanotubes encapsulated metal nickel as high-performance microwave absorber from Ni–Zn metal–organic framework derivative. *Journal of Alloys and Compounds 801*: 609–18.

246. He, P., Ma, W., Xu, J., et al. (2023). Hierarchical and orderly surface conductive networks in yolk–shell Fe_3O_4@C@Co/N-doped C microspheres for enhanced microwave absorption. *Small 19*: 1.

247. Ma, W.J., He, P., Xu, J., et al. (2022). Self-assembly magnetized 3D hierarchical graphite carbon-based heterogeneous yolk–shell nanoboxes

with enhanced microwave absorption. *Journal of Materials Chemistry A 10*: 11405–13.

248. Barrios, E., Fox, D., Li Sip, Y.Y., et al. (2019). Nanomaterials in advanced, high-performance aerogel composites: A review. *Polymers 11*: 726.

249. Zhang, W., Zheng, Y., Zhang, X., et al. (2019). Synthesis and mechanism investigation of wide-bandwidth Ni@MnO$_2$ NS foam microwave absorbent. *Journal of Alloys and Compounds 792*: 945–52.

250. Guo, Y., Su, J., Bian, T., et al. (2023). Construction and application of carbon aerogels in microwave absorption. *Physical Chemistry Chemical Physics 25*: 8244–62.

251. Maleki, H. (2016). Recent advances in aerogels for environmental remediation applications: A review. *Chemical Engineering Journal 300*: 98–118.

252. Li, H., Li, J., Thomas, A., et al. (2019). Ultra-high surface area nitrogen-doped carbon aerogels derived from a Schiff-base porous organic polymer aerogel for CO$_2$ storage and supercapacitors. *Advanced Functional Materials 29*: 1904785.

253. Nardecchia, S., Carriazo, D., Ferrer, M.L., et al. (2013). Three dimensional macroporous architectures and aerogels built of carbon nanotubes and/or graphene: synthesis and applications. *Chemical Society Reviews 42*: 794–830.

254. Meng, F., Wang, H., Wei, et al. (2018). Generation of graphene-based aerogel microspheres for broadband and tunable high-performance microwave absorption by electrospinning-freeze drying process. *Nano Research 11*: 2847–61.

255. Wang, Z., Wei, R., Gu, J., et al. (2018). Ultralight, highly compressible and fire-retardant graphene aerogel with self-adjustable electromagnetic wave absorption. *Carbon 139*: 1126–35.

256. Liu, N., Zhang, X., Dou, Y., et al. (2022). Design of carbon aerogels with variable surface morphology for electromagnetic wave absorption. *Carbon 200*: 271–80.

257. Zhang, Z., Tan, J., Gu, W., et al. (2020). Cellulose-chitosan framework/polyailine hybrid aerogel toward thermal insulation and microwave absorbing application. *Chemical Engineering Journal 395*: 125190.

258. Zhao, H.B., Cheng, J.B., Zhu, J.Y., et al. (2019). Ultralight CoNi/rGO aerogels toward excellent microwave absorption at ultrathin thickness. *Journal of Materials Chemistry C 7*: 441–48.

259. Qin, M., Zhang, L., Zhao, X., et al. (2021). Lightweight Ni foam-based ultra-broadband electromagnetic wave absorber. *Advanced Functional Materials 31*: 2103436.

260. Han, M., Yin, X., Hou, Z., et al. (2017). Flexible and thermostable graphene/SiC nanowire foam composites with tunable electromagnetic wave absorption properties. *ACS Applied Materials & Interfaces 9*: 11803–10.
261. Sun, N., Li, W., Wei, S., et al. (2021). Facile synthesis of lightweight 3D hierarchical $NiCo_2O_4$ nanoflowers/reduced graphene oxide composite foams with excellent electromagnetic wave absorption performance. *Journal of Materials Science & Technology 91*: 187–99.

CHAPTER 4

Next-generation electromagnetic wave (EMW) absorption materials

IN THE DETAILED COLLATION and analysis of the compositional and structural design and data analysis of different types of materials in the previous section, many electromagnetic wave (EMW)-absorbing materials with strong absorption were introduced. While the first generation of EMW-absorbing materials exhibits promising EMW absorption capabilities in theoretical analyses, meeting current market demands and practical applications presents challenges. Real-world applications require environments more complex than theoretical data alone, emphasizing the need for diverse structural morphologies and a shift toward practical applications in future research. For example, for applications in outdoor high-temperature environments, it is essential to design heat-resistant, self-healing, electronically controllable, and superhydrophobic materials. It is also possible for these new-generation materials to achieve strong absorption through rational component optimization and structural design. However, it is difficult for these traditional design methods to effectively balance EMW

DOI: 10.1201/9781003490098-4

absorption performance and additional capabilities. Therefore, some researchers have attempted to incorporate auxiliary strategies to provide better solutions. In this chapter, we summarize the new generation of electromagnetic absorbing materials, organizing their design concepts and principles.

4.1 SUPERHYDROPHOBIC MATERIALS

Hydrophobicity is defined as the phenomenon where a liquid is not able to wet the solid surface, indicated by a contact angle $\theta > 90°$ on the material surface. Materials exhibiting a stable contact angle $\theta > 150°$ and a rolling contact angle $\alpha < 10°$ on their surface are called superhydrophobic. Based on the difference between droplets and surface adhesion in the superhydrophobic state, the morphology of the superhydrophobic material is divided into five forms: Wenzel state, Cassie state, lotus leaf state, Wenzel-Cassie state, and gecko state. Among these five forms, the adhesion of water droplets to the surface is extremely small, and the "lotus leaf state" with a rolling angle of $\alpha < 10°$ is the most widely studied state of a superhydrophobic material. In daily life, superhydrophobic organisms can be found everywhere, such as lotus leaves, water phosphorus, rose petals, butterfly wings, rice leaves, taro leaves, and so on[1-11]. These superhydrophobic materials are often used in self-cleaning, corrosion protection, oil-water separation, medicine and water pollution. Importantly, researchers have found that superhydrophobic materials can protect electronic devices from corrosion in humid environments, thereby extending their usefulness. In general, traditional inorganic materials show poor superhydrophobicity due to the absence of hydrophobic bonds, while hydrophobic materials are mainly found in organic matter and polymers.

However, these hydrophobic materials show weak microwave absorption properties[12]. Therefore, in order to achieve high absorption performance and good hydrophobic properties, researchers have attempted to compound inorganic materials with superhydrophobic materials. For example, Du et al. prepared $FeNi/NiFe_2O_4/NiO/C$ nanofibers by an electrospinning technique and a in situ calcination process, and introducing $Ni(acac)_2$, $Fe(acac)_3$ and DMF/PAN solution as a precursor, as shown in Fig. 4.1(a)[13]. The influence of different structural characteristics formed by different calcination temperatures on the absorption performance of the composites was investigated. The $FeNi/NiFe_2O_4/NiO/CNFs$ exhibited good hydrophobic properties because magnetic nanoparticles were deposited on the surface of the CNFs to form a rough nanostructure. When the calcination temperature

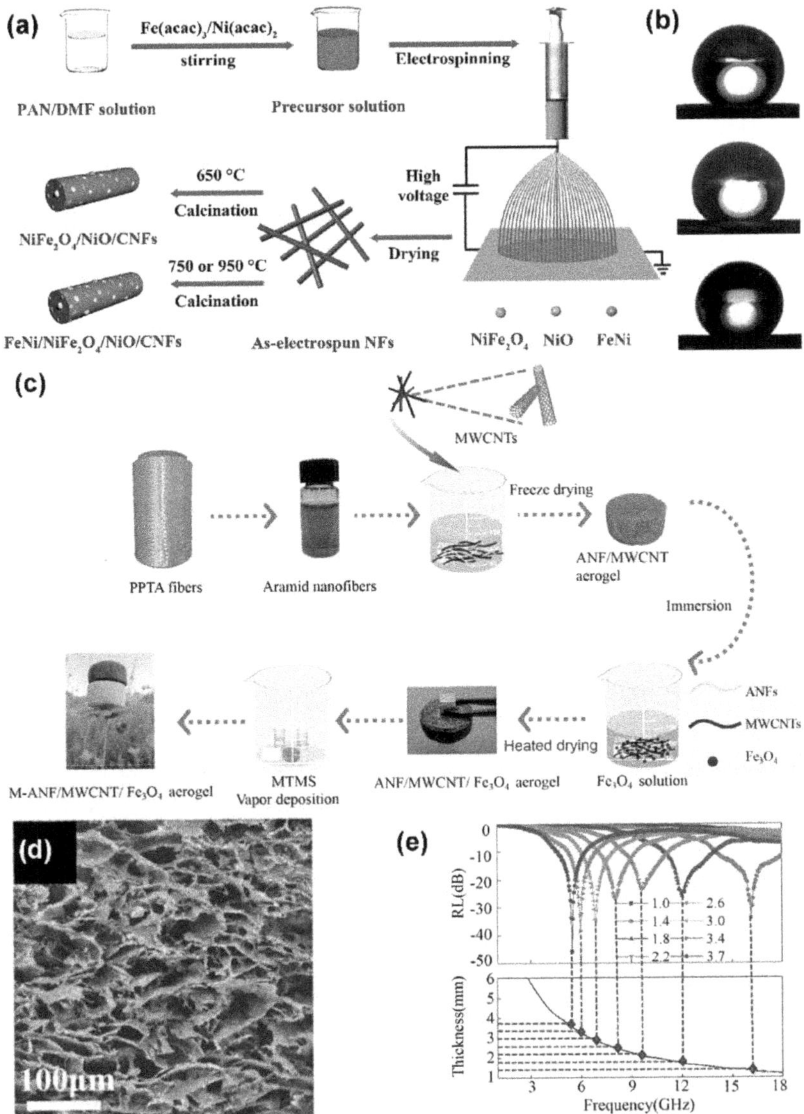

FIG. 4.1   The synthesis process of FeNi/NiFe₂O₄/NiO/C nanofibers (a) and optical photographs for CA measurement with different calcination temperatures of 650, 750 and 950°C (b). (Reproduced with permission from Ref.[13] Copyright 2020, Elsevier.) (c) Schematic of the preparation process M-ANF/MWCNTs/Fe₃O₄ aerogels; (d, e) SEM image of M-AMF32 and its *RL* curves under different thickness and corresponding thickness calculated according to $\lambda/4$ model as a function of microwave frequency. (Reproduced with permission from Ref.[14] Copyright 2022, Elsevier.)

was 750 °C, the FeNi/NiFe$_2$O$_4$/NiO/CNFs displayed a strong absorption with an RL_{min} of −40.9 dB with 5.0 mm and a narrower effective absorption bandwidth (EAB) of 1.5 GHz (3.8–5.3 GHz). Furthermore, when the calcination temperature was 650, 750 and 950 °C, the contact angles (CA) values were 143°, 141° and 144°, respectively (Fig. 4.1b). It can be seen from the CA value that these materials all exhibited strong hydrophobic properties, suggesting that the calcination temperature had an insignificant impact on the CA. Similarly, Wen et al. fabricated super-hydrophobic M-ANF/MWCNTs/Fe$_3$O$_4$ aerogels using a freeze-drying technique and modified the M-AMF with methyl trimethoxy silane during immersion to achieve superhydrophobicity, as shown in Figs. 4.1(c and d)[14]. The M-AMF32 aerogel composite (32 represents the mass fraction of CNT in aerogels) had an RL_{min} of −45.83 dB with 3.7 mm at 5.46 GHz and a maximum EAB of 4.0 GHz, as illustrated in Fig. 4.1(e). The contact angle of M-AMF32 gradually decreased with time and remained above 130° after 600s. More importantly, M-AMF aerogel composites not only showed excellent EMW absorption and superhydrophobic properties, but also excellent thermal insulation properties and selective absorption of non-polar liquids. This further proves that M-AMF aerogel composites can be used in complex environments.

A hierarchical Ti$_3$C$_2$T$_x$ MXene/Ni chain/ZnO array hybrid nanostructure was designed on cotton fabric, which exhibited good EMW absorption performance and self-cleaning function, as displayed in Fig. 4.2(a)[15]. Studies have demonstrated that MXene and ZnO exhibit high dielectric properties, so optimizing impedance matching by controlling the number of Ni chains to enhance the magnetic loss mechanism is a suitable way to improve the EMW absorption performance. The superhydrophobic cotton fabric MXene/Ni chain/ZnO (S-CF-MNZ-9) composite with dip-coated Ni chains exhibited an RL_{min} of −35.1 dB at 8.3 GHz with a thickness of 2.8 mm, and an EAB covering the entire X-band at 2.2 mm. The superhydrophobic coating was constructed using a material with low surface energy and a rough surface (Fig. 4.2b). Thanks to the nanostructure and superhydrophobic coating, the S-CF-MNZ-9 fabric composite exhibited excellent liquid repellency and self-cleaning ability. In another example, a PDMS/CeFe$_2$O$_4$/GO hybrid was synthesized by hydrothermal and mechanical mixing methods (Fig. 4.2c)[16]. The PDMS/CeFe$_2$O$_4$/GO hybrid displayed a preeminent EMW absorption performance (an RL_{min} of −52 dB with an absorption bandwidth of 2.1 GHz) was and superior superhydrophobicity was as shown in Fig. 4.2(d). When the

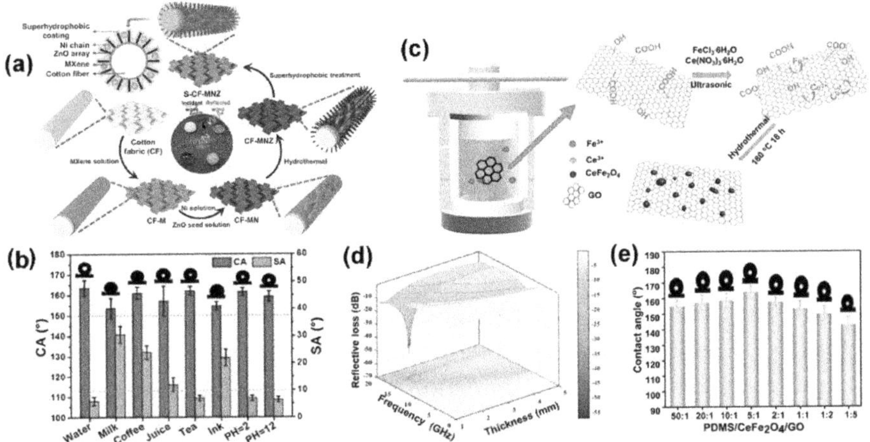

FIG. 4.2 (a) Schematic illustration of preparation process of MXene/Ni chain/ ZnO array cotton fabrics; (b) CAs and SAs of various liquid droplets on the SCF-MNZ-9 fabric surface. (Reproduced with permission from Ref.[15] Copyright 2020, American Chemical Society.) (c) Schematic illustration of the preparation of $CeFe_2O_4$/GO hybrid nanocomposites; (d) 3D RL curve of PDMS/$CeFe_2O_4$/ GO; (e) The contact angle of PDMS/$CeFe_2O_4$/GO with different mass ratios. (Reproduced with permission from Ref.[16] Copyright 2022, American Chemical Society.)

ratio of PDMS/$CeFe_2O_4$/GO was 5:1, the water contact angle can reach 164° (Fig. 4.2e). Although PDMS has little microwave attenuation capability, the introduction of PDMS can provide more interfaces between $CeFe_2O_4$/GO, GO/PDMS, and $CeFe_2O_4$/PDMS, thereby promoting the EMW absorption and superhydrophobicity. Wang et al. successfully prepared a novel N-doped hollow carbon microsphere (N-HCMs) with a hierarchically micro-nano surface and a multiporous shell, and excellent EMW absorption and superhydrophobic properties via a micro-encapsulation and carbonization process[17]. When the filling loading was 15 wt.%, the N-HCMs-700-1 displayed an RL_{min} of −61.2 dB with 2.24 mm and an EAB of 6.8 GHz at 2.5 mm. The CA value of N-HCMs-700-1 can reach 162.1°. Importantly, these N-HCMs were superhydrophobic and can withstand a variety of water environments, acids and alkalis. This shows that the microwave-absorbing coating has universal waterproofness and attractive self-cleaning capabilities.

In a word, for the design of hydrophobic EMW absorbing materials, researchers often focus on selecting the combination of inorganic components and superhydrophobic materials to evaluate their EMW absorption performance, hydrophobic characteristics, corrosion resistance and self-cleaning function. Therefore, future research should select appropriate hydrophobic components with a high specific surface area, a high proportion of hydrophobic bonds in the outermost layer of the composite material, and select hydrophobic components with excellent EMW absorption characteristics and dissipation capacity.

4.2 SELF-HEALING MATERIALS

Self-healing materials, as the name suggests, are materials that can heal and return to their original state after being damaged by some external forces, such as heat, mechanical or light stimulation[18]. Moreover, self-healing materials are mainly concentrated on macropolymers that can produce reversible chemical covalent bonds, such as acylhydrazone bonds, disulfide bonds, thiuram disulfide bonds and imine bonds[19]. In the field of EMW absorption, the mechanical collision, friction or fracture are often encountered by the absorbing materials, which affects their practical application and has received great attention. In order to address this, it is important to develop materials with self-healing capabilities. The self-healing ability of a material is characterized by recovery efficiency. The recovery efficiency is the ratio of the EMW absorption performance of the healed material to that of the original material. It has been found that the EMW absorption capacity of self-healing polymers is relatively weak. Therefore, in order for an absorbing material to have excellent self-healing ability and excellent EMW absorption performance at the same time, the most common measure is to combine a material with excellent EMW absorption performance with a polymer precursor with strong self-healing ability. However, studies have found that the weight ratio between the polymer component and the absorbing component of a absorbing material with strong self-healing ability is generally greater than the filling ratios of absorbing materials. Therefore, developing absorbing materials with ultra-light feature, high-absorption performance, and self-healing capabilities is currently facing great challenges.

In 2020, Fu et al. developed a ZnO@MWCNTs/DA-PDMS nanocomposite[20]. In detail, the first step was to support zinc oxide nanoparticles as absorbents on the surface of multi-walled carbon nanotubes (MWCNTs). Then, polydimethylsiloxane (PDMS) was used as a raw material to synthesize reversible self-healing DA-PDMS, which was reproduced by Diels-Alder (DA) reaction, as shown in Figs. 4.3 (a and b). Finally, ZnO@ MWCNTs particles were evenly distributed in the DA-PDMS. When held

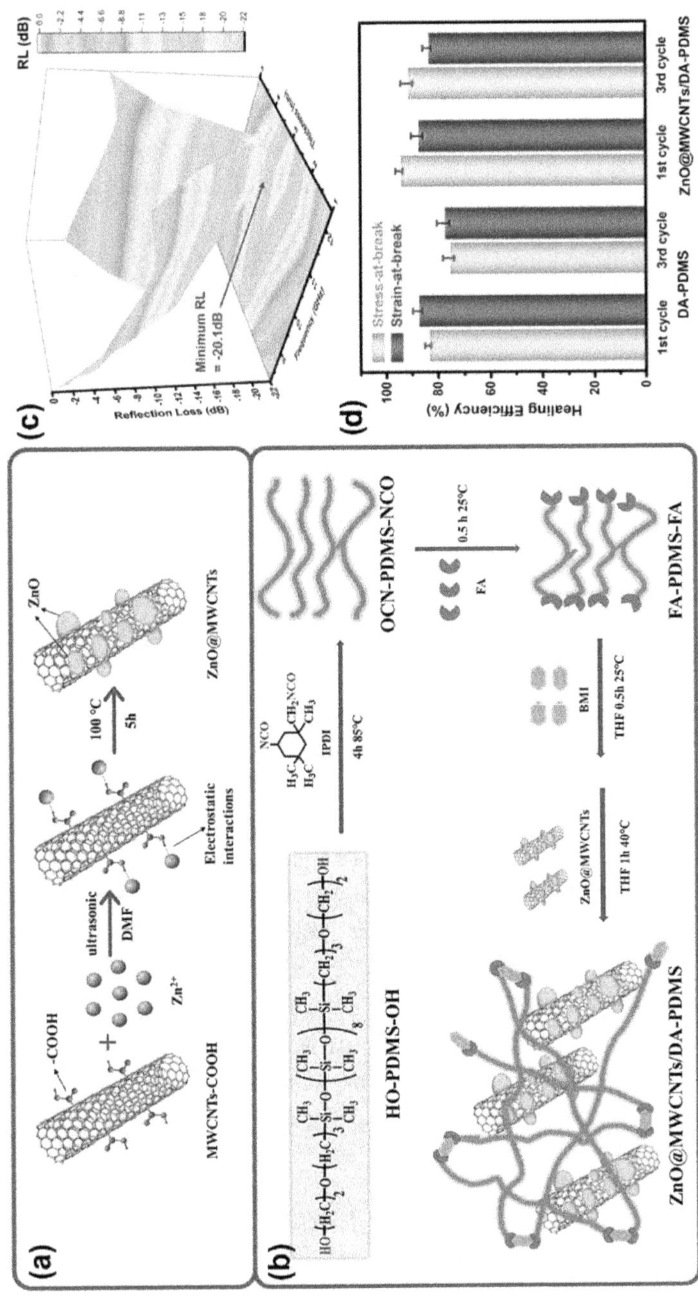

FIG. 4.3   The synthesis process (a) and healing efficiency (b) of ZnO@MWCNTs/DA-PDMS nanocomposite; (c) the 3D RL plots of ZnO@MWCNTs/DA-PDMS (10 wt.%); (d) healing efficiency of DA-PDMS and 5 wt%-ZnO@MWCNTs/DA-PDMS. (Reproduced with permission from Ref.[20] Copyright 2020, Elsevier.)

at 120°C for 10 minutes, the nanocracked ZnO@MWCNTs/DA-PDMS composites can recover 90% of their mechanical strength. When the filling was 10 wt.%, the ZnO@MWCNTs/DA-PDMS nanocomposites exhibited a reflection loss of −20.1 dB, corresponding to an effective absorption bandwidth of 8.2 to 12.4 GHz (full X-band) at 3.0 mm, as shown in Fig. 4.3(c). Moreover, after three cycles, the healing efficiency of the composite still reached 90% (Fig. 4.3d), and the composite still exhibited a reflection loss of −20.0 dB at 3.0 mm.

Chen et al. designed a 3D hollow $NiCo_2O_4$ structure via a precipitation hydrothermal method[21]. Subsequently, benzotriazoles (BTA) were loaded in the hollow $NiCo_2O_4$ material to form epoxy/BTA@$NiCo_2O_4$ coating. The BTA@$NiCo_2O_4$ exhibited an RL_{min} of −35.39 dB at 16.01 GHz at 2.0 mm, corresponding to an EAB of 4.64 GHz. In addition, the epoxy/BTA@$NiCo_2O_4$ coating not only displayed an excellent MA performance but also excellent anticorrosion and self-healing properties. A novel self-healing carbonyl iron powder (CIP)/polydopamine(PDA)@MWCNTs was fabricated based on an imine reversible covalent bond[22]. The core-shell structure formed by coating the outside of CIP with a layer of PDA is conducive to improving the impedance matching and dispersion of the composite. When the filling ratio is 20 wt.%, the CIP/PDA@MWCNTs composites displayed an excellent EMW absorption performance with an RL_{min} of −30.69 dB at 9.44 GHz, covering the whole X-band, which also possessed a superior self-healing property. The self-healing mechanism of the composite material can be attributed to the reversible covalent imine bond and hydrogen bond originating from the urea group.

In self-healing materials, low EMW absorption of the filler is a major bottleneck. Therefore, in order to better balance the self-healing ability and EMW absorption capacity, surface functionalization should be more efficiently integrated into the design of the self-healing materials, thereby enhancing conductive loss and polarization ability. In addition, multi-component structures can be designed to optimize multiple loss mechanisms to achieve excellent self-healing ability and absorption performance.

4.3 THERMAL-RESISTANCE MATERIALS

Traditional EMW-absorbing materials are mainly utilized in common environments and at room temperature. Therefore, they could fail to meet the demands of certain practical applications, such as those in high-temperature environments. For example, when applied to certain parts of an aircraft or weapon (head cones, engine air intakes and nozzles, etc.), they need to resist high temperatures and heat flows. As the temperature requirements for real-world EMW applications increase, the EMW absorption performance is often lower than the estimated level. Therefore, it is urgent to develop EMW-absorbing materials that are resistant to high temperatures.

High-temperature resistant EMW absorbing materials can be divided into ultra-high-, high-, medium- and low-temperature resistant materials. Ultra-high temperature means above 1000 °C, high temperature refers to 500–1000 °C, and medium and low temperature refers to ~500 °C[23]. To improve absorption performance at high temperatures, one should first determine the temperature range. For example, the electromagnetic dissipation capacity, the Curie temperature point of the magnetic alloy, is effective only within medium- and low-temperature region[24]. In addition, the magnetic loss ability of magnetic nanoparticles will be lost in high-temperature environments. Secondly, heat-resistant materials should exhibit excellent stability in the selected temperature region to prevent oxidation, and melting, and thermal decomposition[23].

4.3.1 Medium low-temperature region materials

For EMW absorption materials in the medium- and low-temperature region, they mainly include carbonaceous materials, transition metal oxide materials, and transition metal phosphides (TMPs). Carbonaceous materials are dipole-dominated EMW absorbers because their conductivity loss capacity is limited by low graphitization and wide bandgap electronic structure[23]. There are many carbonaceous materials such as carbon nanotubes, carbon spheres (solid, hollow and hollow porous, etc.), graphene, MXene, biomass-derived carbon, MOFs-derived porous carbon, phenol-formaldehyde resin (PF) derived carbon. They have different temperature response mechanisms and they are widely studied for their absorbing properties. Generally, TMPs were investigated in various electrochemistry fields such as supercapacitance and battery, due to their outstanding low charge-discharge potentials and good thermal stability. TMPs have also been applied in EMW absorption for their magnetic loss and conductive loss ability.

For instance, Liu et al. designed 2D Ni_2P nanosheets anchored on 1D silk-derived carbon fiber via solvothermal reactions and low-temperature phosphating processes at 300 °C[23]. The EMW absorption performance of silk carbon fiber (SCF) was investigated by regulating its carbonization temperature (600, 700, 800 °C) at a fixed phosphating temperature of 300 °C. The high-density 2D Ni_2P-700 nanosheets exhibited a splendid EMW absorption performance with an RL_{min} of –56.9 dB at 2.32 mm and an EAB of 7.2 GHz at 1.93 mm, covering the entire Ku-band. This layered structure enables EMWs to undergo multiple reflections and scattering within the material, enhancing the attenuation of EMWs. In addition, the effect of phosphating temperature on the EMW absorption performance was also explored. Wang et al. fabricated 3D flower-like CoNi-P/C composites by a facile solvothermal self-assembled strategy and a carbonization phosphorization process under different temperatures (300, 400, 500 °C), as shown in Figs. 4.4 (a and b)[23,25]. When the filling ratio was 15 wt.%, the CoNi-P/C-300 composite displayed an optimal RL_{min} of –65.5 dB at 2.1 mm, which is better than CoNi-P/C-400/500 (Figs. 4.4 (c and d)). The CoNi-P/C-300 with the highest ε'' values displayed the strongest dielectric loss and benefited EMW absorption. $Ni_3P@Ni_x(PO_y)_z$ core–shell heterostructures were synthesized via an annealing process and a thermal phosphating strategy at 300°C[26]. The $Ni_3P@Ni_x(PO_y)_z$ heterostructures exhibited an excellent absorption performance with an RL_{min} of –43.1 dB at 1.3 mm and corrosion resistance in acid, alkaline, and salt environment, which was attributed to the surface amorphous $Ni_x(PO_y)_z$.

In summary, the magnetic and dielectric loss characteristics of TMPs render them promising materials for electromagnetic absorbing. Transition metal oxides with narrow band gaps (0–3.0 eV) and good chemical stability are also considered to be low-temperature EMW absorbers. In materials with a narrow band gap, conductivity losses are also enhanced with increasing temperature, thus sustaining the optimal electromagnetic absorption. However, it is challenging to maintain low-temperature EMW absorbers with consistently excellent performance. This is likely attributed to the dominance of conduction loss mechanisms in transition metal oxides. The real permittivity (ε') typically rises with temperature, leading to diminished impedance matching ability and consequently impacting both the EMW absorption performance and its stability. Therefore, in future designs and applications of low-temperature EMW absorbing materials, it is crucial to build multi-component composite materials. Additionally, optimizing impedance matching ability by adjusting electromagnetic parameters is essential.

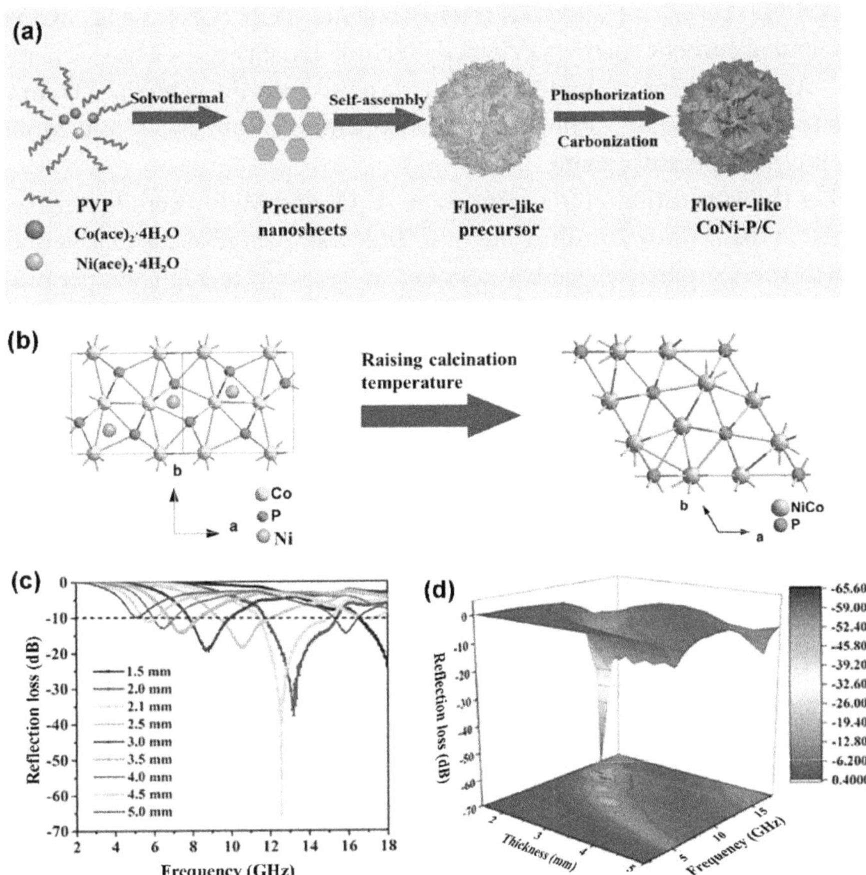

FIG. 4.4 The synthesis route (a), crystal pattern transformation process with calcination temperature raised (b) and their *RL* curves and 3D plots (c, d) of CoNi-P/C composites. (Reproduced with permission from Ref.[25] Copyright 2019, American Chemical Society.)

4.3.2 High-temperature materials

As is well known, ceramic-based materials exhibit excellent thermal stability, with operating temperature often exceeding 1000 °C, making them a primary focus in high-temperature EMW materials. In addition, they have excellent dielectric loss capability. Many researchers are increasingly focusing on the development of high-temperature EMW-absorbing materials, and notable progress has been made in this field. High-temperature absorbing dielectric materials mainly include Ti_3SiC_2, SiC_f, and Al_2O_3-TiC, and magnetic materials primarily include Fe_3Si, Cr_2AlB_2, and

Fe@MoS$_2$. Absorbent materials mainly include SiCN (CNTs) and SiCO@ BN, and absorbent coatings primarily include La$_{0.6}$SrFeO$_{3-\delta}$/MgAl$_2$O$_4$, and TiC-Al$_2$O$_3$/Si[27]. For example, Cai et al. designed an Al$_2$O$_3$-MoSi$_2$/Cu composite coating via atmospheric plasma spraying[28]. Both Al$_2$O$_3$ and MoSi$_2$ are high-temperature stable.

For the fabrication of the composite, Al$_2$O$_3$ and MoSi$_2$ were first mixed together using the ball milling method. They then served as the matrix onto which the Cu particles were subsequently sprayed. In high-temperature environments, Cu nanoparticles can be converted to CuO and Cu$_2$O, which can improve the absorption performance of the Al$_2$O$_3$-MoSi$_2$/Cu composite coating. When the coating temperature reached 700 °C, Al$_2$O$_3$-MoSi$_2$/Cu composite coating exhibited a super absorption performance with an RL_{min} of −19.09 dB at 1.4 mm and 10.51 GHz, demonstrating that the Al$_2$O$_3$-MoSi$_2$/Cu composite coating has an excellent high-temperature EMW absorbing performance. As the coating thickness increased from 1 to 1.4 mm, the absorption capacity of the Al$_2$O$_3$-MoSi$_2$/Cu composite coating also gradually increased. In addition, the absorption performance of the Al$_2$O$_3$-MoSi$_2$/Cu composite coating at 700 °C is higher than the performance at 25 °C. This further shows that Al$_2$O$_3$-MoSi$_2$/Cu composite coating has a strong absorbing capability in high-temperature environments. Similarly, La$_{0.6}$Sr$_{0.4}$FeO$_{3-\delta}$/MgAl$_2$O$_4$ (LSF/MAS) composite ceramic coatings were synthesized by a plasma spraying technology[29]. When the ceramic thickness was 1.5 mm, at a temperature of 673–1173 K, the coatings demonstrated considerable microwave absorption in the 8–18 GHz range. Also, the absorption bandwidth of the LSF30 (the mole ratio of 3:7) was 3.5 GHz for $RL < -10$ dB and the absorption bandwidth was 8 GHz for $RL < -5$ dB. From the above examples, it can be seen that although perovskite-based ceramic materials can work effectively in high-temperature, there is room for improvements in their EMW absorption performance.

Yuan et al. fabricated a SiC/SiO$_2$ monolithic composite via a combined process of nanocasting, cold-pressing, and pyrolysis at different temperatures of 1300 °C and 1400 °C[30]. At 500 °C, SiC/SiO$_2$-1300 °C exhibited an RL_{min} value of −45.5 dB at a thickness of 2.6 mm and an reflection loss of less than −10 dB at 2.8 mm, which means that the EAB covered the entire X-band. Moreover, it was also found that the microwave absorption of the SiC/SiO$_2$-1300 °C was better than that of SiC/SiO$_2$-1400 °C, which may be due to the fact that impedance matching is

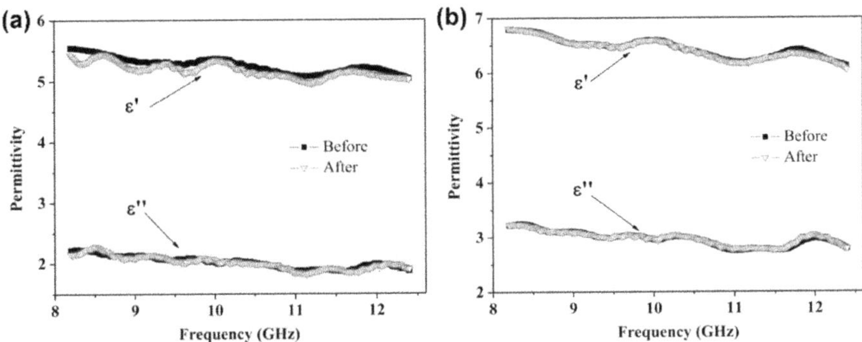

FIG. 4.5 The ε' and ε'' of the SiC/SiO$_2$-1300 (a) and SiC/SiO$_2$-1400 (b) before and after high-temperature pyrolysis. (Reproduced with permission from Ref.[30] Copyright 2017, Elsevier.)

different for samples pyrolized at different temperatures. As shown in Figs. 4.5 (a and b), there is a little change in permittivity values before and after the high-temperature pyrolysis for SiC/SiO$_2$-1300 °C/1400 °C. In addition, TiB$_2$-Al$_2$O$_3$/MgAl$_2$O$_4$ composites were synthesized via spark plasma sintering. In the composite, the TiB$_2$ was the conductive filler, MgAl$_2$O$_4$ was the matrix, and Al$_2$O$_3$ was the third phase used to adjust high-temperature conductivity[31]. The TiB$_2$ was obtained by heating at 1400 °C for 180 min by a carbothermal method. When the content of Al$_2$O$_3$ increased from 70.6 wt.% to 80.6 wt.%, at 25 °C–700 °C, the reflection loss range of the TiB$_2$-Al$_2$O$_3$/MgAl$_2$O$_4$ composites was –19.4 to –14.3 dB at 8.2–12.4 GHz, and the corresponding effective absorption bandwidth with $RL < -5$ dB was 3.19–3.55 GHz, demonstrating temperature insensitivity and good EMW absorption performance. Gong et al. constructed a titanium nitride/boron nitride (TiN/BN) composite material by in-situ synthesis, using TiN as the loss unit and BN as the impedance matching unit. The TiN/BN composite exhibited an excellent absorption performance with an $RL_{\min}$ of –16.74 dB at 873 K and an EAB of 3.26 GHz in the temperature range of 293–873 K[32] confirming that the TiN/BN/SiO$_2$ composites have good high-temperature absorption performance.

In summary, for high-temperature EMW absorbing materials, although there are many innovative studies, numerous bottlenecks remain difficult to solve, such as their EMW absorption performance is generally inferior to that of traditional EMW absorption material. Moreover, the pyrolysis of ceramic-based high-temperature materials under relatively high-temperature conditions will produce a variety of complex composites, making it difficult to conduct mechanistic studies. Secondly, when ceramic matrix composites are calcined at high temperatures, the occurrence of defects or vacancies is inevitable, necessitating quantitative analysis. Therefore, it is still a great challenge to develop high-temperature EMW absorbing materials with excellent absorption performance, high-temperature stability, and oxidation resistance.

Researching ultra-high temperature EMW-absorbing materials presents greater challenges compared to those for high-temperature and medium-high-temperature materials. This is primarily due to the complexities of ultra-high temperature environments, making it difficult to determine electromagnetic parameters, thus posing a significant technical challenge. At ultra-high temperatures, the crystalline phases, lattices, and defects of the absorbing materials could undergo irreversible changes[23]. Therefore, future research on ultra-high temperature absorbing materials should focus more on their mechanical strength and adhesive viscosity. At present, while high-temperature absorbing materials have been extensively studied, the challenge of developing them for use under even higher temperatures persists.

4.4 ELECTRICALLY CONTROLLABLE MATERIALS

Traditional EMW absorbing materials can achieve excellent EMW absorption performance through components and structural design. However, their own physical properties make it difficult to independently optimize electromagnetic parameters or the real and imaginary parts of magnetic permeability[23,33]. Also, changes in the real part will cause alterations in the imaginary part. Based on the quarter-length theory, an increase in thickness will cause gradual shift of the absorption peak to a lower frequency[23]. Therefore, low-frequency electromagnetic absorbers were developed at the expense of increasing the thickness[34]. However, even with increased thickness, effective absorption in the low-frequency region remains challenging. Considering the aforementioned issues, the introduction of an external electric field may enable effective absorption in the low-frequency region or facilitate the development of frequency-selective EMW absorber[23].

4.4.1 Low-frequency EMW absorbers

Traditional EMW absorbers have been widely used in the high-frequency or intermediate-frequency region. However, achieving more effective absorption in the low-frequency region, especially in 1.0–2.0 GHz (L-band) range, remains challenging[35]. At present, there is still a significant bottleneck in the design of absorbers intended for use in the low-frequency region. Exploring low-frequency absorbers holds great significance in addressing electromagnetic pollution more comprehensively, rather than relying solely on separate shielding methods. In 2018, Lv et al. proposed a strategic scheme to develop materials with high ε' value and low ε'' value, and control the dielectric value by applying an applied electric field[36]. Flexible $SnS/SnO_2@C$ composites with a sandwich structure were constructed by using a carbon film electrode to conduct external current. When the applied voltage was 16 V and the thickness was reduced to about 5 mm, an absorption value of 85% within 1.5–2.0 GHz was observed. Moreover, it exhibited good absorption characteristics in the temperature range of 25–150 °C[36]. However, the increase of ε' led to an increase in ε'', which limits their application in EMW absorption. Xu and co-workers fabricated $\gamma\text{-}Fe_2O_3$ nanocubes/graphene (GFC) composites via a solvothermal method. An RL_{min} value of –59.3 dB at 2.4 GHz was obtained for the composite, as shown in Fig. 4.6[35]. Importantly, the composites exhibited outstanding EMW absorption performance in the range of 2–18 GHz, especially in the S-band (2–4 GHz) (< –30.0 dB) and C-band (4–8 GHz) (< –26.0 dB) under specific frequency and thickness matching (Fig. 4.6c). It was found that high impedance matching is a key factor in promoting the EMW absorption performance of the $\gamma\text{-}Fe_2O_3$ nanocubes/graphene composites in the S band and C band.

From the discussion above, it is evident that although certain EMW absorption performance has been attained in the low-frequency regions such as L, S, and C band, achieving sufficient absorption remains challenging. Moreover, the thickness of the absorbers is high, making their commercial utilization impractical. Therefore, to achieve excellent EMW absorption performance in the low-frequency region, it is important to identify more suitable ε' and ε''. In the meantime, the use of magnetic fields to adjust the magnetic permeability of absorbers is also of vital importance.

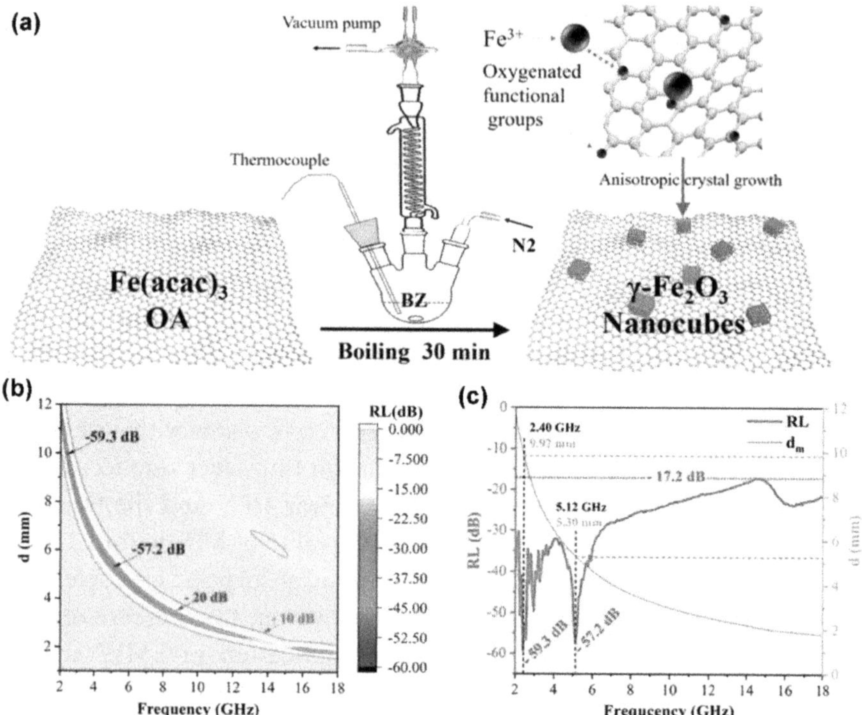

FIG. 4.6 The synthesis process (a) and *RL* map (b) of γ-Fe₂O₃ nanocubes/ graphene materials. (c) The *RL* values at matching frequency and thickness of γ-Fe₂O₃ nanocubes/graphene in 2–18 GHz. (Reproduced with permission from Ref.[35] Copyright 2021, Elsevier.)

4.4.2 Frequency-selective EMW absorbers

Frequency-selective EMW absorbers have important application prospects in low-energy rapid charge and discharge. In our daily lives, radio technology has brought many conveniences to human beings and also improved the quality of life. However, the instability of signal caused by electromagnetic interference necessitates operation in a specific frequency range. Therefore, it is essential to develop EMW absorbers that selectively target a specific frequency range while excluding microwaves in unrelated frequency ranges. This also helps to minimize signal loss between transmitters and receivers. However, achieving this is challenging as most electromagnetic absorbing materials fail to achieve effective absorption in the low-frequency region, and their dielectric permittivity and magnetic permeability remain largely unchanged in the microwave region.

Therefore, it is difficult to rely solely on EMW-absorbing materials to achieve absorption in a specific frequency range. Some researchers have designed an electrically switchable flexible electromagnetic absorber in order to explore practical applications in the low and medium-frequency regions. For instance, Wu et al. designed core–shell $Sn/SnS/SnO_2$@C composite absorbers via reducing thioacetamide (TAA), and then annealing at different temperatures, as shown in Fig. 4.7(a)[37]. Through electrically tunable calculation and electromagnetic simulation, a qualified $\varepsilon_r-\varepsilon''$ was obtained, corresponding to $0 <$ effective absorption $(f_E) < 2.0$, a thickness of 1.0–2.0 mm and a specific frequency region of 2–8 GHz (Fig. 4.7b). The $Sn/SnS/SnO_2$@C composite absorber with an operation voltage of ≤ 18 V displayed an exceptional electrically switchable capability, and narrow $(< 2.0$ GHz), selective, and multiple f_E regions (corresponding tunable bands $N = 7$) were achieved in 90% of C-band (Fig. 4.7c). In addition, the results show that the Sn/SnS@C-based composite absorber can achieve the largest six frequency bands, the widest Δf, and switchability at 4–8 GHz.

CONCLUSION

The development of next-generation EMW absorbing materials has been driven by the need for advanced solutions that offer superior performance, durability, and versatility. Among the next-generation materials that have emerged in recent years, the typical representatives are superhydrophobic materials, self-healing materials, thermal-resistance materials, and electrically controllable materials. These materials hold great promise for revolutionizing the field of EMW absorption and opening up new possibilities for various applications. In this chapter, we provided a comprehensive summary and analysis of these advanced materials. Nevertheless, there still exist some problems and challenges.

(1) Superhydrophobic materials have better absorption of EMWs in a specific frequency range, but have poor absorption performance at other frequencies. Therefore, it is necessary to develop absorbent materials that exhibit effective absorption in a wider frequency range to meet practical applications.

(2) Self-healing materials have garnered significant attention for their ability to repair damage autonomously, thereby extending the lifespan and reliability of absorbing materials. However, the self-healing efficiency, durability, and repair speed of self-healing

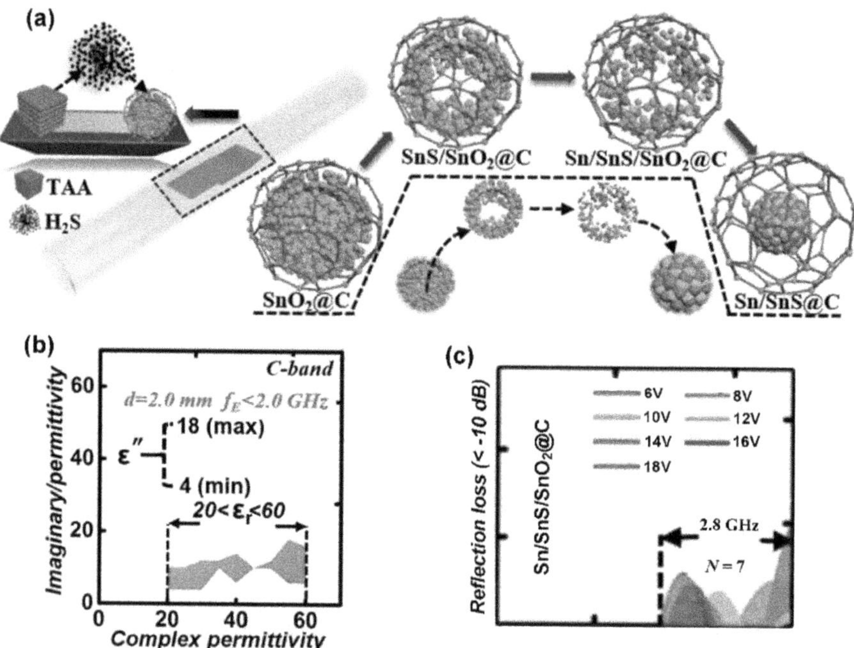

FIG. 4.7   (a) The schematic illustration of the formation of SnS/SnO$_2$@C; (b) the optimized ε_r–ε'' pairs (f$_E$ < 2.0 GHz in C-band, thickness = 2.0 mm); (c) the qualified f_E regions for the optimal Sn/SnS/SnO$_2$@C sample. (Reproduced with permission from Ref.[37] Copyright 2019, Wiley & Sons.)

materials need to be improved. In addition, it should exhibit a robust self-healing capability under complex environmental conditions, such as high temperature, high humidity and other challenging scenarios. Further optimization of the EMW absorption performance is also necessary.

(3) Thermal-resistance materials exhibit high thermal stability and resistance to temperature fluctuations, making them suitable for applications in high-temperature environments. However, limitations in flexibility, weight, and thermal conductivity may restrict their applicability in certain scenarios. Future research will focus on optimizing the thermal properties of materials to improve absorption performance and heat dissipation capabilities. Lightweight, flexible, and thermal-resistance materials will be developed to meet the demands of space/weight-constrained environments.

(4) Electrically controllable materials enable dynamic adjustment of EMW absorption properties, offering enhanced tunability and adaptability for diverse operating conditions. These materials have the potential for active control and modulation of electromagnetic signals, opening up new possibilities for applications requiring real-time adjustments. However, complex control mechanisms, power requirements, and stability challenges may hinder their widespread adoption. Future research will focus on developing efficient and low-power consumption control systems for electrically controllable materials.

REFERENCES

1. Wang, S., Liu, K., Yao, X., et al. (2015). Bioinspired surfaces with superwettability: New insight on theory, design, and applications. *Chemical Reviews 115*: 8230–93.

2. Guo, Z., and Liu, W. (2007). Biomimic from the superhydrophobic plant leaves in nature: Binary structure and unitary structure. *Plant Science 172*: 1103–12.

3. Mehanna, Y.A., Sadler, E., Upton, R.L., et al. (2021). The challenges, achievements and applications of submersible superhydrophobic materials. *Chemical Society Review 50*: 6569–612.

4. Gao, X.F., Jiang, L. (2004). Water-repellent legs of water striders. *Nature 432*: 36–36.

5. Yin, W., Zheng, Y.L., Lu, H.Y., et al. (2016). Three-dimensional topographies of water surface dimples formed by superhydrophobic water strider legs. *Applied Physics Letters 109*: 163701.

6. Bhushan, B., and Her, E.K. (2010). Fabrication of superhydrophobic surfaces with high and low adhesion inspired from rose petal. *Langmuir 26*: 8207–17.

7. Koch, K., Bhushan, B., and Barthlott, W. (2009). Multifunctional surface structures of plants: An inspiration for biomimetics. *Progress in Materials Science 54*: 137–78.

8. Yuan, J., Lv, Z., Lu, Q., et al. (2015). First-principles study of the phonon vibrational spectra and thermal properties of hexagonal MoS2. *Solid State Sciences 40*: 1–6.

9. Favret, E.A., and Andrada, A.N. (2009). Studies on lotus and rice leaf surfaces by using RIMAPS technique. *Microscopy and Microanalysis 15*: 946–47.

10. Huang, J.A., Zhang, Y.L., Zhao, Y., et al. (2016). Superhydrophobic SERS chip based on a Ag coated natural taro-leaf. Nanoscale 8: 11487–93.

11. Peng, J., Wu, L., Zhang, H., et al. (2022). Research progress on eco-friendly superhydrophobic materials in environment, energy and biology. Chemical Communications 58: 11201–19.

12. Quan, Y.Y., Chen, Z., Lai, Y., et al. (2021). Recent advances in fabricating durable superhydrophobic surfaces: a review in the aspects of structures and materials. Materials Chemistry Frontiers 5: 1655–82.

13. Shen, Y., Wei, Y., Ma, J., et al. (2020). Self-cleaning functionalized FeNi/NiFe$_2$O$_4$/NiO/C nanofibers with enhanced microwave absorption performance. Ceramics International 46: 13397–406.

14. Ma, Y., Li, Y., Zhao, X., et al. (2022). Lightweight and multifunctional super-hydrophobic aramid nanofiber/multiwalled carbon nanotubes/Fe$_3$O$_4$ aerogel for microwave absorption, thermal insulation and pollutants adsorption. Journal of Alloys and Compounds 919: 165792.

15. Wang, S., Li, D., Zhou, Y., et al. (2020). Hierarchical Ti$_3$C$_2$T$_x$ MXene/Ni chain/ZnO array hybrid nanostructures on cotton fabric for durable self-cleaning and enhanced microwave absorption. ACS Nano 14: 8634–45.

16. Guo, Y., Chen, X., Ma, W., et al. (2022). CeFe$_2$O$_4$ nanoparticle/graphene oxide composites with synergistic superhydrophobicity and microwave absorption. ACS Applied Nano Materials 5: 6513–22.

17. Yang, N., Luo, Z.X., Wu, G., et al. (2023). Superhydrophobic hierarchical hollow carbon microspheres for microwave-absorbing and self-cleaning two-in-one applications. Chemical Engineering Journal 454: 140132.

18. Kang, J., Tok, J.B.H., and Bao, Z. (2019). Self-healing soft electronics. Nature Electronics 2: 144–50.

19. Wang, S., and Urban, M.W. (2020). Self-healing polymers. Nature Reviews Materials 5: 562–83.

20. Zhou, M., Yan, Q., Fu, Q., et al. (2020). Self-healable ZnO@ multiwalled carbon nanotubes (MWCNTs) /DA-PDMS nanocomposite via Diels-Alder chemistry as microwave absorber: A novel multifunctional material. Carbon 169: 235–47.

21. Ma, C., Wang, W., Wang, Q., et al. (2021). Facile synthesis of BTA@NiCo$_2$O$_4$ hollow structure for excellent microwave absorption and anticorrosion performance. Journal of Colloid and Interface Science 594: 604–20.

22. Zheng, J., Zhou, Z., Zhu, L., et al. (2022). Room temperature self-healing CIP/PDA/MWCNTs composites based on imine reversible covalent bond as microwave absorber. Reactive and Functional Polymers 172: 105179.

23. Lv H.L., Yang, Z.H., Pan H.G., et al. (2022). Electromagnetic absorption materials: Current progress and new frontiers. Progress in Materials Science *127*: 100946.

24. Pajda, M., Kudrnovský, J., Turek, I., et al. (2001). Ab initiocalculations of exchange interactions, spin-wave stiffness constants, and Curie temperatures of Fe, Co, and Ni. Physical Review B *64*: 174402.

25. Wang, R., He, M., Zhou, Y., et al. (2019). Self-assembled 3D flower-like composites of heterobimetallic phosphides and carbon for temperature-tailored electromagnetic wave absorption. ACS Applied Materials & Interfaces *11*: 38361–71.

26. Liang, L.L., Song, G., Chen, J.P., et al. (2021). Crystalline-amorphous $Ni_3P@Nix(PO_y)_z$ core–shell heterostructures as corrosion-resistant and high-efficiency microwave absorbents. Applied Surface Science *542*: 148608.

27. Gao, C., Jiang, Y., Cai, D., et al. (2021). High-temperature dielectric and microwave absorption property of atmospheric plasma sprayed Al_2O_3-$MoSi_2$-Cu composite coating. Coatings *11*: 1029.

28. Gao, C., Jiang, Y., Cai, D., et al. (2021). Effect of temperature on the microwave-absorbing properties of an Al_2O_3–$MoSi_2$ coating mixed with copper. Coatings *11*: 940.

29. Jia, H., Zhou, W., Nan, H., et al. (2020). High temperature microwave absorbing properties of plasma sprayed $La_{0.6}Sr_{0.4}FeO_{3-\delta}$/$MgAl_2O_4$ composite ceramic coatings. Ceramics International *46*: 6168–73.

30. Yuan, X., Cheng, L., Guo, S., et al. (2017). High-temperature microwave absorbing properties of ordered mesoporous inter-filled SiC/SiO_2 composites. Ceramics International *43*: 282–88.

31. Liu, X., Zhang, S., Luo, H., et al. (2021). Temperature-insensitive microwave absorption of TiB_2-Al_2O_3/$MgAl_2O_4$ ceramics based on controllable electrical conductivity. Science China Technological Sciences *64*: 1250–63.

32. Shi, Y., Li, D., Si, H., et al. (2022). TiN/BN composite with excellent thermal stability for efficiency microwave absorption in wide temperature spectrum. Journal of Materials Science & Technology *130*: 249–55.

33. Quan, B., Liang, X., Ji, G., et al. (2017). Dielectric polarization in electromagnetic wave absorption: Review and perspective. Journal of Alloys and Compounds *728*: 1065–75.

34. Shu, J.C., Huang, X.Y., and Cao, M.S. (2021). Assembling 3D flower-like Co_3O_4-MWCNT architecture for optimizing low-frequency microwave absorption. Carbon *174*: 638–46.

35. Wang, F., Li, X., Chen, Z., et al. (2021). Efficient low-frequency microwave absorption and solar evaporation properties of γ-Fe_2O_3 nanocubes/graphene composites. Chemical Engineering Journal *405*: 126676.

36. Lv, H., Yang, Z., Wang, P.L., et al. (2018). A voltage-boosting strategy enabling a low-frequency, flexible electromagnetic wave absorption device. Advanced Materials *30*: 1706343.

37. Lv, H., Yang, Z., Xu, H., et al. (2019). An electrical switch-driven flexible electromagnetic absorber. Advanced Functional Materials *30*: 1907251.

CHAPTER 5

Component manipulation of electromagnetic interference (EMI) shielding materials

INTRODUCTION

Materials commonly used in the production of electromagnetic wave (EMW) shields include carbon-based materials, ceramics, cement, metals, conducting polymers, and various composites. Among these, graphene/carbon-based materials and metals are widely utilized due to their high conductivity and the presence of itinerant electrons that can interact with the electrical field in a radiation environment. Ceramics and cement-based materials are less effective, but their functional groups can still respond to the electrical field within a radiation region. Polymeric matrices generally exhibit less success, unless they possess conductive properties. The inclusion of magnetic constituents in these materials enhances EM wave shielding by collaborating with the magnetic field in a radiation zone[1].

DOI: 10.1201/9781003490098-5

Electromagnetic interference (EMI) shielding materials can be classified into two categories: structural and functional. Functional shielding materials, which are essential in this field, are often used as shielding components in devices like mobile phones, where they are integrated into or added onto a given structure. Research on EMI shields primarily focuses on functional shielding materials rather than their structural counterparts. Structural materials refer to those that not only provide shielding but also serve as load-bearing elements in a structure. For example, continuous-carbon fiber polymeric-matrix nanocomposites are utilized in lightweight structures such as frames. Another example is cement-based structural shields, which are crucial for civil engineering infrastructure, such as the concrete cover of a large transformer dome. Shielding is also necessary for buildings to prevent electromagnetic (EM) spying and provide EM pulse protection. Structural EMI shields, like graphene-based composites, are often multifunctional.

Compared to incorporating functional shields onto a structure to achieve the desired shielding, researchers are increasingly interested in multifunctional structural EMI shields due to their durability, cost-effectiveness, high functional surface area (large functional volume), and improved mechanical performance. In the case of compact devices like mobile phones, functional EMI shields must be effective even at low thicknesses. Additionally, functional EMI shields with low density are appealing to reduce the overall weight of the device. However, being effective at low density alone is not sufficient unless they also perform well at low thicknesses. On the other hand, for larger structures such as infrastructures, multifunctional EMI shielding materials do not need to be effective at low thicknesses. Similarly, except for lightweight structures like airplanes, multifunctional EMI shielding materials do not require effectiveness at low density. Low density is generally not a requirement for civil engineering structures where cement-based shields are commonly employed.

5.1 METAL-BASED EMI SHIELDING MATERIALS

For a considerable period, metals in bulk or glazed forms have been commonly used for EMI shielding. Film Al is a typical example of a metal used in custom-made electronic enclosures for bulk shielding. In the case of metal coatings, electroplated Ni is a common choice for electronic components that require EMI shielding, given its conductivity and ferromagnetic properties. However, these coatings face challenges such as susceptibility to scratching, although these scratches are somewhat tolerable due to the long wavelength of radiation. Most of these materials also face challenges related to their large mass and volume. Enclosures, in particular, encounter difficulties in maintaining EMI shielding integrity at joints and seams. For example, an EMI-shielded room door may experience a deficit in EMI shielding at the joint, which can be addressed by using EMI gaskets. These gaskets need to be robust and springy while effectively providing EMI shielding. Metals and alloys with magnetic properties are especially suitable for absorption-based EMI shielding[1]. Some examples in this domain include permalloy (an alloy of Ni (80 wt%) and Fe (20 wt%), with a comparative magnetic permeability (MP) of 20000 at 1 kHz), mu-metal (a ferromagnetic alloy of Ni and Fe, with a comparative MP of 100000 at 1 kHz), and stainless steel (#430, ferritic, with a comparative MP of 500). However, these alloys tend to have high weight and mass due to their high densities. Nickel, though possessing some magnetic properties, has a relatively low relative permeability of 100[2]. Nonetheless, it is intriguing to researchers and industrialists due to its ease of coating onto predetermined surfaces through electroplating.

The practical approach to achieving EMI shielding involves the combined use of magnetic alloys and conducting but nonmagnetic materials like carbon. For example, FeNi micro/nanoparticle CNTs are more effective in enhancing the EMI shielding of polymeric-matrix nanocomposites compared to using magnetic alloys or CNTs alone as reinforcement[3]. Another approach is blending mu-metal magnetic alloy with graphite flakes in EMI shielding materials. By using a lower magnetic constituent in the polymeric-matrix composite compared to the conductive constituent, the EMI SE can be improved compared to composites with either the magnetic constituent or the conductive constituent alone as reinforcement[4]. It has been found that for enhanced EMI shielding, the material's magnetic constituent is effective as long as the composite's electrical resistivity is less than 10 Ω cm[5].

The common belief is that high conductivity in EMI shields leads to SE primarily through reflection rather than absorption[5,6]. However, high conductivity can also contribute to SE through absorption, as seen in the case of Al which is widely used in EMI with good SE. To understand the factors influencing the EMI shielding of these materials, it is important to compare their SE with their nano/microstructural parameters, magnetic, electrical, and dimensional characteristics. This relationship has been studied for nickel, which is both conductive and ferromagnetic. The parameters include SE, magnetic properties, electrical resistivity, and nano/microstructure of fibrous nickel with three different diameters (20 μm, 2 μm, and 0.4 μm). The EMI SE at 1-2 GHz, hysteresis energy loss, and coercive force consistently decrease with increasing diameter of nickel, while the DC resistivity is lowest at around 2 μm diameter. This is due to the competing influences of improved contact between nickel filament units (resulting in reduced resistivity) and an increased number of touch points (resulting in increased resistivity) with decreasing diameter. Therefore, a compromise between these two influences, as indicated by the transitional thickness, yields the lowest resistivity.

The grain size of 0.016 μm and 0.018 μm corresponds to the widths of 0.4 and 2 μm, but is higher for the 20 μm diameter. The number of grains per unit or grain size is not correlated with the diameter and/or resistivity. Higher EMI SE is associated with higher hysteresis energy loss and coercive force values, as well as smaller diameters. However, EMI SE is not related to resistivity or grain size, including the number of grains per unit diameter. At the same filler concentration, a smaller diameter is linked to a higher specific interfacial area, which enhances EMI SE due to the skin effect that influences and governs conductivity. Similarly, higher specific interfacial area contributes to increased energy loss due to hysteresis and coercive force. As absorption occurs within the shield's interior, greater material thickness leads to an increased contribution from absorption. Additionally, higher conductivity causes a significant impedance mismatch between air and the specimen, resulting in substantial reflection of incident EM radiation. Aluminum EMI shields tend to be dominated by the absorption mechanism, where the shielding effectiveness ratio (SE_R) is lower than the SE_A. For example, at 1 GHz, SE_A was 43 dB, while SE_R was reported to be 5 dB.

Despite its high conductivity and low thickness, aluminum foil still exhibits a dominance of absorption. This is partially attributed to its relatively high electric permittivity value of 54,800 at a frequency of 2 kHz[7]. The higher permittivity arises from the presence of free electrons within the aluminum atoms, and the polarization aided by permittivity enhances absorption. Recent research has shown promising results with materials such as 12% Al-doped MoS_2/rGO nanohybrids and 3D porous nickel metal foam/polyaniline heterostructures, which exhibit high AC electrical conductivity and EMI SE values. Furthermore, there have been reports of a remarkable

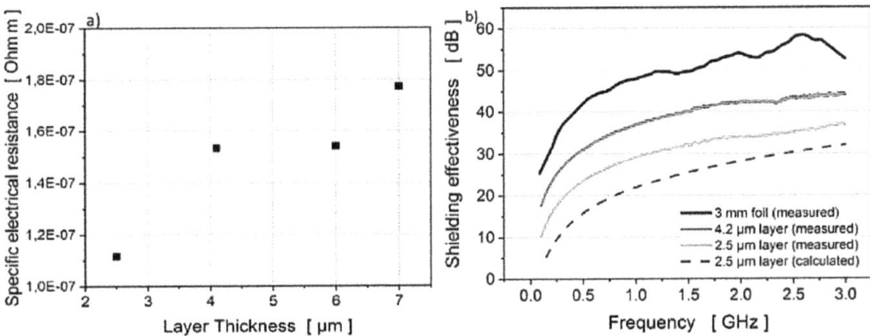

FIG. 5.1   (a) Specific electrical resistance as a function of layer thickness. (b) Shielding effectiveness in magnetic near field configuration as a function of frequency for an aluminum foil (smoothed) and two aluminum layers of different thickness. The reference dashed line was calculated according to the impedance concept (used parameters: source/sample distance = 0.25 mm, aluminum layer thickness = 2.5 μm, electrical conductivity = 7.106 S m^{-1}). (Reproduced with permission from Ref.[8] Copyright 2022, Elsevier Science Ltd.)

EMI SE and absorption from a 3D porous nickel metal foam/polyaniline heterostructure. This is achieved through the deliberate construction of a macroscopic conductive network. Heinß and Fietzke[8] demonstrated in their research that the thickness of the layered composite EMI shield and specific electrical resistance increased as the samples became thicker, as illustrated in Fig. 5.1.

5.2 CARBON-BASED EMI SHIELDING MATERIALS

Various carbon materials, including graphene, graphite, coke, carbon nanofibers (CNFs), carbon fibers (CFs), and carbon nanotubes (CNTs), exhibit not only electrical conductivity but also efficient EMW absorption across a wide frequency range[9,10]. Researchers have noted significant differences in terms of conductivity, structure, morphology, and cost among different types of carbon materials, CNFs, for example, have also been referred to as C-filament. Among the notable forms of graphitic material is exfoliated graphite, which can be obtained by intercalating graphite flakes followed by rapid heating, resulting in the exfoliation of graphite[11]. Each exfoliated carbon piece takes a worm-like shape due to significant enlargement along the c-axis and subsequent elongation in the same direction. When these worm-like materials are compressed together, mechanical interlockings occur among neighboring worms, leading to the formation of a sheet without a polymeric matrix. This highly flexible sheet is referred to as "flexible graphite." Due to the extensive compression during its preparation, the sheet exhibits a strong preferred crystallographic alignment of the carbon layers along the sheet axis. As a result of this preferred orientation, the sheets display high anisotropy, with significantly higher electrical conductivity within the sheet plane compared to the perpendicular direction.

Furthermore, the exfoliation process of flexible graphite results in the formation of open pores and blocked pores within its structure, leading to an increased surface area. A higher surface area is desirable for improved EMI shielding due to the skin effect. In this regard, flexible graphite exhibits an EMI SE of approximately 130 dB[12], which represents the upper limit of power loss measurement. Additionally, due to its cellular structure and favorable alignment, the sheet possesses strong resilience perpendicular to its plane, allowing flexible graphite to function as an effective EMI seal. The exfoliated flexible graphite also offers high thermal conductivity, chemical inertness, and a low thermal expansion coefficient, making it highly valuable in microelectronics fabrication. With respect to flexible carbon materials, both SE transmission loss and SE absorption are enhanced as the frequency increases. Specifically, SE transmission loss surpassing 49 dB and SE absorption exceeding 44 dB can be achieved. The frequency dependence of both SE absorption and SE reflection is weak, with SE absorption being significantly larger than SE reflection, indicating a greater absorption effect at higher frequencies. Furthermore, the high density of flexible graphite, attributed to its cellular structure, contributes to its effectiveness as a shielding material. The SE absorption relative to the unit thickness of the specimen increases with increasing frequency, approaching the thickness of the carbon itself[3]. This highlights the importance of considering the material's density in EMI shielding applications.

Compared to Al, the SE absorption relative to unit thickness is significantly lower for flexible graphite. This is partly due to the fact that aluminum is much thinner compared to flexible graphite. Moreover, both the electrical conductivity and permittivity of flexible graphite are lower than those of aluminum. Activation of CNFs involves a chemical reaction that results in increased surface porosity. Activated CNFs with higher surface areas and utilizing the skin effect are more effective as shielding materials compared to unactivated CFs[13]. However, the activation process does lead to a reduction in the mechanical properties of CFs, particularly with higher degrees of activation. Therefore, if CFs are intended for structural applications, a lighter activation process is recommended. However, for non-structural purposes, there is no need to limit the degree of activation. Additionally, nanocarbons, including graphene, CNFs, and fullerenes, are considered effective reinforcing fillers in EMI shields due to their higher surface area and skin effect. Instead of dispersing nCs within a matrix, it is possible to create nC paper, such as fullerene paper, which can serve as reinforcing fillers in nanocomposite systems for EMI shielding.

Hybrid nCs and porous nCs are primarily synthesized to enhance their surface area and reduce impedance mismatch between the shielding material and the surrounding air. This is crucial for EMI shielding in carbon materials as it helps decrease the SE_R and increase absorption loss. The combination of continuous CNFs and fullerenes has received significant attention due to the alignment of nanotubes along the CF axis. Hybrid shields can be composed of carbon and/or metal fillers, such as CNFs, nickel nanoparticles, and graphene[14]. When it comes to polymeric matrices with transparency to radiation, nanofiber-reinforced composites exhibit superior properties compared to microfiber counterparts for EMI shielding. The phenomenon of higher interfacial interaction within NF-reinforced composites, along with the skin effect, occurs when NFs and MFs are at equal volume fractions. Metal-coated CNFs, particularly nickel-glazed CNFs (Ni-glazed C-filament or Ni-filament), have been found to be more effective in EMI shielding than unglazed CNFs. Nickel-coated CNFs demonstrate increased conductivity and magnetic behavior compared to uncoated CNFs. In applications where nickel-coated CNFs are used as functional reinforcements in a matrix, the resulting composite EMI shield exhibits an EMI SE of 87 dB (1–2 GHz) with a filler volume fraction of 7%. In EMI seals and gaskets, the filler should be continuous and its volume fraction should be low to maintain resilience provided by the polymer matrix[15]. Therefore, even at low volume fractions, the reinforcing filler is expected to be effective in EMI seals.

The arrangement of fiber-like platelike structures and stacking patterns at the macroscopic scale significantly affects the EMI SE. In Fig. 5.2, the SE of three planar arrangements of a continuous carbon fiber chain comprising approximately 12,000 fibers is depicted[16]. In arrangement 1, the chain forms a coil-like structure where it is continuous throughout the coil. In arrangement 2, the chain is arranged in parallel straight lines, with continuity between adjacent lines. Fig. 5.2 shows two variations of arrangement

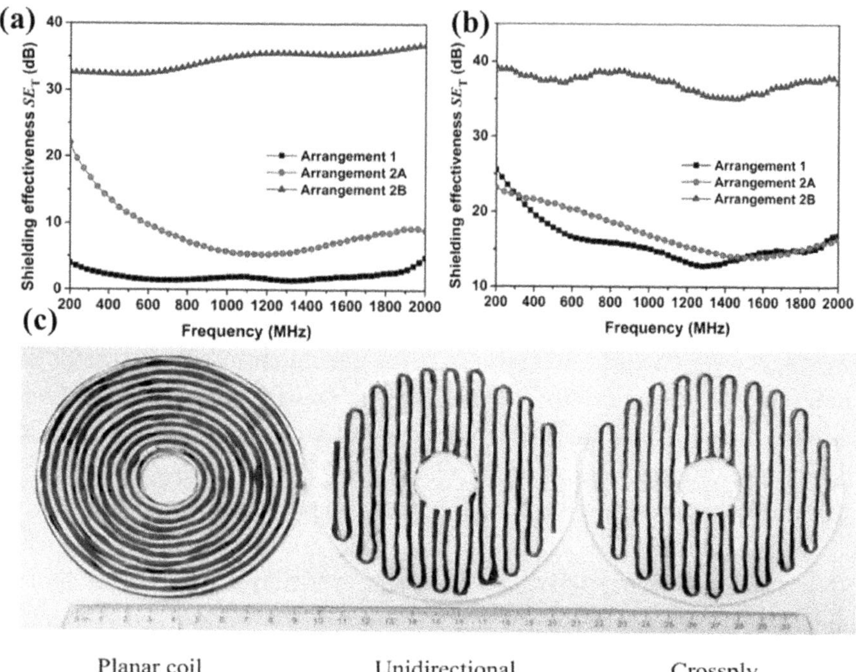

FIG. 5.2 (a) Absorption loss of continuous carbon fiber (uncoated) with three fiber chain planar arrangements. (b) Absorption loss of continuous carbon fiber (nickel-coated) with three fiber chain planar arrangements. (c) Optical photograph of the specimens corresponding to the 3 platelike arrangements, together with a ruler with major divisions in centimeters. The backside of the cross-ply arrangement is hidden from the reading in panel c, such that the photos seem to be identical for the cross-ply (Arrangement 2B) and unidirectional (Arrangement 2A) arrangements in panel c. (Reproduced with permission from Ref.[16] Copyright 2019, Elsevier Science Ltd.)

2, labeled as 2A and 2B, which correspond to chain directions that are 0° and 90° apart on opposite sides of the paper. Thus, arrangement 2A is referred to as unidirectional, while arrangement 2B is defined as cross-ply. Ni is an effective coating material for enhancing the SE of CFs due to its physical properties and magnetism. Fig. 5.2c compares unglazed and nickel-glazed continuous CFs among the three planar arrangements[17]. Arrangement 2B consistently demonstrates higher SE compared to arrangement 2A, regardless of fiber coating. This aligns with the fact that EM waves are unpolarized, meaning the electric field of the waves is randomly distributed within the sample plane. Unpolarized EM waves are more commonly encountered than polarized EM radiation. Arrangement 1 exhibits a lower EMI SE than arrangements 2A and 2B for uncoated fibers but shows a similar EMI SE to arrangement 2A when fibers are coated with Ni. The inferiority of arrangement 1 compared to arrangement 2 becomes more apparent when no glaze is applied. From a magnetic attraction perspective, the planar coil arrangement (as in arrangement 1) is more favorable in terms of magnetic interaction compared to the linear configuration (as in arrangement 2). However, the contribution of the magnetic interface to EM wave shielding is less significant than that of the electrical interface, even though the magnetic interface is enhanced when fibers are coated with Ni.

In terms of the unidirectional arrangement 2A, the EMI SE is primarily determined by the interaction of the electrical field, which results in better SE compared to arrangement 1 for uncoated fibers. On the other hand, the cross-ply arrangement 2B exhibits significantly higher interaction with the electrical field compared to arrangement 2A. Consequently, arrangement 2B offers a much higher SE and shielding effectiveness attenuation (SE_A) than its counterpart (arrangement 2A) for both unglazed and Ni-glazed fibers. Hybrid-based (nano)composites that incorporate two or more types of fillers are expected to demonstrate superior EMI shielding properties compared to composites with only a single type of reinforcing material, such as fibers or particles. For instance, a combination of fibers and/ or particles as dual reinforcing fillers within the composite can lead to enhanced EMI shielding. However, due to the significant ratio of fibers compared to particles[18], the presence of fibers improves the electrical connectivity within the (nano)composite.

Researchers worldwide have been exploring the use of multiscale (nano) composites, which involve incorporating various fillers of different scales. For example, fillers like microcarbon "mC" (carbon fibers) and nC (CNTs) have been widely employed. While microcarbon provides continuity, nC does not, and adding nC to mC may not substantially improve the EMI SE due to the dominant role of continuous fibers[19]. Moreover, continuous C-nanocomposites exhibit superior mechanical properties compared to their noncontinuous counterparts. However, incorporating irregularly shaped nC into continuous mC can result in a reduced weight percentage of well-connected microcarbon, leading to a decrease in the mechanical properties of the composites and potentially affecting the EMI SE. The effect of the reinforcing filler should be carefully considered in the design of multifunctional hybrid/composite structures.

5.3 CERAMIC-BASED EMI SHIELDING MATERIALS

Compared to carbon or metals, ceramic materials are less commonly used for fabricating EMI shields due to their relatively low conductivity. However, among ceramic materials, metal carbides are often chosen for their conductive properties. Ceramic magnetic materials, such as ferrite (Fe_3O_4)[20,21] and nickel ferrite $(NiFe_2O_4)$[22], which exhibit ferromagnetic properties, are highly effective in shielding EMWs through absorption. Nonconductive ceramics with magnetic properties are preferred over metallic materials because they are resistant to corrosion. Therefore, combining ceramics like Fe_3O_4 with conductive elements like metals or carbon-based materials (such as rGO, CNFs or CNTs)[20] and inherently conducting polymers like PEDOT can be effective. Ceramics alone, with their magnetic, polarizability, and nonconductive properties, are not sufficient for EMI shielding. Thus, research on ceramic materials for EM wave shielding often involves the use of carbon-ceramic[23] or clay-metal composites. Examples of such systems include silver-ceramic[24], ceramic-FeSiAl, and Al_2O_3-metal composites[25]. These systems are also suitable for high-temperature shielding applications, such as SiC-C and Ti_3AlC_2 nanocomposites. Ceramics with high permittivity, such as perovskite-based materials, are also of interest for EMI shielding. Reported structural-based EMI shields in the literature include SiC-C, SiC nanowires, $SiC-Si_3N_4$, and SiC-SiCN composites[26]. MXenes, which are 2D inorganic materials, are a promising category of ceramic materials for EMI shield fabrication. They exhibit good conductivity (4000 S m^{-1}), polarizability, and large surface area. MXenes are composed of a small number of atomically thin layers of transition metal carbides, carbonitrides, nitrides, etc. and are considered excellent materials for EMI shield applications[27]. According to the literature, MXene-based materials have demonstrated shielding effectiveness (SE) values ranging from 24 dB to 70 dB within the frequency range of 8.2–12.4 GHz. These SE values are lower compared to carbon-based materials. Inexpensive continuous-glass fibers, as reported in the literature, are nonconductive and have poor EMI shielding properties. As a result, glass fiber-based polymeric matrix composites are widely used in cost-effective structures such as boats, concrete buildings, wind turbines, etc. To enhance the EMI shielding properties of glass fibers, conductive entities like CNTs and CNFs are added or coated onto the polymeric matrices[28]. There are reports of coating glass fibers with conductive components like Ni and rGO[29], which have also been applied to microsphere-based hollow glass fibers[30]. Additionally, glass fibers provide a radiation-transmitting coating to mitigate impedance mismatch issues.

When considering solid aggregates, cement is referred to as cement paste in natural processes. Mortar, on the other hand, is a fine particle combination that does not contain granular materials. Sand is commonly used as the fine combination, while stone or gravel is used as the coarse combination. These combinations act as reinforcing materials and help reduce material shrinkage caused by drying. Concrete admixtures are additional constituents, apart from water, cement and aggregates, that are used to enhance specific properties of cured cement. Continuous fibers cannot be fully incorporated in their aligned form during the cement mixing process. Therefore, cement shielding systems are not as effective as EMI shields unless the admixtures have the ability to enhance the EMI SE. EMI shield-enhancing mixtures primarily consist of conductive fibers such as CFs, CNTs, and steel fibers, as well as semiconductive fillers such as coke, carbon, graphene, graphite, and nickel. Carbon is a cost-effective nanoparticle aggregate compared to CFs and CNTs, making it necessary for the production of concrete materials. Currently, EMI shielding components are predominantly led by C-based materials, particularly CFs, CNTs, C, and coke, etc.

Ceramic nano/microparticles, such as ash containing Fe_2O_3, are another type of EMI shielding-enhancing compounds. However, semiconductive blends are much more effective in providing enhanced EMI SE compared to ash[31]. The use of discontinuous fibers/particles to create continuous conducting paths enhances the SE of these materials. However, achieving a dispersed distribution of reinforcing fibers/particles in a cement admixture can be challenging, especially when the fiber/particle size is in the nano-scale range. For example, CNFs tend to get entangled, making their dispersion difficult[32].

Composite amalgamations based on nanoscale materials result in a reduction of air voids within the final product, which is necessary for improving mechanical properties. While decreasing the percentage of voids is important for enhancing mechanical properties, using an innovative nanofiller like CNTs for this purpose is not economically feasible due to the higher cost compared to carbon. The inclusion of nanofillers to improve SE may also contribute to the reduction of air voids in the composite system. Consequently, the improvement in the EMI SE of nanofiller-reinforced nanocomposite systems primarily relies on the interfacial interaction between fillers and themselves, as well as with EMWs.

To address the aforementioned challenge of nanoparticle dispersion, one possible solution is to incorporate $nSiO_2$ (size 0.1 µm) into the cement admixture. $nSiO_2$ can be obtained as a low-cost waste product. On the other hand, steel microfibers with a large diameter (8 mm) and length (6 µm) have a high surface area, leading to strong interaction with EM waves and thus achieving a good SE of approximately 70 dB at a volume fraction of 0.72% in a cement-based system. In comparison, flat stainless steel 304 with a thickness of 4.12 mm achieves an SE of approximately 78 dB[33]. Therefore, steel fiber cement is inferior to plate-like stainless steel of the same thickness as an EMI shield. The superior SE-enhancing capability of steel microfibers compared to carbon microfibers in cement materials is attributed to their higher conductivity. CFs are typically coated with appropriate organic compounds to improve their processability in cement systems, while unsized CMFs are commonly used in this manner. The EMI SE tends to increase with increasing fiber volume percentage. At a constant CF volume percentage, the EMI SE is comparable between mortar and cement paste.

5.4 POLYMER-BASED EMI SHIELDING MATERIALS

Metal-encapsulated composite fibers are known for their efficient EMI shielding properties compared to unencapsulated CNFs, as conventional polymeric materials are nonconductive and transparent to EM radiation. Inherently conductive polymers, either produced by modifying nonconductive polymers or synthesizing directly from standard monomers, exhibit conductivity and are valuable candidates for EMI shielding materials[31]. Examples include polypyrrole (PPy), polyacetylene (PA), polyindole (PIn), polyaniline (PANI), and their copolymers. However, these polymers often have mechanical properties inferior to standard polymeric matrices and are more expensive. Polymer matrices, mainly used as binding materials in nanocomposites and composites for EMI shielding, can enhance thermal and electrical conductivity along with electrical continuity in the composite structure[34]. Conducting polymer matrices are particularly beneficial for this purpose. Additionally, nonconducting nanocomposites/composites can incorporate conductive polymer matrices as supplementary components to improve their EMI SE. The ability of a polymeric matrix to effectively disperse fillers is crucial for enhanced EMI shielding, as filler dispersion affects the conduction of the nanocomposite. Resilient polymer matrices are essential for EMI seals.

Researchers worldwide have extensively utilized composite materials in fabricating EMI shielding materials. For instance, studies on CNTs in PLA/PVDF composites have examined their impact on EMI SE, electric conductivity, and rheological performance. These studies found that processing time, PVDF viscosity, CNT aspect ratio, and processing technique significantly influenced the EMI SE of the final composite materials. The inclusion of CNTs improved the SE, dielectric permittivity (ε'), dielectric loss (tan δ), and EC (σ; S·cm−1) of the composites/blends. In their research, Zhang et al.[5] introduced an innovative method to convert a guest material into a host, resulting in the creation of a unique mountain-like structural wall composed of WS_2 and rGO. This structure exhibited efficient and environmentally friendly EMI SE. The reported EMI SE was greater than or equal to 20 dB in the frequency range of 2–18 GHz, with a maximum SE of 32 dB and a green index of approximately 1.0. The success of achieving effective and environmentally friendly EMI SE in their study was attributed to the multilevel architecture, inherent dielectric properties of the WS_2-rGO structure, relaxation and conduction synergy, multi-scattering among voids and interfaces, as well as the corresponding wedge effects. In another study by Jia et al.[10], highly porous high-performance 3D graphene/CNT/polydimethylsiloxane nanocomposites were developed using a simple process (as depicted in Fig. 5.3). These nanocomposites exhibited SE at the X-band as high as 54.43 dB and a specific shielding effectiveness (SSE) of 87.86 dB·cm^3 g^{-1} at a filler loading of ≤0.98 wt%.

According to the Schelkunoff theory, the structural architecture of EMW shielding materials, particularly those with an isolated and foam-based structure, enhances multireflection and attenuation of EM waves within the material. This characteristic provides a limitless advantage in enhancing the SE of structural materials[35]. Different structural orders have been considered in polymer-based composites for EMI shielding, including uniform structure, segregated structure, foamed structure, and layered structure.

When aiming to enhance the SE ratio and SE attenuation of EMI shields, the architectural design of these materials becomes crucial. It affects their macrostructure, microstructure, nanostructure, and overall composition, playing a vital role in achieving high SE. Therefore, it is not only the choice of an effective conductive filler that should be considered but also the structural design, which greatly influences the EMI shielding ability of the shield. Exceptional conductive properties and well-designed conductive network

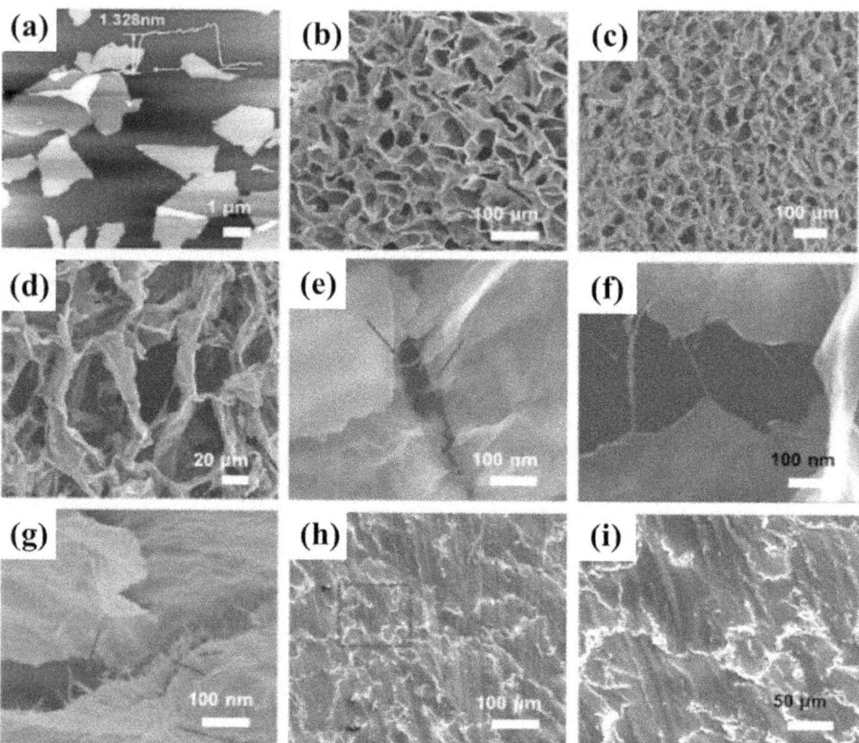

FIG. 5.3 (a) AFM image of initial GO sheets. (b) SEM image of initial GO/C foam. (c, d) SEM images of GC foam at different magnifications. (e, f) SEM images of MWCNTs' different distribution state in the 3D skeleton. (h, i) Cross-sectional SEM images of GC/PDMS composites at different magnification (bright parts are graphene sheets and MWCNTs, dark parts are PDMS matrix). (Reproduced with permission from Ref.[10] Copyright 2020 Elsevier Science Ltd.)

structures are essential requirements in the design, fabrication, and application of EM shielding nanocomposites and composites to achieve high SE and efficient absorption and reflection at multiple interfaces. Additionally, factors such as hysteresis loss and controllable shielding performance are imperative considerations. Architectural design tuning of EMI composite shields has gained increasing attention, considering various forms such as uniform, segregated, foamed, and layered structures[36]. Polymer materials, known for their excellent plasticity and ease of processing, play a progressively significant role in the structural design of EMI shielding composites. These different structural forms offer diverse properties and advantages in EMI shielding applications.

CONCLUSION

Functional, structural, and multi-structural shielding materials are covered, with small-scale utilization of functional shielding materials recently gaining attention from researchers and industrialists. Structural shielding materials enable multifunctional structures in extensive applications, with continuous carbon fiber hybrids/composites and their cement-based counterparts being dominant among these materials. The reviewed shielding materials include polymers, cement, ceramics, carbons, metals, hybrids, nanocomposites, and others. Researchers typically fabricate these structures through hybrids/nanocomposites to enhance their SE. Carbons and metals are the most effective materials for offering efficient SE, while nonconductive polymers, ceramics, and cement are generally not effective unless combined with functional components/materials. However, certain ceramics and intrinsically conducting polymers are conductive and effective as shielding materials. With various forms of microcarbons and nanocarbons available, shielding materials such as cement-carbon, ceramic-carbon, metal-carbon, and/or polymer-carbon mixtures have recently received much attention.

REFERENCES

1. Chung, D.D.L. (2020). Materials for electromagnetic interference shielding. Materials Chemistry and Physics *255*: 123587.
2. Sushmita, K., Madras, G., and Bose, S. (2020). Polymer nanocomposites containing semiconductors as advanced materials for EMI shielding. ACS Omega *5*: 4705–18.
3. Menon, A.V., Madras, G., and Bose, S. (2017). Magnetic alloy-MWNT heterostructure as efficient electromagnetic wave suppressors in soft nanocomposites. Chemistry Select *2*: 7831–44.
4. Wu, J., and Chung, D.D.L. (2008). Combined use of magnetic and electrically conductive fillers in a polymer matrix for electromagnetic interference shielding. Journal of Electronic Materials *37*: 1088–94.
5. Zhang, D.-Q., Liu, T.-T., Shu, J.-C., et al. (2019). Self-assembly construction of WS2–rGO architecture with green EMI shielding. ACS Applied Materials & Interfaces *11*: 26807–16.
6. Joshi, A., and Datar, S. (2015). Carbon nanostructure composite for electromagnetic interference shielding. Pramana *84*: 1099–116.
7. Du, Y., Liu, T., Yu, B., et al. (2012). The electromagnetic properties and microwave absorption of mesoporous carbon. Materials Chemistry and Physics *135*: 884–91.

8. Heinß, J.-P., and Fietzke, F. (2020). High-rate deposition of thick aluminum coatings on plastic parts for electromagnetic shielding. Surface and Coatings Technology 385: 125134.

9. Xu, Y.L., Uddin, A., Estevez, D., et al. (2020). Lightweight microwire/graphene/silicone rubber composites for efficient electromagnetic interference shielding and low microwave reflectivity. Composites Science and Technology 189: 108022.

10. Jia, H., Kong, Q.-Q., Liu, Z., et al. (2020). 3D graphene/ carbon nanotubes/polydimethylsiloxane composites as high-performance electromagnetic shielding material in X-band. Composites Part A: Applied Science and Manufacturing 129: 105712.

11. Sun, Z., Chen, J., Jia, X., et al. (2021). Humidification of high-performance and multifunctional polyimide/carbon nanotube composite foams for enhanced electromagnetic shielding. Materials Today Physics 21: 100521.

12. Zhang, D., Yang, H., Pan, J., et al. (2020). Multi-functional CNT nanopaper polyurethane nanocomposite fabricated by ultrasonic infiltration and dip soaking processes. Composites Part B: Engineering 182: 107646.

13. Lan, C., Guo, M., Li, C., et al. (2020). Axial alignment of carbon nanotubes on fibers to enable highly conductive fabrics for electromagnetic interference shielding. ACS Applied Materials & Interfaces 12: 7477–85.

14. Kim, T., and Chung, D.D.L. (2006). Mats and fabrics for electromagnetic interference shielding. Journal of Materials Engineering and Performance 15: 295–98.

15. Xiangcheng Luo, D.D.L.C. (1999). Electromagnetic interference shielding using continuous carbon-fiber carbon-matrix and polymer-matrix composites. Composites Part B: Engineering 30: 227–31.

16. Guan, H., and Chung, D.D.L. (2019). Effect of the planar coil and linear arrangements of continuous carbon fiber tow on the electromagnetic interference shielding effectiveness, with comparison of carbon fibers with and without nickel coating. Carbon 152: 898–908.

17. Chung, D.D.L., and Eddib, A.A. (2019). Effect of fiber lay-up configuration on the electromagnetic interference shielding effectiveness of continuous carbon fiber polymer-matrix composite. Carbon 141: 685–91.

18. Micheli, D., Vricella, A., Pastore, R., et al. (2016). Ballistic and electromagnetic shielding behaviour of multifunctional Kevlar fiber reinforced epoxy composites modified by carbon nanotubes. Carbon 104: 141–56.

19. Yao, K., Gong, J., Tian, N., et al. (2015). Flammability properties and electromagnetic interference shielding of PVC/graphene composites containing Fe3O4 nanoparticles. RSC Advances 5: 31910–19.

20. Yu, K., Zeng, Y., Wang, G., et al. (2019). rGO/Fe3O4 hybrid induced ultra-efficient EMI shielding performance of phenolic-based carbon foam. RSC Advances 9: 20643–51.

21. Yadav, R.S., Kuřitka, I., Vilcakova, J., et al. (2019). NiFe2O4 nanoparticles synthesized by dextrin from corn-mediated sol–gel combustion method and its polypropylene nanocomposites engineered with reduced graphene oxide for the reduction of electromagnetic pollution. ACS Omega 4: 22069–81.

22. Pan, H., Yin, X., Xue, J., et al. (2017). Microstructures and EMI shielding properties of composite ceramics reinforced with carbon nanowires and nanowires-nanotubes hybrid. Ceramics International 43: 12221–31.

23. Ru, J., Fan, Y., Zhou, W., et al. (2018). Electrically conductive and mechanically strong graphene/mullite ceramic composites for high-performance electromagnetic interference shielding. ACS Applied Materials & Interfaces 10: 39245–56.

24. Qing, Y., Su, J., Wen, Q., et al. (2015). Enhanced dielectric and electromagnetic interference shielding properties of FeSiAl/Al2O3 ceramics by plasma spraying. Journal of Alloys and Compounds 651: 259–65.

25. Tan, Y., Luo, H., Zhou, X., et al. (2018). Dependences of microstructure on electromagnetic interference shielding properties of nano-layered Ti3AlC2 ceramics. Scientific Reports 8: 7935.

26. He, P., Cao, M.-S., Cai, Y.-Z., et al. (2020). Self-assembling flexible 2D carbide MXene film with tunable integrated electron migration and group relaxation toward energy storage and green EMI shielding. Carbon 157: 80–89.

27. Wu, X., Han, B., Zhang, H.-B., et al. (2020). Compressible, durable and conductive polydimethylsiloxane-coated MXene foams for high-performance electromagnetic interference shielding. Chemical Engineering Journal 381: 122622.

28. Zhai, J., Cui, C., Ren, E., et al. (2020). Facile synthesis of nickel/reduced graphene oxide-coated glass fabric for highly efficient electromagnetic interference shielding. Journal of Materials Science: Materials in Electronics 31: 8910–22.

29. Yang, J., Liao, X., Wang, G., et al. (2020). Gradient structure design of lightweight and flexible silicone rubber nanocomposite foam for efficient electromagnetic interference shielding. Chemical Engineering Journal 390: 124589.

30. Zhang, H., Guo, Y., Zhang, X., et al. (2020). Enhanced shielding performance of layered carbon fiber composites filled with carbonyl iron and carbon nanotubes in the Koch curve fractal method. Molecules 25: 969. doi: 10.3390/molecules25040969.

31. Cao, J., and Chung, D.D.L. (2003). Coke powder as an admixture in cement for electromagnetic interference shielding. Carbon *41*: 2433–36.
32. Hong, X., and Chung, D.D.L. (2017). Carbon nanofiber mats for electromagnetic interference shielding. Carbon *111*: 529–37.
33. Hosseini, E., Arjmand, M., Sundararaj, U., et al. (2020). Filler-free conducting polymers as a new class of transparent electromagnetic interference shields. ACS Applied Materials & Interfaces *12*: 28596–606.
34. Shakir, H.M.F., Tariq, A., Afzal, A., et al. (2019). Mechanical, thermal and EMI shielding study of electrically conductive polymeric hybrid nano-composites. Journal of Materials Science: Materials in Electronics *30*: 17382–92.
35. Lee, T.-W., Lee, S.-E., and Jeong, Y.G. (2016). Carbon nanotube/cellulose papers with high performance in electric heating and electromagnetic interference shielding. Composites Science and Technology *131*: 77–87.
36. Yu, W.-C., Xu, J.-Z., Wang, Z.-G., et al. (2018). Constructing highly oriented segregated structure towards high-strength carbon nanotube/ultrahigh-molecular-weight polyethylene composites for electromagnetic interference shielding. Composites Part A: Applied Science and Manufacturing *110*: 237–45.

CHAPTER 6

Macrostructure design of electromagnetic interference (EMI) shielding materials

INTRODUCTION

As a member of porous functional materials, hydrogels, composed of a network of cross-linked hydrophilic building blocks surrounded by water, not only maintain the porous structure but also have a water-enriched interior environment.[1,2] On the one hand, most of the engineering nanomaterials can be easily and uniformly dispersed in the hydrogels to form the conductive paths, leading to the high conductive loss of incident EM waves[3]. On the other hand, the water can produce polarization loss to consume the EM wave energy owing to the change of hydrogen-bond networks as well as the polarization of water molecules in gigahertz and terahertz bands. Benefiting from the water-rich nature and the soft polymer-based building blocks, the broad application of hydrogels includes, but is not limited to, tissue engineering, soft electronics, and electromagnetic interference (EMI) shielding. Due to the water being made of polar molecules, the polarization of H_2O molecules and variations of hydrogen-bond networks can be evoked under the applied electromagnetic field. Combined with the efficient

DOI: 10.1201/9781003490098-6

155

multiple reflection capability derived from the existing porous structure, the hydrogels show great potential for novel and high-performance EMI shielding materials.

The network of cross-linked polymer chains is crucial for forming the hydrogels with a stable 3D structure. Through forming the dynamic ionic/hydrogen bonds between polymers or creating covalently cross-linked networks, the polymer chains dispersed in an aqueous solution are assembled to construct the hydrogels with outstanding elasticity and flexibility[4]. Although the interactions between inorganic nanoparticles are weak, the cross-linked polymer networks can allow the conductive fillers to be embedded into the polymer matrix to prepare the freestanding and conductive hydrogels[5,6].

With the rapid development of high-precision and high-sensitivity electronic devices, there is an increasingly urgent need for ultra-high-performance EMI shielding materials. Among various forms of EMI shielding materials, electromagnetic shielding fabrics have received widespread attention due to their advantages such as light weight, good flexibility, and strong adaptability in protecting human bodies and sensitive electronic devices from electromagnetic radiation. The efficient EMI shielding performance depends on the material composition and structure of the coating on the fabric surface.

With the advancement of modern high technologies, the problems related to EMI and electromagnetic (EM) radiation pollution caused by EM waves have become increasingly severe. EM pollution, in addition to noise pollution, air pollution, and water pollution, has emerged as a significant public health threat. Particularly in the era of 5G and miniaturized, multifunctional electronic packaging, the mitigation of undesired EM energy generated by electronic components has become an urgent and challenging task[7,8]. Therefore, it is imperative to address the extremely severe issue of EM pollution.

EMI shielding is considered the most effective and convenient approach to eliminate EMI. It involves utilizing the inherent properties of electrically conductive or magnetic materials to block the propagation path of EMI sources, thereby reflecting or confining EM waves within the shield and regulating EMI at its root cause. Among various EMI shielding materials, elemental metals such as gold, aluminum, copper, and silver, which possess freely moving electrons, are commonly used due to their high electrical conductivity. However, these metals have limitations in their application for EMI shielding due to their high density, poor resistance to acid, alkali, oxidation, and vulnerability to secondary reflection pollution[9,10].

In recent years, there has been an increasing demand for EMI shielding materials that are lightweight, have high absorption loss, low cost, wide absorption band, and good processability, particularly in aircraft, aerospace, and fast-growing electronics[11,12]. To meet these requirements, researchers have been exploring the use of nanomaterials, such as zero-dimensional (0D) metal nanoparticles[13], one-dimensional (1D) carbon nanotubes (CNTs)[14,15], nanowires[16], nanofibers[17], and two-dimensional (2D) graphene[18] and MXene[19]. These nanomaterials are being incorporated into three-dimensional (3D) free-standing porous materials with porous network structures, which offer superior absorption and dissipation of EM waves compared to thin-layered paper, film or textile-based EMI shields[20,21].

Free-standing porous composites for EMI shielding can be categorized into three main types: foams, aerogels, and sponges. The aforementioned advanced nanomaterials can serve as fillers or reinforcement phases embedded in the foam, aerogel or sponge matrices. This improves the physicochemical properties of composite foams, aerogels, and sponges for EMI shielding, including electrical conductivity, magnetic permeability, specific surface area, dispersity, density, porosity, mechanical strength, and flexibility[22,23]. Consequently, these composite materials have shown great potential for efficient shielding of EM waves.

Over the past few decades, there has been significant progress in the exploration of various materials for EMI shielding and electromagnetic wave absorption (EMWA)[24]. In academia, there has been a growing interest in carbon-based materials, ranging from fundamental theoretical studies to experimental design and analysis. This interest stems from the pursuit of high-performance EMW absorbing and shielding materials that possess strong absorption/shielding capabilities, are lightweight, have low thickness, and operate within a wide bandwidth range.

Carbon nanomaterials such as CNTs[25,26], carbon nanofibers[27], reduced graphene oxide (rGO)[28], and their composites[29] have shown immense potential due to their good conductivity, high aspect ratio, specific surface area, and excellent chemical stability. However, the complex and relatively expensive manufacturing and processing methods pose challenges for mass production of these materials. Additionally, these carbon materials heavily rely on the petrochemical and coal chemical industry, which is not sustainable in the long term due to resource depletion and environmental pollution. Therefore, the search for alternative materials that can serve as substitutes for these carbon-based products is underway, with the goal of creating a greener and more sustainable society.

Wood, which includes 3D self-supporting natural wood and its primary components such as cellulose and lignin, has attracted increasing attention as an ideal renewable precursor for carbon materials due to its abundant reserves[30]. Although hemicellulose is an important component of wood biomass, there is limited research on hemicellulose-derived carbons in the field of EMWA and EMI shielding, and thus it is not covered in this review. Cellulose and lignin, the two most abundant natural polymers on Earth[31], along with widely available wood, offer advantages in terms of low cost and accessibility. Compared to high-cost carbon materials like graphene and CNT, wood-based carbons demonstrate clear cost advantages. Moreover, relying on sustainable forestry resources, wood biomass ensures consistent yield and stable quality, enabling mass production of products based on wood biomass-derived carbon materials. Wood-based carbons possess desirable properties such as good electrical conductivity, lightweight, abundant pores, and a stable structure[32], making them and their composites strong contenders in the fields of EMWA and EMI shielding. Carbonized cellulose and lignin, in particular, exhibit excellent dispersibility in water, facilitating the uniform loading of functional fillers. Additionally, their strong gelling ability allows for the creation of tailored porous structures that aid in the dissipation of EMWs, a feature not commonly found in other carbon materials. Carbonized wood monoliths with regular pore structures offer ample space for loading functional fillers and enhancing multiple reflections of EMWs within the micro-level structures of the materials. Regarding EMWA, wood biomass-derived carbons can effectively suppress incident EMWs through conductive losses caused by the conductive carbon network, as well as polarization and relaxation losses within the materials[33]. The high porosity of these carbons promotes impedance matching and multiple reflections of EMWs[34]. Furthermore, the abundant functional groups on the surfaces of wood biomass materials simplify the loading of functional fillers. In summary, wood biomass-derived carbons play crucial roles in EMW absorption, with their microstructures and compositions being key factors[35]. In terms of EMI shielding, wood-based carbons provide an electron transmission path that enhances EMW reflection and absorption, along with unique porous architectures that facilitate multiple reflections and scattering of EMWs. Moreover, these carbons can act as supportive hosts for functional fillers, enabling the preparation of high-performance EMI shielding composites. Consequently, wood-based carbons show great promise for EMWA and EMI shielding applications, underscoring the importance of reviewing related research on these promising carbon materials.

6.1 HYDROGEL MATERIALS FOR EMI SHIELDING

6.1.1 Conductive hydrogels

Generally, the conductive hydrogel is composed of an intrinsically conductive material and a cross-linked hydrogel network, which combines the unique advantages of the conductive material and the hydrogel, and has the advantages of good conductivity, adjustable mechanical flexibility, and easy processing, which can be used to prepare strain sensors with different requirements[36]. So far, various electronic and ionic conductive materials have been developed, such as conductive polymers[37], metal-based materials[38], MXenes[39], carbon-based materials[40], and ionic materials[41]. They can be added into the hydrogel matrix to prepare various conducting hydrogels. For electrically conductive hydrogel materials, electron transport in hydrogel mainly depends on tunnel effect at low concentrations, so hydrogel has low conductivity. When the concentration increases to the seepage threshold, the conductive material forms a continuous conductive network inside the gel, thus forming an electron transmission path with relatively low resistance. Therefore, the hydrogel exhibits high conductivity. For ion-conducting hydrogels, the network structure inside the gel and the water-rich environment provides the necessary channels for ion transport[42]. Therefore, the uniform porous structure, high concentration of electrolyte solution, and high-water content are conducive to improving the conductivity of the gel materials. Both electron and ion-conducting hydrogels can be given appropriate conductivity by optimizing the loading amount of conductive materials and designing appropriate network structure of hydrogels. In this section, conductive hydrogels are named and summarized according to conductive materials.

6.1.1.1 Conductive polymer-based conductive hydrogels

Conductive polymer is a kind of synthetic polymer with electronic conductivity, which is widely used to prepare various conductive hydrogels due to its adjustable conductivity and easy synthesis[43,44]. Conductive polymers are used as fillers or polymer scaffolds to construct conductive hydrogels. When the conductive polymer is used as the filler, the rigid and hydrophobic chain segments of the conductive polymer tend to twist or tangle in the solution during the polymerization process, thus forming conductive polymer "islands" in the porous hydrogel matrix, resulting in low conductivity of the hydrogel. Therefore, the most effective way to prepare highly conductive polymeric hydrogels is through doping/crosslinking conductive polymers.

Zhou et al.[45] prepared conductive polymer hydrogels using tannic acid (TA) doping and cross-linked in-situ polymerized polypyrrole (PPy) (Fig. 6.1). As Ppy is the hydrogel skeleton material, hydrogels show better electrical conductivity performance. It was found that the conductivity and mechanical properties of hydrogel could be regulated by the concentration of TA. With the increase of TA, the hydrogel crosslinking density increases, and then the elastic modulus of hydrogel changes from 394 ± 15 Pa to 2260 ± 20 Pa. And the conductivity increased from 0.05 S cm^{-1} to 0.18 S cm^{-1}.

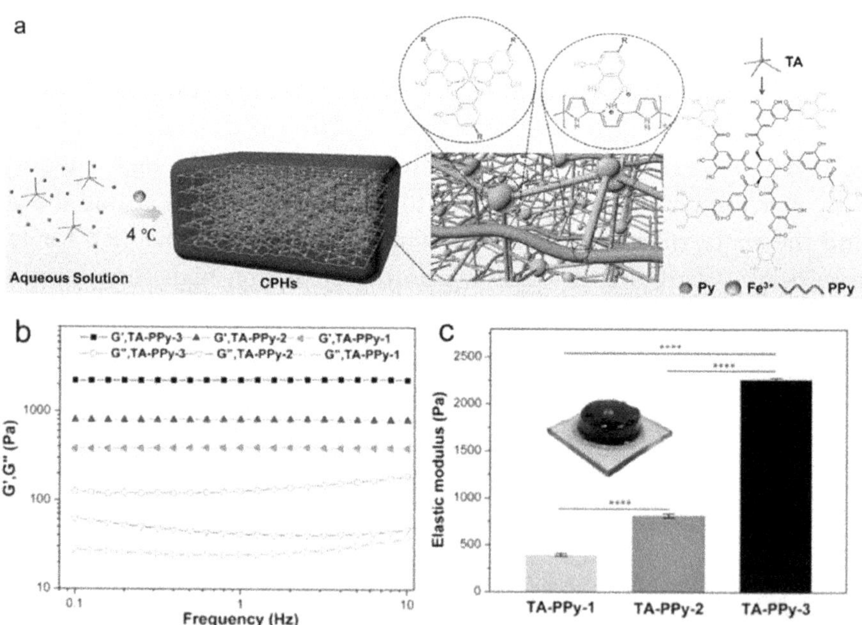

FIG. 6.1   (a–c) Fabrication and properties of CPHs conductive polymer hydrogel. (Reproduced with permission from Ref.[45] Copyright 2018, American Chemical Society.)

6.1.1.2 Carbon-based conductive hydrogels

Carbon materials, especially 1D and 2D carbon materials, such as CNTs and graphene (GO), are widely used as conductive fillers to prepare conductive hydrocondensates due to their high conductivity, excellent stability, and mechanical strength[46]. CNTs and GO are carbon nanomaterials with large specific surface area, which can construct a conductive network at a lower concentration, and give hydrogels good electrical conductivity. However, CNTs and GO are difficult to directly combine with hydrophilic hydrocondensates due to their inherent hydrophobicity and poor dispersion, which make them easy to self-aggregate in an aqueous medium. In addition, their weak interfacial interactions with hydrophilic polymers cause hydrogels to exhibit poor electrical conductivity. To solve these problems, cellulose nanofibers (CNF) and hydrophilic polymers or compounds are often used to improve their compatibility with hydrogels[47].

Chen et al.[48] mixed CNF and CNT in an acrylamide (AM) solution, and then used N, n-methylene diacrylamide as a crosslinking agent to initiate AM polymerization in situ by the initiator potassium persulfate, effectively doping CNTs into polyacrylamide (PAM) hydrogel (Fig. 6.2). Because of the excellent dispersion of CNFs, when the content of CNFs was 1 wt%, the conductivity of the PAM/CNF/CNTs composite hydrogel was reduced from $0.041 S\ m^{-1}$ increased to $0.085\ S\ m^{-1}$. At the same time, the appropriate addition of CNTs can also improve the conductivity and mechanical properties of hydrogels. However, when the CNTs content increased to 2 wt%, the conductivity and tensile property of the hydrogel decreased due to the poor dispersion of CNTs, which restricted the transport of electrons in the hydrogel and resulted in stress concentration.

6.1.1.3 MXene-based conducting hydrogels

MXene is a 2D transition metal carbide or carbon-nitride material. Because of its large aspect ratio, excellent mechanical strength, electrical conductivity, photothermal effect, hydrophilicity, and abundant surface functional groups, MXene has been widely used in drug delivery, seawater desalination, flexible sensors, EMI shielding, and other fields[47]. Because of the abundant hydrophilic groups on the surface of MXene, it has good dispersion in water and strong interaction with the polymer network, so that it can be evenly dispersed in the hydrogel and form a stable conductive network. Therefore, MXene is often used to construct high-performance conductive hydrogels. MXene can be doped into hydrogel polymer as

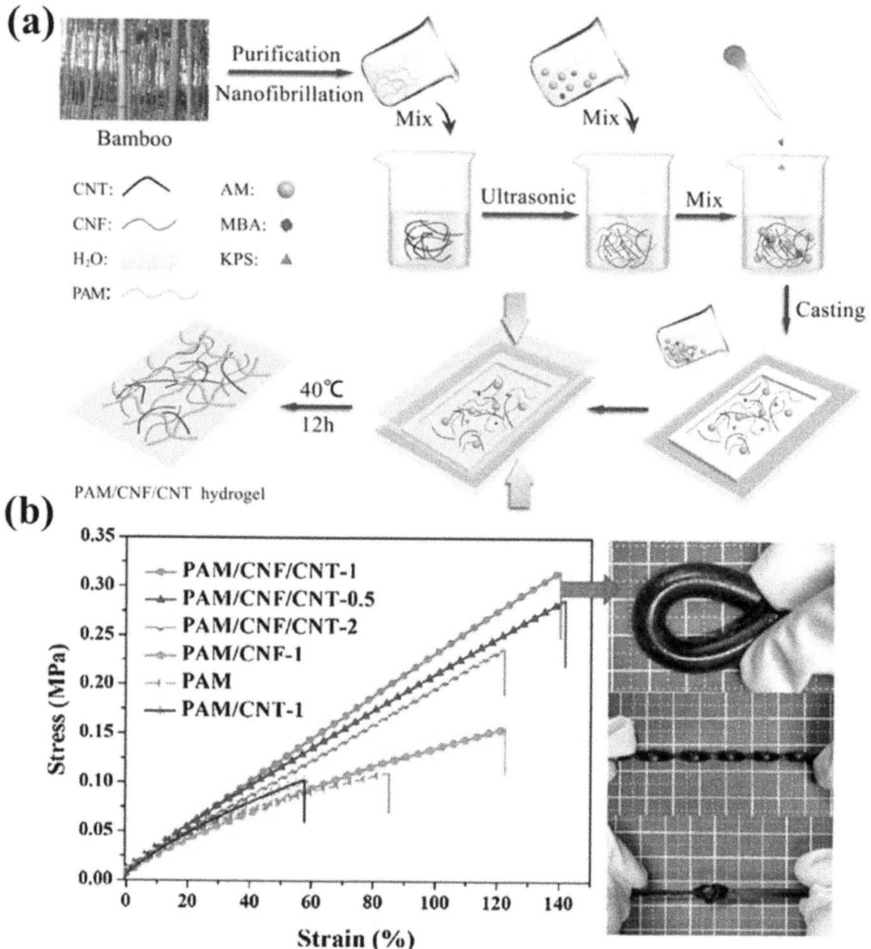

FIG. 6.2 (a, b) Preparation and properties of the PAM/CNF/CNT conductive hydrogel[48]. (Reproduced with permission from Ref.[22] Copyright 2019, Springer Nature.)

nanometer conductive filler in the matrix, the skeleton material that can also be used to construct the hydrogel enhancing the hydrogel electrical conductivity[49–51].

Inspired by pearls with excellent mechanical properties, Shi et al.[52] proposed a high-performance MXene-based flexible strain sensor with a structure similar to "brick-mortar" (Fig. 6.3). The prepared composites exhibit ultra-high sensitivity, high tensile strength, and long-term durability. Two-dimensional MXene nanosheets and 1D silver nanowires were

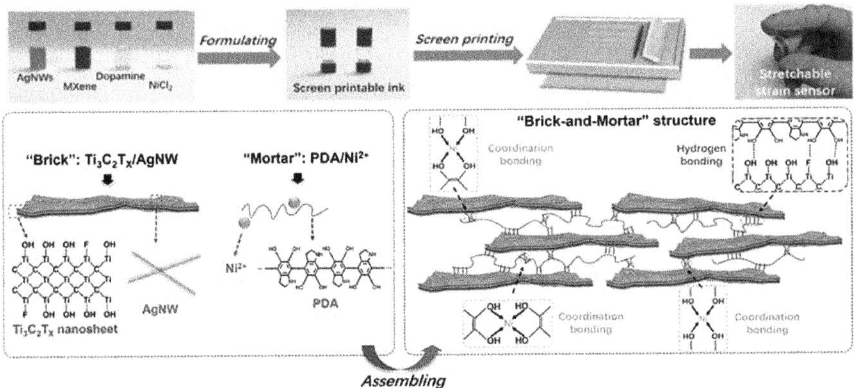

FIG. 6.3   Preparation the MXene-based conductive hydrogel with "brick-and-mortar" structure.

used as "bricks" and PDA/Ni^{2+} was used as "mortar" to form a "brick-mortar" structure that mimics the nacre layer. The "brick" component gives the hydrogel high electronic conductivity, while the "mortar" structure gives the hydrogel high toughness and excellent transmission. The results showed that the composite conductive hydrogel exhibited high conductivity (10500 S cm^{-1}), but also had good tensile properties and ultra-high sensitivity (GF=8767.4). The sensor made of hydrogel can be used for human movement monitoring and health monitoring because of its ultra-high sensitivity.

6.1.1.4 Metal-based conductive hydrogels

In recent years, metal-based conductive materials (such as metal nanoparticles and low melting point metal alloys (LMs)) have been used to construct conductive hydrogels, and a series of high-performance flexible strain/stress sensors have been developed. Metallic nanoparticles (such as silver nanoparticles and gold nanoparticles) are often used as conductive fillers to prepare conductive hydrogels because of their excellent conductive properties[53]. However, inorganic metal particles tend to aggregate and settle in the polymer matrix, resulting in the inability to disperse evenly in the hydrogel, which ultimately affects the properties of the hydrogel (such as electrical and mechanical properties)[54]. In order to solve these problems, some effective methods have been reported, including in-situ synthesis of metal nanomaterials on hydrophilic cellulose nanocrystals (CNCs), in-situ synthesis and fixation of metal nanomaterials using some hydrophilic polymers or compounds with reducing ability, and finally incorporation into hydrogel network to achieve the purpose of improving the performance of hydrogels.

In order to load silver nanoparticles (Ag NPs) on CNCs, Lin et al.[55] coated TA on CNCs for the first time. Ag^+ was reduced to Ag NPs by the phenol hydroxyl group of TA. Reduced Ag NPs are fixed to CNCs by coupling with the catechol group on TA. Finally, complex Ag/TA@CNC was successfully introduced into polyvinyl alcohol (PVA) hydrogel. The hydrogels exhibit excellent electrical conductivity, tensile properties, and antibacterial properties. When the content of Ag/TA@CNC increased from 0 wt% to 5 wt%, the conductivity of hydrogel increased from 0.16 S m^{-1} to 4.61 S m^{-1}, and the elongation at break increased from about 1700% to about 4000%.

As soft conductive materials, LMs (such as gallium-based eutectic alloy) can be better used to prepare stretchable conductive hydrogels by fixed addition of hydrophilic polymers or polar monomers into the hydrogel network. In addition, LMs can catalyze the in-situ polymerization of monomers to form hydrogels. Liao et al.[56] used PVA to fix LM particles (LMP) and crosslinked boric acid to form PVA-LMP composite hydrogel (Fig. 6.4). The interaction between PVA's hydroxyl group and LMPs immobilized LMPs and improved the compatibility of LMPs with hydrogels. Therefore, the composite hydrogel showed excellent electrical conductivity (3.75× 10-3 S m^{-1}). In addition, due to the dynamic interaction between the PVA hydroxyl group and the borate ion, the composite hydrogel showed excellent self-healing properties.

6.1.1.5 Ionic-based conducting hydrogels

Gel is a soft material composed of a 3D polymer frame structure and continuous aqueous phase, with abundant pore structure inside, providing a way for ion migration, which provides technical support for the synthesis of high-performance ionic conductive hydrogels[57,58]. The free ions in the gel may come from polyelectrolytes, electrolytes or even ionic liquids dissolved in the aqueous phase that form hydrogel networks. Because ionic hydrogels genereally do not use dark and hydrophobic electronic conductive materials, they usually have excellent transparency and good mechanical properties[59]. In addition, ion-based conductive hydrogels also have excellent electrical conductivity and are widely used in electronic skin, personalized medical monitoring, motion monitoring, artificial muscles, and ion touchpads[60].

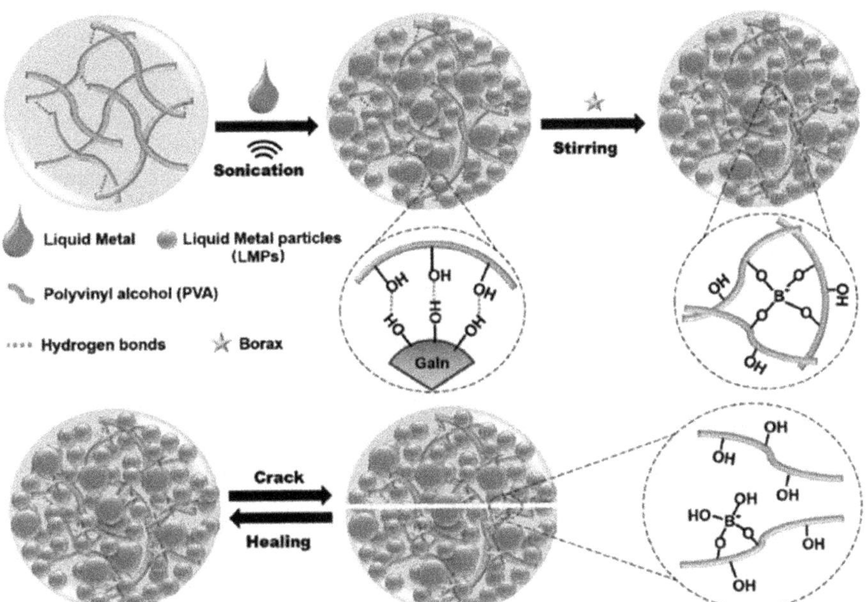

FIG. 6.4   Schematic illustration of the preparation of PVA-LMP composite hydrogel. (Reproduced with permission from Ref.[56] Copyright 2022, American Chemical Society.)

Wang et al.[61] reported the preparation of PVA/sodium chloride (PVA/NaCl) conductive hydrogels with anisotropic microstructures by thermal stretching and directional freezing (Fig. 6.5). Thermal stretching and directional freezing make the PVA chains aligned along the tensile direction, thus forming anisotropic microstructures inside the gel. The hydrogel shows anisotropy in mechanical properties and fatigue resistance. In addition, NaCl was introduced in the preparation process of hydrogel to give the hydrogel good electrical conductivity. Due to the anisotropy of the gel, the ionic conducting hydrogel prepared has different sensing sensitivity in the tensile direction and the vertical direction, so as to achieve accurate anisotropy sensing capability, and the monitoring accuracy of the sensor for strain can reach 0.5%. The PVA/NaCl hydrogel sensor can also realize real-time monitoring of human body joint motion and display anisotropic motion information, providing technical inspiration for the study of anisotropic flexible material.

Conductive hydrogels are usually black or dark, and ionic conducting hydrogels are generally transparent or light-colored. Compared with

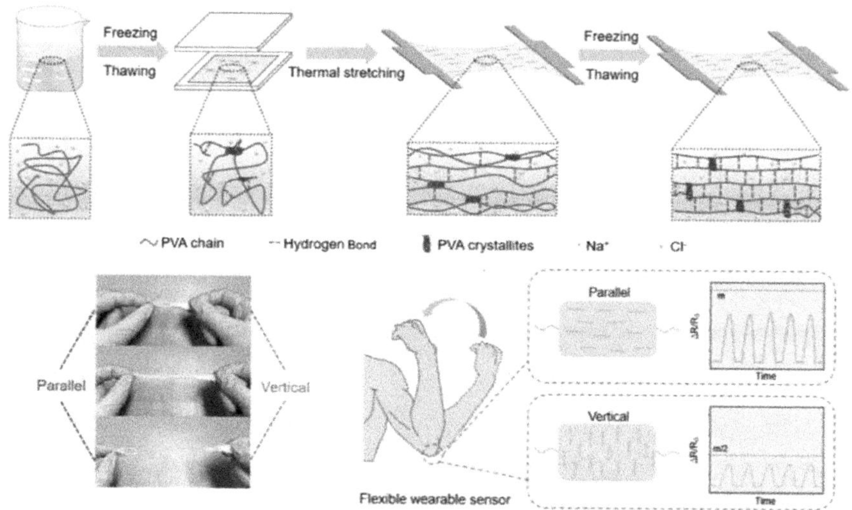

FIG. 6.5 Preparation, properties and applications of anisotropic PVA/NaCl hydrogels. (Reproduced with permission from Ref.[61] Copyright 2022, Springer Nature.)

other conductive hydrogels, the hydrogels with higher conductivity can be prepared by using conductive polymers as the polymer framework and using efficient admixtures. However, the inherent brittleness and hydrophobicity of conductive polymers destroy the tensile properties of hydrogels. In contrast, the more dispersed graphene/RGO and CNTs used as conductive fillers for hydrogels can not only provide good electrical conductivity for hydrogels, but also ensure excellent mechanical properties of hydrogels. The excellent dispersion of MXene in water is conducive to the formation of a continuous conductive network. Therefore, MXene-based hydrogels show better electrical conductivity and tensile properties. In addition, MXene can form a hydrogel network skeleton through crosslinking, and the resulting gel shows better electrical conductivity. However, due to the lack of elastic polymer network structure, hydrogels exhibit low mechanical properties. The conductive hydrocondensates prepared by metal nanoparticles, low melting point metal alloys (LMs), and other metal materials usually show higher elongation at break and lower electrical conductivity. The use of 1D nanomaterials (such as CNCs) to fix metal nanoparticles is conducive to the formation of a continuous conductive network, so as to effectively improve the conductivity of metal-based conductive hydrocondensates. In addition, LMs have catalytic properties and can catalyze monomers and prepare conductive hydrogels by in-situ polymerization, which is conducive to improving the mechanical properties and sensing properties of LMs-based conductive hydrogels. Ion-based conducting hydrogels constructed from ions of electrolytes or polyelectrolytes have excellent mechanical properties. However, ionic conductive hydrogels exhibit low conductivity due to the balance between ionic concentration and water content. The conductive properties of the hydrogel can be improved by combining electronic conductive materials such as conductive polymer with ionic conductive hydrogel. Because of different conductive materials, conductive hydrogels have different properties, which will provide an important reference for the construction of electronic skin and wearable electronic devices in the future.

6.1.1.6 Ferrogel hydrogels

Magnetic gels are generically called as ferrogel or magnetoelastic gels. Magnetoelastic gels are composed of a flexible 3D network of polymeric chains having a viscoelastic nature and magneto sensitivity. Magnetoelastic gels are composed of finely distributed colloidal magnetic nanoparticles or nano wires inside gel matrix with swellability[62].

Among magnetic loss materials, magnetic metal nanoparticles with high saturation magnetization and Snoek's limit value, involving Fe, Co, Ni, and their alloys, are compact to achieve stronger magnetic loss. Magnetic chains can be thought of many 0D magnetic nanoparticles arranged in 1D splicing. The domain motion in the magnetic chains is faster than that in the 0D nanoparticles. It is well known that the structure and motion of magnetic domains affect the magnetization behavior of materials, which is related to magnetic loss. The relative complex permeability is determined by the domain migration confined inside the magnetic chains and the magnetic flux field surrounded outside the magnetic nanochains. Compared with dispersed magnetic particles, the magnetic coupling in the chains is more obvious. Besides, the magnetic response shows strong correlation with the chain length, causing an entirely different performance of electromagnetic parameters. These findings make magnetic chains a promising candidate for EMWA[63].

6.1.2 Research progress of hydrogel-based EMI shielding materials

Generally, high-performance EMI shields often possess high electrical conductivity. Nevertheless, the polymer-based building blocks of hydrogels, such as PVA, PAAM or PAM, and cellulose, are typically non-conductive, leading to insufficient EMI shielding capability. The factors affecting the shielding performance of hydrogels include conductivity, multiple reflections of incident EM waves, and conductive and polarization loss capabilities. Therefore, filling conductive nanomaterials to achieve considerable conductivity is the key point for developing hydrogel-based EMI shielding materials. Meanwhile, retaining the internal water-rich environment also contributes to the potent polarization loss capability derived from the changed hydrogen-bond networks and molecule polarization under the EM fields. Probing the effects of the components in hydrogels on EMI shielding performance is warranted. Thus, this discussion can be divided into three main categories according to the components and functionalities, including the hydrogel-based EMI shields embedded with inorganic conductive fillers as well as the hydrogel-based EMI shields integrated with conductive polymers.

6.1.2.1 Hydrogel-based EMI shields embedded with inorganic conductive fillers

Recently, owing to the large aspect ratio and high conductivity, 1D nanomaterials such as silver nanowires (AgNWs)[64,65] and CNTs[66,67] are most popularly employed for constructing 3D conductive networks and improving the conductivity and EMI shielding performance of hydrogels. Through vacuum-assisted filtration, the AgNWs were attached to the PVA/aramid nanofibers (ANFs) mats fabricated by electrospinning. Furthermore, Zhou et al. fabricated the sandwich structured hydrogels combined with the ANF-reinforced PVA hydrogels layer and AgNWs/PVA layer[68]. The crystalline structure of PVA formed during electrospinning acts as physical crosslinking sites to form a 3D network with ANFs, and then the hydrogels were constructed by swelling in the water. In addition, the ANFs further reinforced the mechanical properties of hybrid hydrogels, exhibiting the modulus of 10.7–15.4 MPa, which were superior to those of pure PVA hydrogel. With the addition of AgNWs increasing from 0.02 vol.% to 0.23 vol.%, the conductivity of the hybrid hydrogels reached 16,000 S·m^{-1}, and the EMI SE in X-band increased from 32 to 52 dB at the thickness of approximately 0.3 mm. It was noticed that the enhancement of more than 20 dB mainly originated from SE_A as the SE_R increased slightly. The addition of highly conductive AgNWs mainly contributed to the formation of the AgNWs layer with potent conductive loss capability, lengthening the propagation path of incident EMWs and thus dissipating the EMWs more. This mainly promoted the absorption loss-dominated EMI SE. Considering the high density, the metal-based fillers were expected to be replaced by lightweight, highly conductive, and chemically stable carbon-based 1D conductive nanomaterials, such as CNTs[3,69]. More importantly, the CNTs with huge aspect ratios are able to form conductive networks, increasing the conductivity of hydrogels. Yang et al. fabricated the hydrogels by incorporating multiwalled CNTs (MWCNTs) into hydrophobically associated PAM hydrogels by using cellulose nanofiber as the dispersant[70]. The admirable dispersion of MWCNTs also facilitates the close interconnection between MWCNTs to form a strong and well-established conductive network throughout the hydrogels. As a result, with the increased MWCNT content from 0.1 wt.% to 1 wt.%, the electrical conductivities of these as-prepared hydrogels gradually enhanced to 0.85 S·m^{-1}, and the EMI SE increased from 15 to 28.5 dB at a thickness of 2 mm. The enhancement of EMI SE was mainly ascribed to the SEA derived from the increased charge carriers with greater dissipation capacity with increased MWCNT contents. Notably, benefiting from the reversible crosslink bonds such as dynamic hydrogen bonding and ionic interactions, this conductive hydrogel also possessed the autonomously self-healing capability. After self-healed from two sections of hydrogel contacting for a week, the mechanical, electrical and EMI shielding properties remained unchanged, which cannot be achieved by bulk and foam/aerogel-based EMI shields.

Yu et al. reported the MXene-based PVA/PAAM hydrogel via polymerizing acrylamide (AAm) and connecting the polymers by hydrogen bonds (Fig. 6.6)[71]. With the increased solvent displacement time, the pore size and thickness of the cell wall of MXene hydrogels decreased, and the EMI SE firstly increased from 12 to ~ 35 dB with an addition of 1.1 wt.% MXene, and then decreased to ~ 30 dB at a MXene content of 2.2 wt.%. The nonmonotonic change of SE was attributed to the competition between increased electron conduction from MXene and decreased ion conduction. In addition, the decreased cell walls derived from high MXene content were also against the electrical conductivity of networks and multiple reflections of incident EM waves, decreasing the EMI SE. In the practical application, the low-temperature tolerance and anti-drying performance are vital for the hydrogel-based EMI shields. To achieve the long-term stability of hydrogel-based EMI shields, the solvent displacement of MXene-based hydrogel was carried out, and the MXene organic hydrogel containing binary solvent of

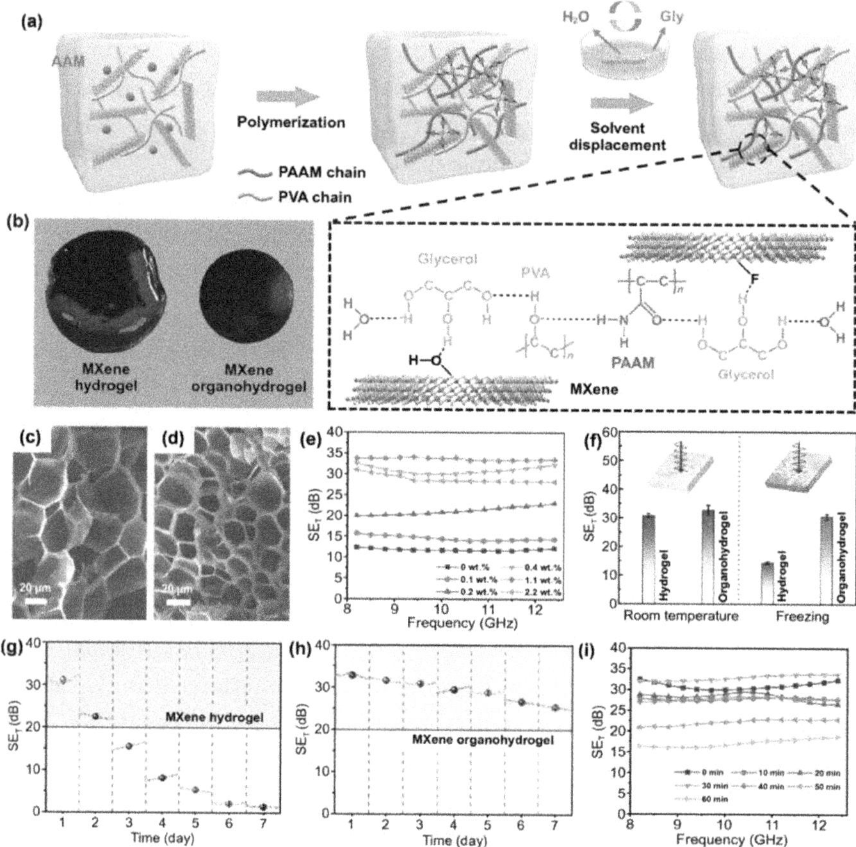

FIG. 6.6   (a) Schematic of preparing MXene-based organic hydrogels, (b) photography of hydrogel and organic hydrogel, (c) and (d) SEM images of porous structures with different displacing time, (e) EMI SE of MXene-based hydrogels, (f) average SE of hydrogel and organic hydrogel before and after freezing, (g) and (h) the stability of hydrogels with storage time, and (i) EMI SE of MXene-based organic hydrogel with water-Gly displacement time. (Reproduced with permission from Ref.[71] Copyright 2022, Springer Nature.)

glycerol (Gly) and water was obtained. After freezing at -25 °C, the EMI SE of MXene organic hydrogel was maintained at more than 30 dB, while the SE of MXene hydrogel declined to 14 dB. More importantly, during 7-day storage at room temperature, the SE of MXene hydrogel decreased from 30.8 to 1.54 dB, while the SE of MXene organic hydrogel only decreased to 25.3 dB. However, after the solvent displacement of 60 min, the content of Gly in organic hydrogels further increased, and the EMI SE gradually declined to below 20 dB due to the weak polarization loss of Gly molecules in the GHz band.

Through a facile, scalable unidirectional freezing followed by a salting-out approach, a type of hydrogels composed of "trashed" MXene sediment (MS) and biomimetic pores is manufactured by Yang et al.[72] Upon the synergistic effects of the MS-based conductive network, PVA chains, water, and porous structure, the MS-based hydrogels exhibit good EMI shielding performance. Our MXene-based hydrogels possess the EMI SE of 31 to 91 dB at a thickness of 1.0 to 7.5 mm, respectively, and an SE of more than 40 dB at a thickness of 2.0 mm in the ultra-broad band GHz band (8.2–40 GHz). Particularly, the influences of water hydrogels on EMI shielding performance via a resumable approach to control the water fraction was quantitatively identified. With the addition of AgNWs, the shielding performance of the MS-based hydrogels significantly increases, which is derived from the enhanced multiple reflection loss (RL) caused by the biomimetic-aligned porous structure. This contributes to the implementation of an anisotropic EMI shielding performance for the hydrogel-based EMI shields, achieving the preparation of hydrogels with controllable EMI shielding performance derived from interior porous structures beyond the material constituent.

6.1.2.2 *Hydrogel-based EMI shields integrated with conductive polymers*

Although the conductive fillers lift the shielding performance of hydrogels, due to the weak interactions and biocompatibility of inorganic nanomaterials, the compliance, deformability and biodegradability of hydrogels are compromised[42]. To solve this, employing conductive polymers seems to be an effective strategy to balance the contradiction between conductivity and mechanical properties of hydrogels. Thus, the conducting polymer-based hydrogels have been explored for high-performance EMI shielding.

The PEDOT: PSS, one of the most promising conductive polymers, holds great potential for enabling new EMI-shielding applications due to the inherent electrical conductivity[73,74]. In PEDOT: PSS molecules, the PEDOT contributes to the high electrical conductivity and PSS holds the structure of the polymer as well as improves solubility. To achieve accurate customizability, additive manufacturing was employed to prepare the hydrogel-based EMI shields.

To further lift the conductivity of PEDOT: PSS, Wang et al. utilized the ionic liquid to modify the PEDOT: PSS polymers for constructing the conductive hydrogels[75]. Through the dry-annealing film formation followed by the rehydration method, the conductive polymer hydrogels without conductive inorganic fillers were fabricated. After the modification and rehydration in Gly/water solution, the 50 wt.% ethyl-3- methylimidazolium bis(trifluoromethylsulfonyl)imide (EMIMTFSI) doped PEDOT: PSS hydrogels (PEDOT: PSS hydrogel/EMIMTFSI) exhibited the conductivity of 30,600 $S \cdot m^{-1}$, which was far higher than that of PEDOT: PSS hydrogels. This brought about the EMI SE of 53 dB at a thickness of merely 0.045 mm and a thicknesses-normalized SE of 1,182.9 $dB \cdot mm^{-1}$ in the X band. Liu et al. developed the Ti_3C_2-MXene-modified PEDOT: PSS ink for 3D printing, and a freeze-thawing method was designed to transform the printed objects directly into conductive and robust hydrogels with high shape fidelity[76]. Dispersing the freeze-dried PEDOT: PSS into MXene dispersion, the inks with optimal printability were simply achieved by adjusting the solid concentration of PEDOT: PSS and MXene to 4 wt.% and 1 wt.%, respectively. The as-printed objects were firstly frozen so that the ice crystals extruded the MXene and PEDOT: PSS to form a framework. Subsequently, the frozen as-printed objects were thawed in the H_2SO_4 solution immediately. During this process, the partial PSS was dissolved and the crystallinity of PEDOT was increased, increasing the conductivity of hydrogel. After this treatment, the Ti_3C_2-MXenemodified PEDOT: PSS hydrogel with flexibility, stretch ability, fatigue resistance and high fidelity was fabricated.

6.2 FILM/FABRIC MATERIALS

Commonly used coating materials for preparing EMI shielding fabrics include conductive materials such as CNTs, graphene, AgNWs and MXene[77,78]; magnetic materials such as Fe, Ni and Co[79]; and conductive polymers such as PANI and PPy[80]. In addition, polymers can be used to enhance the interaction between the coating and the fabric base or as a hydrophobic coating to maintain the performance stability of the shielding fabric in long-term use[81]. Jia et al. prepared an electromagnetic shielding fabric with an ultra-high electroEMI SE of 106 dB by impregnating carbonized non-woven fabric (CFF) in AgNWs dispersion and encapsulating it with polyurethane (PU) (Fig. 6.7a)[82]. AgNWs are uniformly attached to single fibers and gaps in fibers to build a good conductive path on the fabric. The PU layer plays a role in protecting the AgNWs network, so that the PU-AgNW/CFF fabric still exhibits excellent electromagnetic shielding performance even after being subjected to bending, ultrasonic, peeling and acidic and alkaline solutions.

Macroscopic or microscopic structural design of coating materials can improve the electromagnetic shielding performance of fabrics. Fu et al. alternately assembled conductive MXene and insulating polyethyleneimine (PEI) on the fabric base[83]. The dielectric properties of MXene/PEI composite fabric were improved, and the rich heterogeneous interface inside promoted the multiple reflection and absorption of EMW, leading to an increase of 138.95% in EMI SE compared with pure MXene-coated fabric. Ren et al. chemically deposited silver on nonwoven fabric (NWF) and encapsulated it with either waterborne polyurethane (WPU) or WPU containing FeCo@rGO to prepare Ag/NWF/WPU and Ag/NWF/FeCo@rGO/WPU composite fabrics, respectively (Fig. 6.7b)[84]. When the amount of Ag was both 10.5 wt%, the EMI SE_T of the latter was 4.6 dB higher than that of the former, with SE_A increased from 55.8 dB to 70.6 dB and SE_R decreased from 16.7 dB to 6.5 dB. This was because FeCo@rGO not only weakened the reflection caused by impedance mismatch, enhanced dielectric loss and magnetic loss, but also formed a special absorb-reflect-absorb layered structure with the highly conductive Ag/NWF network, thereby enhancing the multiple reflection and absorption loss of EMW.

6.2.1 Conductive film/fabric materials

With the current development demands for EMI shielding materials to be "thin, light, strong, and wide," there is an urgent need for thin film materials that possess ultralight, flexible, freestanding and high mechanical properties. These materials can satisfy the requirements for packaging materials, energy storage devices and composites. Furthermore, the rising popularity of wearable electronics and portable devices has increased the demand for EMI shielding materials that are much thinner, lighter and more flexible. This makes thin film materials unique and indispensable in this field. Therefore, there have been numerous research studies on graphene-based films in the field of electromagnetic shielding, including microstructure, specific properties and theoretical research. These studies have demonstrated the significant advantages and potential for the application of these films.

What's more, EMI shielding fabrics have low density, good flexibility and light weight which make them widely used in manufacturing

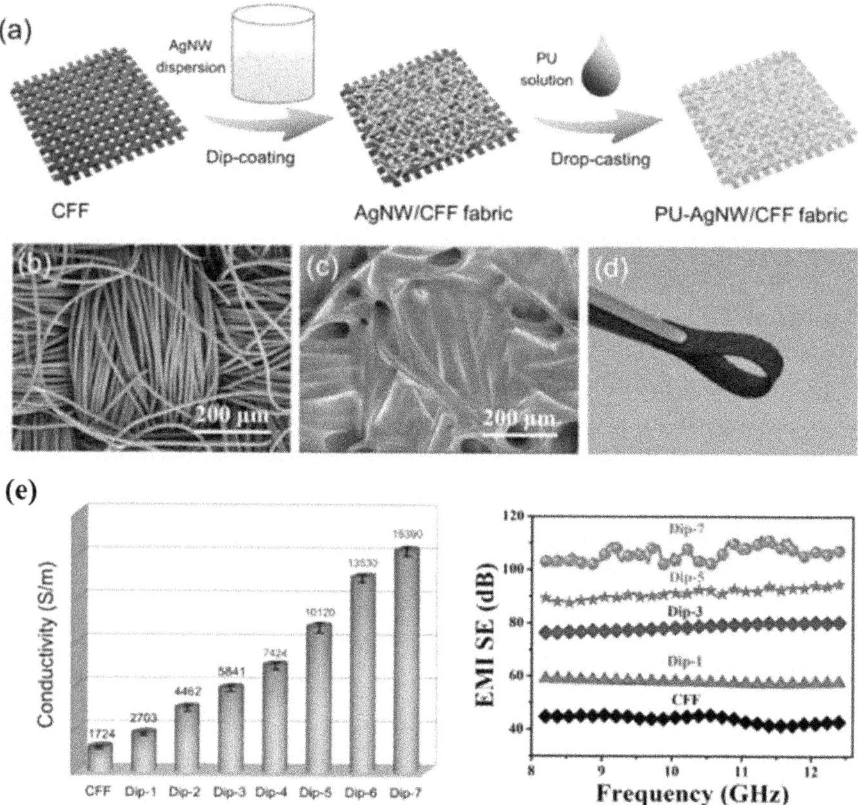

FIG. 6.7 The fabrication and EMI shielding performances of (a) PU-AgNW/CFF. (b) SEM image of CFF. (c) SEM image of the PU-AgNW/CFF fabric. (d) Digital photograph of the PU-AgNW/CFF fabric under bending state. (e) Electrical conductivity of the CFF and the PU-AgNW/CFF fabrics as a function of dip-coating cycles. (f) EMI SE of CFF, Dip-1, Dip-3, Dip-5 and Dip-7 in X-band. (Reproduced with permission from Ref.[82] Copyright 2018, Elsevier Ltd.)

electromagnetic wave protection products like protective clothing, shielding tents and shielding gun-suits. These fabrics can be easily molded, boast excellent designability and have breathable properties; they possess both softness and EMI shielding properties. They can also be turned into different shapes to shield radiation sources, or processed into shielding suits and caps to safeguard staff from electromagnetic wave radiation. Additionally, metal fiber fabrics have other functions including anti-static, antibacterial, and deodorizing. It is clear that EMI shielding fabrics are indeed ideal shielding materials with outstanding properties.

6.2.1.1 Metal-based EMI shielding film/fabric

Metal is a conventional conductive material[85] that is known for its relatively low cost and excellent mechanical properties. With the progress of nanotechnology, it is now possible to obtain metal materials at the nanoscale in various dimensions: 0D, 1D or 2D, through a variety of fabrication processes. In comparison to traditional metal wires or foils, nanoscale metal materials exhibit a significant improvement in conductivity, allowing them to be incredibly lightweight and thin. Ultra-thin metal composites typically involve the combination of metals with other materials such as fiber membranes, gels, aerogels and so on. Fiber membranes can be produced using various methods, and electrospinning stands out for its ability to create ultra-thin fiber membranes. The porous structure resulting from electrospinning plays a crucial role in increasing the number of times electromagnetic (EM) waves are reflected, thus contributing to improved shielding efficiency.

Guo et al. have even conducted a review that outlines the application value of electrospinning in EMI shielding materials[86]. Through the addition of viscous substances like polyacrylonitrile (PAN)[87] and polyvinyl pyrrolidone (PVP)[88], as well as conducting physical hot pressing and applying post-treatments using organic solvents such as acetone and ethanol, the connection between spun fibers can be effectively strengthened, resulting in significant improvements in their mechanical properties. In the case of fabrics produced through other textile methods, factors such as yarn arrangement, yarn diameter, yarn composition, warp and weft yarn density, weaving mode, weaving angle and fabric pattern can all impact the EMI SE[89,90]. For EMI shielding composites, several aspects influence their overall EMI SE, including the volume fraction, structure, shape and size of the conductive filler, the composite structure, polymer matrix type, viscosity and material preparation method. It has also been noted that the weight fraction of the metal in the composite can be determined by the ratio of metal ion concentration to polymer concentration during in-situ polymerization, thereby influencing the total EMI SE. These factors should all be carefully considered during material selection and structural design. Furthermore, the development of advanced coating technologies and materials is crucial for achieving ultra-thin shielding membranes.

6.2.1.1.1 Ag-based EMI shielding film/fabric   Silver is a highly conductive metal that offers excellent cost performance. It is considered a suitable material for EMI shielding and possesses remarkable antibacterial properties. In terms of structural design, silver matrix composites are typically categorized into two aspects: silver nanoparticles and silver nanowires.

For EMI shielding materials consisting of silver nanoparticles wrapped with polymers, achieving uniform deposition of the silver particles is crucial, and this can be influenced by different silver-plating processes. Additionally, ensuring good adhesion between the metal layer and the polymer is critical. Most metal particles have difficulty forming strong connections with both synthetic and natural polymers, making sole reliance on Van der Waals' force unreliable. Moreover, weak interactions can compromise the durability of EMI shielding materials. To address these challenges, researchers have developed several bonding methods, primarily categorized as physical methods and chemical methods. Physical methods include techniques like hot pressing[91], hot rolling[92] and high-temperature sintering[93]. These methods involve elevated temperatures and melting processes, which facilitate cross-linking between the polymer nanofibers, alter the internal structure and transition the glass state to a highly elastic state. This promotes the embedding of metal particles and enhances the interaction between metal nanoparticles and the polymer substrate.

In chemical methods, the selection of an adhesive that can create a strong bond through a chemical reaction is crucial for connecting polymers and metal nanoparticles[94]. Inspired by the adhesive properties of certain proteins secreted by mussels, which exhibit exceptional adhesion to various interfaces, Haeshin et al. proposed that the presence of o-diphenol and amino groups could enable adhesion. They developed dopamine self-polymerization to form a polydopamine (PDA) adhesive, which effectively connects different interfaces. In addition to enhancing adhesion, the presence of PDA also improves the hydrophilic characteristics of composites. This is particularly beneficial when applying a metal coating to the surface of hydrophobic polymers like polylactic acid (PLLA)[81,95].

By leveraging the properties of PDA, an excellent process for coating metal nanoparticles has been developed. Firstly, a polymer fiber membrane is prepared using electrospinning technology and subsequent post-treatment; then, it is immersed in a mildly alkaline PDA solution to mimic a natural sea environment; finally, the PDA-grafted fiber membrane is immersed in a metal ion precursor solution and combined with a reducing agent to form an ultra-thin metal-wrapped membrane. This membrane exhibits ultra-high conductivity and EMI SE. In this process, the concentration of the precursor and the coating time significantly impact the material's conductivity and EMI SE[96], while maintaining a pH around 7 prevents the oxidation of metal nanoparticles[97]. The presence of PDA ensures a more uniform metal layer, and PDA itself affects the surface morphology of the material. Because of its stability, many EM interference shielding composites utilize PDA as an adhesive. For example, Yun Mao et al. utilized PDA to create silver-graphite felt[98], while Zeng et al. developed metal-encased cellular membranes through PDA attachment[99].

However, the exact mechanism behind the adhesion between PDA and metals remains unclear. One possible mode of action[100]. The adhesion of PDA may result from the coexistence of o-diphenol and amino groups. It is worth noting that dopamine can be relatively expensive, prompting researchers to seek more cost-effective adhesive alternatives. Complexes containing both polyphenol and polyamine, such as TA and PEI, have been explored as potential substitutes for PDA, offering a more affordable option[101,102]. Based on this principle, future research may lead to the development of even more suitable adhesives. Additionally, other chemicals like (3-aminopropyl) triethoxysilane and 3-mercaptopropyltriethoxysilane have been utilized to modify the polymer surface and facilitate the deposition of silver[103].

Nanowires or nanorods, which are metal nanostructures, are widely utilized and their formation is dependent on the crystal lattice orientation of metals[97]. Just like silver nanoparticle-polymer composites, it is necessary to establish a strong bond between silver nanowires and polymers[104]. Moreover, for silver nanowires, it is crucial to create interconnected conductive networks within polymer matrices. Despite the gaps between silver nanowires, these networks enable high conductivity and EMI SE while maintaining light transmittance[105]. Yu et al. developed an EMI SE membrane with a thickness of only 20 μm using a silver nanowire/polymer composite film, achieving an EMI SE of 33 dB[106]. Additionally, a composite film of rGO-modified silver nanowires exhibits excellent light transmission, stability and EMI SE[107].

The fabrication process of silver nanowire polymer composites typically involves two steps: synthesis of silver nanowires and their subsequent incorporation into polymers. The polyol method is commonly employed for synthesizing silver nanowires. Initially, polyvinylpyrrolidone (PVP) is completely dissolved in ethylene glycol under continuous stirring. Then, a salt solution and silver nitrate ($AgNO_3$) are added to the solution and stirred until homogeneous. The mixture is then subjected to high-temperature treatment, followed by thorough washing with acetone, ethanol and deionized water before dispersion in deionized water[16]. PVP acts as a surfactant, guiding the directional growth of nanoparticles and facilitating the formation of 1D structures. Cetyltrimethylammonium bromide is another surfactant used for inducing the formation of nanorods, and the concentration of the surfactant plays a crucial role in determining the nanostructure formation. Arjmand et al. also achieved silver nanowires with a higher aspect ratio through the AC electrodeposition approach[108].

6.2.1.1.2 Ni-based EMI shielding film/fabric   Nickel, as a ferromagnetic metal, possesses similar characteristics to other metals such as high hardness and ductility. Its oxidation resistance, high conductivity and affordability make it widely used in the field of EMI shielding. However, nickel is classified as a class I carcinogen by the International Agency for Research on Cancer (IARC), which restricts its use in wearable devices and medical protection.

The process of coating nickel is similar to that of silver and copper. However, when hypophosphorous acid is used as a reducing agent for nickel, phosphorus is incorporated into the nickel coating. Since phosphorus is not conductive, a high concentration of phosphorus reduces the material's conductivity. Thus, controlling the deposition of phosphorus is crucial. Researchers have found that adjusting the pH of the reaction solution effectively regulates the deposition of phosphorus, leading to higher EMI SE in Ni-P composite coating materials[109]. Some researchers have also developed nickel-iron-phosphorus alloy polymer composites specifically for EMI shielding[110]. To connect nickel with polymers, a molecular bridge called APTMS is commonly used to link organic and inorganic substances. Activators such as Sn^{2+} and Pd^{2+} are also utilized. There has been an exploration into using carboxymethyl chitosan complexed with palladium cation as an activation solution for fabrics[111]. Researchers have conducted projects on altering, modifying, and activating fabrics like Tencel and ramie, resulting in Co-Ni-P coatings with high EMI SE[112]. Rare earth elements have been employed as catalysts for the successful plating of the Co-Ni-P layer on fabric surfaces[113].

Recently, a unique uniform closed microporous PVDF/Ni composite has been developed using foaming techniques. This loose structure not only reduces material density but also enhances electromagnetic wave loss during propagation, providing higher conductivity with less volume138.

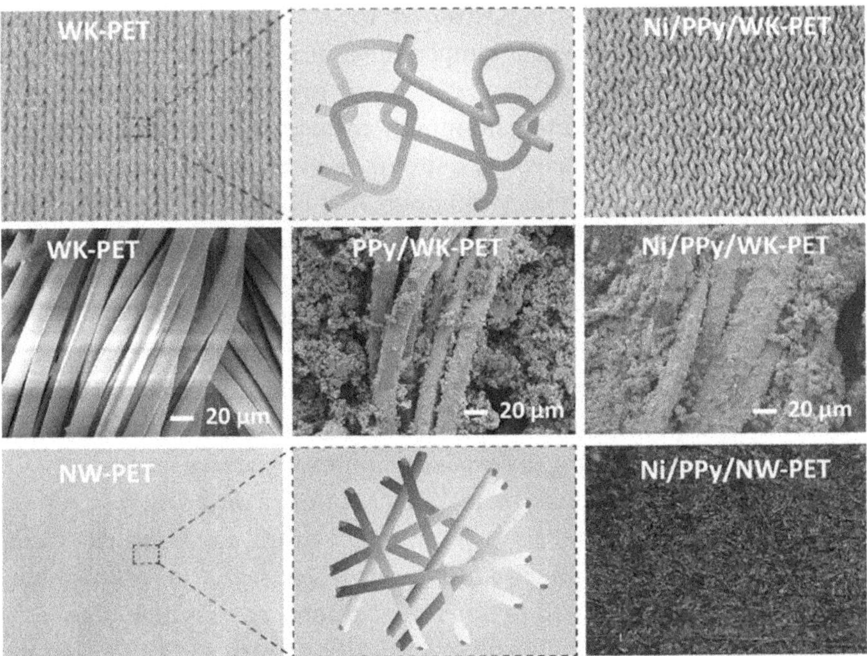

FIG. 6.8   Ni/PPy/WK-PET and Ni/PPy/NW-PET conductive fabrics for EMI SE. (Reproduced with permission from Ref.[115] Copyright 2021, Elsevier Ltd.)

An ultra-thin metal film consisting solely of Ni and Ag, measuring a mere 2 μm in thickness, has also been created and exhibits versatility and high EMI SE[114]. Additionally, flexible EMI SE films with a thickness of less than 500 μm have been produced by combining warp-knitted PET (WK-PET) and non-woven PET (NW-PET) with polypyrrole (PPy) and nickel (Fig. 6.8)[115]. Nickel-coated flaky graphite particles are also utilized as fillers for EMI shielding[116].

Nickel, as both a magnetic and electrically conductive material, plays a significant role in EMI SE. Even a small quantity of nickel can contribute to achieving desirable EMI SE. Compared to copper, the reduction process for nickel requires fewer modifications and activations, making it relatively simpler. However, due to its high density and carcinogenic nature, nickel has limited applications. Therefore, it is frequently combined with other metals and conductive substances in small amounts to enhance EMI SE.

6.2.1.1.3 Other metal-based EMI shielding film/fabric   Metals like aluminum, zinc and magnesium also possess conductivity and can be utilized as materials for EMI shielding. However, their high reactivity makes them prone to oxidation at room temperature, limiting their practical use. Nonetheless, their combination with other materials remains extensive and meaningful. For instance, aluminum-coated paper enhances EM wave reflection[88]. Rare earth elements, a group of 17 elements including scandium, yttrium and lanthanides in Group IIIB of the periodic table, often denoted as R or RE, exhibit distinct magnetic and photoelectric properties. Introducing trace amounts of rare earth elements can adjust the EMI shielding properties of composites, thereby increasing their EMI SE and presenting significant potential for applications. Furthermore, rare earth elements can enhance reaction rates, crystallinity and corrosion resistance of materials. Incorporating rare earth sources into metal chemical deposition processes for EMI shielding materials can improve material quality[113,117]. Gd has been employed in electroless plating to optimize materials[100].

In addition to solid metals at room temperature, some metals exist in a liquid state, such as gallium and indium. These liquid metals also have high conductivity compared to water or solutions, possess higher viscosity and can maintain a relatively stable physical state with good chemical stability. Thus, the development of liquid EMI shielding materials is worth exploring. However, the challenge lies in their packaging due to their fluid nature. Xu et al. proposed a feasible solution and achieved success[118]. They inflated microspheres using the liquid's surface tension and then stabilized them by adding metal powders to create aerogels, which encapsulate the liquid metal. Hydraulic metals possess strong driving force and elasticity but require size optimization when applied to thin-layer EMI shielding materials.

More stable metallic materials offer better durability and longevity, but their widespread usage is limited by price. Cheaper metals, which are more reactive and commonly found in nature, tend to be susceptible to oxidation and corrosion, and have limited EMI shielding properties. Selecting the most cost-effective metal based on application requirements and expected investment is crucial. Additionally, the physical properties of the metal also impact its application to some extent. Factors such as light transmission, wear resistance, ductility, and gloss should be considered when selecting materials. In the case of metallic EM shielding composites, it is important to consider the necessary properties of the material, minimize costs, and incorporate trace elements or modifications to enhance overall performance. Metals possess a significant advantage over other conductive materials due to the movement of numerous free electrons, resulting in excellent electrical conductivity. Even at low concentrations, they can achieve impressive EMI SE. Furthermore, metals are readily available in nature, cost-effective to prepare, and can easily be utilized in building nanostructures, making them superior to carbons and MXenes. Most importantly, the shortcomings of metals can be easily compensated through appropriate combinations, expanding their applicability and enabling mass production, thus ensuring high economic viability.

In conclusion, each metal matrix composite has its own set of advantages and drawbacks. Silver exhibits high conductivity and stability but comes at a relatively higher price. Copper, although cheaper than silver, offers good conductivity but is prone to corrosion. Nickel is corrosion-resistant but poses a potential carcinogenic risk. Liquid metal possesses good fluidity but lacks controllability. In practical applications, it is essential to leverage their respective advantages. While there are numerous types of metals, the formation of ultra-thin metal composite films often follows similar principles. Electrically conductive substances, magnetic substances, and polar molecules are more likely to reflect and absorb EM radiation at the interface, resulting in the loss of EM waves throughout the material. Therefore, selecting suitable materials is an effective strategy for achieving ultra-high EM shielding through structural design. Based on these characteristics, numerous conductive composite films can be prepared.

6.2.2 Carbon-based EMI shielding films/fabrics

6.2.2.1 Carbon fiber-based EMI shielding film/fabric

Carbon fiber composites that are flexible possess most of the physical requirements needed for EMI shielding applications such as high electrical conductivity, large specific surface area, light weight, flexibility, porous structure, and adjustable physical and chemical properties. Carbon fibers-based composites with various morphologies have been applied in EMI shielding. Carbon fiber alone has low absorption efficiency when used for EMI shielding, and it is usually combined with additional EMI filling materials.

6.2.2.1.1 Carbon fiber/metal (oxide)   Several studies have demonstrated that combining carbon fibers (CFs) with metal/metal oxides such as Ag, Ni, Cu, Al_2O_3, Fe, and Fe_3O_4 can effectively enhance EMI shielding performance[119]. Metals like Ag, Ni and Cu possess high conductivity and are capable of dissipating electrostatic charges[81,120]. Magnetic materials like Fe, Ni and Fe_3O_4 exhibit a high saturation magnetization. These metal/metal oxides were incorporated into a flexible CF matrix to create efficient flexible EMI shielding composites[121].

Jiao et al.[122] developed a core-shell composite consisting of cotton-based CFs (CDCFs) and nano-copper. The cotton fibers were converted into conductive and hydrophobic CDCFs, onto which a layer of nano-copper was deposited. The synergistic effect of the nano-copper and CDCF composites resulted in an EMI SE of 29.3 dB. The EMI shielding mechanism of the core-shell nano-Cu/CDCF structure. The conductive structure with a multi-scale network allows for the generation of alternating electromagnetic field-induced currents, leading to rapid attenuation of incident waves.

Furthermore, dielectric relaxation and interfacial polarization loss contribute to the attenuation of EMW. A uniform thin copper layer was electrolessly plated on the surface of CFs[123]. The findings indicate that the EMI SE of Cu@CF is significantly higher than that of pure CFs and Cu foils in the frequency range of approximately 30 MHz to 10 GHz. The porous structure of CFs, along with the core-shell microstructure of Cu@CF, facilitates multiple internal reflection effects at both the micro- and macro-levels, resulting in effective electromagnetic wave attenuation. The core-shell microstructure also promotes interface polarization, increasing the dielectric loss and leading to excellent EMI shielding performance[124,125].

Liu and Kang[126] applied a conductive silver film onto the surface of CFs (shown in Figs. 6.9a–c), resulting in an EMI SE of 80.82 dB with 100 sprays. However, the expensive nature of silver prevents its practical use as an EMI shielding material. These studies aim to establish the groundwork for the development of high-performance EMI shielding materials suitable for large-sized metalized CFFs. Considering that CFs exhibit poor magnetic properties[127], magnetic fillers are introduced to enhance the magnetic loss for EMI shielding. Recently, a new composite coating of magnetic CFs using nickel/ferroferric oxide nanoparticles (Ni/Fe_3O_4-NPs) was prepared[128]. Following Schelkunoff's EMI shielding theory, the magnetic CFs with high magnetic permeability contribute to a coordinated loss mechanism involving both wave reflection and absorption processes, significantly improving the EMI SE of CFs. Yim and Lee[129] utilized electroless plating to coat CF surfaces with soft magnetic metal FeCoNi, resulting in a 40% enhancement in EMI shielding performance compared to pure CFs.

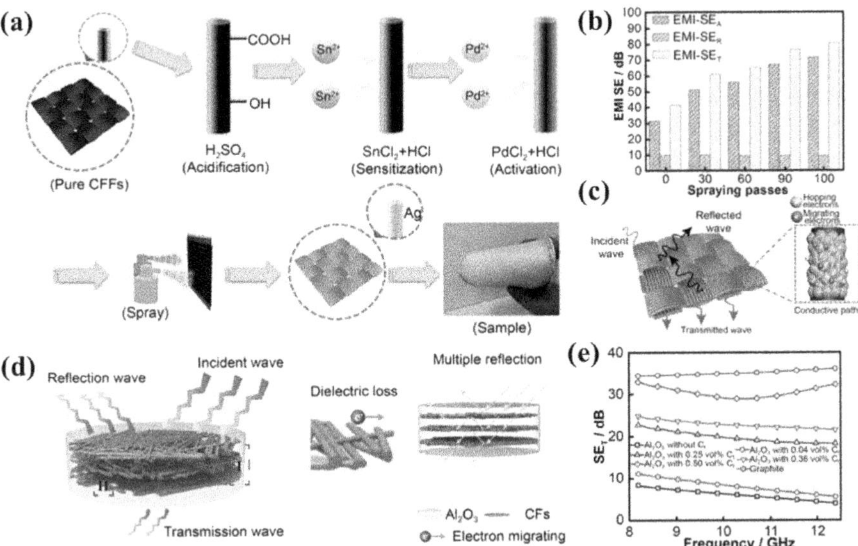

FIG. 6.9   (a) Facile fabrication of conductive silver films on CFFs via spray technique; (b) comparison of SE_A, SE_R, and SE_T of Ag@CFs composites with different spraying passes; (c) schematic illustration of EMI shielding mechanism of Ag@CFFs. (Reproduced with permission from Ref.[126] Copyright 2019, Elsevier Ltd.) (d) Multiple-reflection mechanisms of CF/Al_2O_3 composites; (e) SE_T of CF/Al_2O_3 ceramics. (Reproduced with permission from Ref.[130] Copyright 2019, Elsevier Ltd.)

Ceramic composites also serve as a reliable option for EMI interference-shielding components used in high-temperature applications. Zhu et al.[130] coated nonwoven CFs with Al_2O_3. The outcome demonstrated that seven-layer parallel CF layer/Al_2O_3 composites achieved an EMI shielding effectiveness (SE) as high as 29-32 dB in the X-band frequency range (as shown in Figs. 6.9d–e), effectively transforming Al_2O_3, originally a wave-transmitting material, into an EMI shielding material. The multiple reflections generated by the multilayer structure play a crucial role in enhancing the EMI shielding performance.

6.2.2.1.2 Carbon fiber/(metal)/polymer composites   CF/metal composites face limitations in the field of EMI shielding due to factors such as their high density, difficulty in machining, low corrosion resistance and mechanical properties. However, there are alternative materials that have been widely used for EMI shielding and metal corrosion protection, such as WPU, waterborne polyacrylate, PVA and waterborne epoxy resin[131–133]. These conductive polymers, when combined with a multilayer heterogeneous interface structure, can enhance absorption loss through interface scattering and polarization loss[134]. By forming a heterogeneous layer structure on the CF framework, conductive polymers provide EMI shielding materials with multifunctional characteristics[135].

For instance, Li et al.[27] developed a composite membrane by coating a large area with an electrospun carbon nanofiber felt (CNFF) and a silicone alternating multilayer structure (AMS). The AMS film's resistance can be adjusted by altering the number of alternating layers of CF and silicone, enabling tunability of EMI shielding, electric heating and light-heating behavior. With a thickness of 0.5 μm, the best AMS film achieved an EMI SE of around 100 dB, thanks to the high conductivity, large specific surface area of CNFFs, and multiple reflection and scattering of AMS. Lu et al.[136] prepared CF felt (CFF)/ethylene-co-vinyl acetate (EVA) composites with an EVA-CF-EVA sandwich structure. Through a prefabricated film method and a hot-pressing composite process, they achieved a highly conductive network within the composite. The CFF20/EVA composites, with an area density of 20 g/m^2, achieved a maximum EMI SE of 45 dB and SE$_R$ of 99%. These composites also exhibited shape-memory characteristics controlled by electric heating and found application in smart components.

To construct flexible corrosion-resistant CFF-based EMI shielding composites, conductive fillers like silver and conductive polymers are usually incorporated into the flexible CF matrix. Highly conductive silver has been recognized as a potential excellent EMI shielding material[137]. Jia et al.[82] deposited silver nanowires (AgNWs) on a CFF and coated it with PU, resulting in high-performance and durable EMI shielding materials. The PU-AgNW coating on the CFF's surface bridged adjacent conductive fibers and increased electron transmission, leading to PU-AgNW/CFF fabrics with excellent conductivity and high EMI SE. The conductivity of the PU-AgNW-coated sample with seven dip cycles was even higher than previously reported, with an EMI SE of 106 dB. The PU-AgNW/CFF fabric demonstrated stable and reliable EMI shielding performance, even under extreme conditions such as bending deformation, ultrasonic irradiation, tape peeling and exposure to strong acidic and alkaline solutions.

Zhu et al.[138] enhanced the conductivity of CFF by electroless plating a thin layer of nickel (Ni) on its surface (Figs. 6.10a–b). To achieve a strong interfacial interaction between CF and epoxy matrix, dopamine was self-polymerized on the surface of the Ni coating. The CF-Ni-PDA composite, produced through vacuum-assisted resin infusion (VARI) processing,

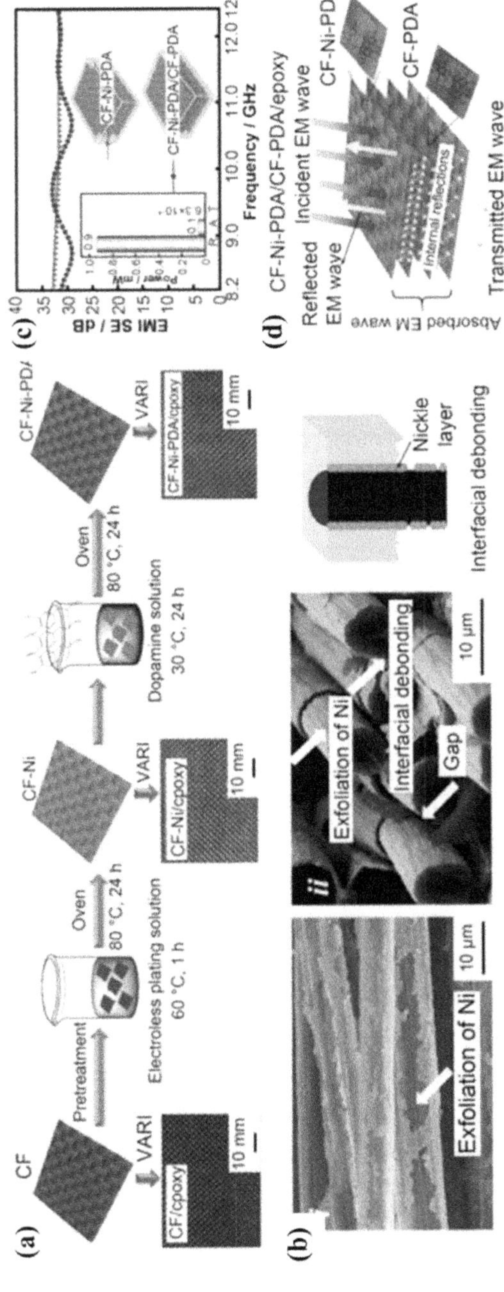

FIG. 6.10 (a) Schematic illustration of surface treatment of CF and their composites; (b) SEM images of fracture surfaces and schematic mechanisms of CF-Ni/epoxy; (c) EMI SE of CF-Ni-PDA/epoxy and CF-Ni-PDA/CF-PDA/epoxy; (d) EMI shielding mechanisms schematic illustration of multilayer hetero-structure composites. (Reproduced with permission from Ref.[138] Copyright 2021, Elsevier Ltd.)

exhibited an EMI SE of 32 dB in the X-band. To reduce the usage of Ni element, composites with a multilayer structure were designed, whose EMI SE was approximately equivalent to that of CF-Ni-PDA/epoxy (Fig. 6.10c). The EMI shielding mechanism of the multilayer heterostructure composites is shown in Fig. 6.10(d), which shows that the main shielding action of CF-Ni-PDA/epoxy is by reflection. This study optimizes the structure of composites to help guide the future design of composite materials to expand the application of CFs.

6.2.2.1.3 Carbon fiber/nano carbon composites Conductive nanocarbon fillers have been extensively utilized for EMI shielding[28,139]. The EMI shielding capabilities of carbon fiber composite materials with nanocarbon fillers such as CNTs and graphene and their derivatives are being discussed[140]. CF/CNT composites hold potential value for flexible EMI shielding due to the high conductivity, lightweight nature, large specific surface area, and excellent mechanical properties of CNTs. Mei et al.[141] reported an enhancement in the EMI shielding performance of CF fabrics by introducing CNTs and densifying them through chemical vapor infiltration. This improvement was indicated by a 56.5% increase in SE attenuation, which was attributed to the main loss mechanism.

In addition, multi-walled carbon nanotube (MWCNT)-coated CF fabrics have been investigated for their EMI shielding performance[142]. MWCNTs cover the surface of CF and fill the pores of the CF fabric, resulting in a carbon nanostructure with nanovoids. This enhances the interaction of electric and magnetic fields with the magnetic/dielectric properties of the filler, thereby increasing the reflection and absorption properties of the shielding material. Ramírez-Herrera et al.[143] studied MWCNT-filled carbon nanofiber composites, which exhibited an EMI SE of 52 dB at 900 MHz for an eight-layer flexible carbon composite with a thickness of 0.65 mm. These findings suggest potential applications in the communication field, particularly for low-frequency bands.

Graphene, possessing high electron mobility (around 15,000 $cm^2 V^{-1} s^{-1}$), a large aspect ratio, adjustable electrical properties, excellent mechanical properties (~1.0 TPa) and abundant functional groups, also meets the requirements of EMI shielding materials[144]. Chen et al.[145] found that rGO improved the EMI shielding performance of unsaturated polyester (UP)/CF composites. With a mass fraction of 0.75% rGO-CF, the SE of rGO/CF/UP reached 37.8 dB at 12.4 GHz, which was 16.3% higher than that of pure CF/UP. Wu et al.[146] utilized rGO-coated CFs (GCFs) and magnetic Fe_3O_4 nanoparticles deposited with rGO (MG) for reinforcing an epoxy resin. The resulting GCF/MG/EP composites exhibited EMI SE values ranging from 31.3 to 51.1 dB in the frequency band of 8.2-26.5 GHz, indicating wide-band EMI shielding characteristics. Lin et al.[147] prepared flexible CF/graphene (CFFG) composite films, where graphene was fused with PVDF to form a porous structure, creating an electromagnetic energy dissipation network and a heat conduction path. The CFFG composite film not only demonstrated significant EMI shielding capabilities compared to other materials but also exhibited a notable photoelectric heating effect, ensuring stable performance even in extremely low-temperature environments.

6.2.2.2 Graphene-based EMI shielding film/fabric

Graphene oxide (GO) is a widely used precursor for graphene and has garnered significant interest as a building block for producing macroscopic materials due to its excellent dispersion in water and polar organic solvents[148]. For instance, GO sheets can be easily assembled into ultrathin films with a layered structure using solution processes. Although most oxygen can be removed from thermally reduced graphene films following annealing at elevated temperatures (1000 °C), structural defects may still remain[149]. Recent work has suggested post-treatment of the reduced graphene film under graphitization conditions to fully repair these defects, resulting in graphitized films with continuous oriented graphene layers and very few defects[150]. Furthermore, the preparation of graphene sheets with few layers is complicated and challenging, and significant amounts of solvents are wasted during the process due to their apparent hydrophobicity and difficult dispersion[151]. Therefore, GO is the primary raw material for the production of electromagnetic shielding graphene films.

6.2.2.2.1 Pure graphene film/fabric Achieving a high level of electromagnetic shielding performance in pure graphene films requires maintaining the integrity, orientation and connectivity of graphene sheets to form a robust conductive network[152]. This not only ensures superior conductivity of graphene thin films but also facilitates uninterrupted reflection of EMW during internal transmission. Currently, researchers chiefly rely on selecting suitable preparation methods, primarily porous film and multilayer film based on structural characteristics, to construct pure graphene EMI shielding films.

Porous films are well-suited for achieving impedance matching and increased multiple reflections of EM waves to a certain extent due to their internal porosity. The porous structure also extends the transmission path of EM waves, causing a further attenuation of EM waves. However, multiple interfaces and defects within the porous structure can also increase direct EMWA, providing an opportunity to prepare electromagnetic shielding materials with enhanced absorption. Along with preparing porous materials, it is crucial to ensure that they possess desirable mechanical and flexibility properties to meet the current comprehensive development requirements of "thin, light, strong and wide" for electromagnetic shielding materials.

Jiabin Xi et al.[153] developed layered graphene film with a particular space (GAF) by reducing GO film prepared through bar casting with hydroiodic acid and high-temperature expansion treatment at 3000 °C. By controlling the casting concentration of GO and the coating height of the rods, researchers could control the lamination spacing of the final graphene membranes. The prepared films exhibit impedance matching and expansion enhancement, resulting in excellent EMI performance. Multiple incidences, transmissions, reflections and absorptions can occur at various layers, porous interfaces, holes and other positions of the material, ultimately improving the loss of EM waves[154]. This provides a useful idea for the subsequent preparation of alternating conductive and non-conductive (air, insulator) materials.

Although self-expandable porous graphene films have excellent EMI properties, their application is limited due to poor mechanical properties caused by irregular expansion. Therefore, some researchers have improved their expansion process. Dengguo Lai et al.[155] created graphene films with controllable thickness (PGFs) by reducing GO film with a spatial confinement strategy (Fig. 6.11a). The prepared PGFs exhibit excellent elastic behavior and strong repeatability stability. Additionally, the annealed PGF shows a very high EMI SE of 63.0 dB, with a maximum SE/thickness ratio of 49750 dB/cm^2/g^1 (Fig. 6.11b) due to multiple reflection and absorption (Fig. 6.11c).

Inspired by the irreversibility of hydrogel deformation, Jian-Jun Jiang et al.[156] mechanically compressed the prepared graphene hydrogels to varying degrees, gradually transformed the original cellular structure into a layered

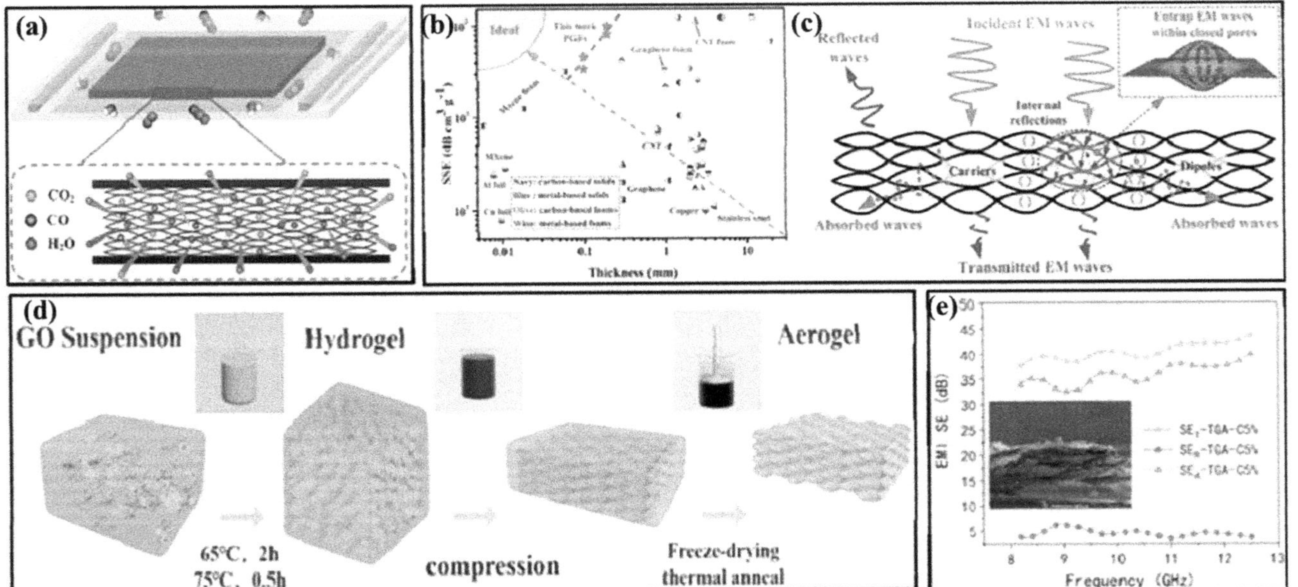

FIG. 6.11 (a) Schematic representation illustrating the spatial confinement strategy and mechanism forming of GO films. (b) Comparison of the specific SE as a function of thickness for PGF. (c) Proposed EMI shielding mechanism for PCFS. (Reproduced with permission[155]. Copyright 2020, Elsevier Ltd.) (d) Schematic illustration of the TGA fabrication procedure. (e) EMI shielding performance of TGA. (Reproduced with permission from Ref.[156] Copyright 2020, American Chemical Society.)

structure via freeze-drying and annealing at 900 °C, resulting in dense porous aerogel films with a layered structure (Fig. 6.11d). Because of the uniform dispersion of graphene sheets, good 3D network charge carriers are formed, and the porous multi-layer structure creates multiple interfaces conducive for multiple reflections and absorption of incident EM waves. Finally, the optimal EMI shielding performance can reach 43.29 dB (Fig. 6.11e). Multilayer films provide multiple mechanisms for the dissipation of EM waves, including multiple reflections of incident waves, ohmic loss due to a conductive network, and polarization loss at multiple interfaces and grain boundaries. The selection of appropriate preparation methods is crucial for achieving good graphene nanosheet orientation and constructing a complete structure. Li Peng et al.[157] prepared multi-layer graphene films with microfolding by casting debris-free, giant GO aqueous dispersions followed by thermal annealing with 300 MPa static pressure up to 3000 °C. The resulting films showed ultra-high flexibility, thermal conductivity and electrical conductivity. Similarly, Chi Fan et al.[158] prepared a flexible multilayer graphene film with high conductivity and small thickness achieved by high-temperature thermal treatment of GO films and subsequent compression rolling. In both cases, the films demonstrated excellent electromagnetic shielding properties and potential applications in electromagnetic conduction and radiation devices. The size and content of the raw material graphene sheets also play a significant role in determining the final properties of the films.

6.2.2.2.2 Composite graphene film/fabric   While pure graphene film has good EMI performance, its performance can be further improved by various methods such as pressure-assisted thermal reduction and a space limitation strategy to construct porous, oriented and other internal structures that enhance EMI performance. However, there are still challenges, including low EMI performance, insufficient mechanical properties, limited preparation methods and single shielding mechanisms. Compounding graphene with other components can effectively address these issues and improve the comprehensive properties of graphene films. Composite inorganic nanomaterials, organic conductive/non-conductive polymers and multi-component composites are common compound forms. We will summarize several composite forms and highlight their advantages and disadvantages below.

Inorganic materials used in composites can be categorized into two types based on their electromagnetic loss mechanism: magnetic loss and electric loss. When EMW travel through mutually perpendicular electric and magnetic fields, only materials with electric or magnetic losses can dissipate the waves, while the rest will disappear eventually. By combining magnetic loss-type inorganic materials with electric loss-type graphene, the film can achieve better impedance matching, leading to multiple and synergistic electromagnetic wave dissipation and loss under both mechanisms. Common magnetic loss materials include nickel, ferrite, carbonyl iron, polycrystalline iron fiber, and ferromagnetic alloy, with hysteresis loss and ferromagnetic resonance loss as the main loss mechanisms.

When graphene is combined with magnetic nanoparticles, it serves as a nucleation site for heterogeneous nucleation of the magnetic nanoparticles. This combination can increase the layer spacing of graphene and prevent graphene from agglomerating. Among magnetic nanoparticles, Fe_3O_4 is the most commonly used material and can absorb EM waves via hysteresis loss, domain wall resonance, natural resonance, and eddy current loss[158]. Cong Chen et al.[159] prepared materials with various loss mechanisms, such as dipole polarization, interfacial polarization, magnetic loss and electrical loss, by loading Fe_3O_4 nanoparticles on sulfur-doped graphene sheets. Other magnetic materials, such as $MnFe_2O_4$ and FeCoNi, are also widely used in composites.

Graphene/electrical loss-type composites may experience impedance mismatch at the interface due to different material impedances, but they can bridge the gaps between the graphene nanocrystals and compensate for defects, eventually forming a "brick-mud" structure that facilitates the formation of a good conductive network. These composites can cause multiple reflections, scattering and absorption of EMW, resulting in the dissipation of EMW and excellent electromagnetic shielding effects. Common electrical dissipation materials used in composites include metal nanowires, carbon nanofibers, CNTs and MXene[160–162].

Huili Fu et al.[163] developed a lightweight film made of single-walled carbon nanotubes (SWCNTs) and graphene, which had a unique interconnected porous layered sandwich structure. They achieved this by welding the SWCNTs between graphene layers to form a skeleton (Fig. 6.12a). This special structure meant that the resulting film, known as SGF, was highly resistant to repeated folding, even after being folded over 1000 times, without any significant changes in its structure (Fig. 6.12b), mechanical or electrical properties. The SGF also displayed exceptional EMI SE of approximately 80 dB (Fig. 6.12c), due to the high conductivity of SWCNTs

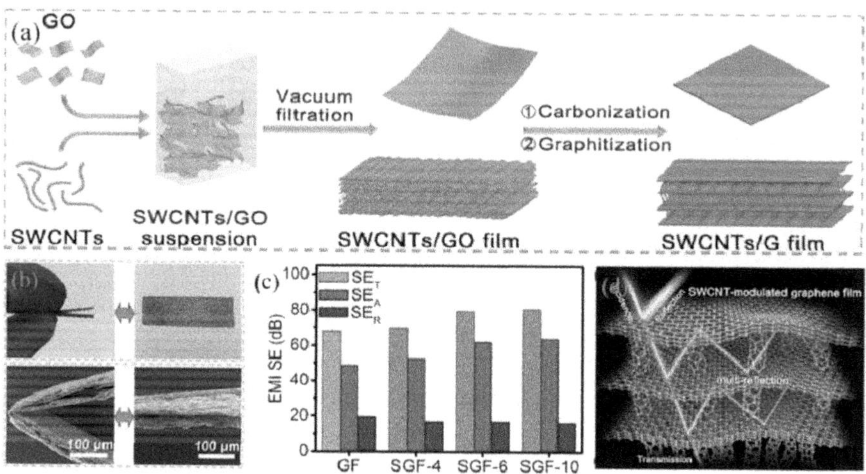

FIG. 6.12 (a) Schematic representation for the preparation process of SGF. (b) Digital photos (up) and SEM images (down) during folding. (c) Comparison of SE_T, SE_R and SE_A of GF and SGF samples annealed at 2800 °C. (d) Proposed EMI shielding mechanism. (Reproduced with permission from Ref.[163] Copyright 2020, Elsevier Ltd.)

and graphene network, as well as multi-reflections that took place in the interlayers of SGF (Fig. 6.12d). The "brick-mud" structure, with graphene as the brick, and the densification effect, was mainly due to the synergetic effect of the two components, which ultimately resulted in excellent EMI shielding performance. These findings could serve as a reference point for preparing future electromagnetic shielding materials with high thermal conductivity. Recently, Qiang Song et al.[164] first reported a simple method of self-sacrificing template plasmon enhanced chemical vapor deposition (PECVD) and thermal reduction to prepare a lightweight and flexible and free-standing 3DGMTS which assembled with SiC nanocrystals decorated edge-rich vertical graphene nanosheets (VGNs) exhibits superb EMI SE of around 38 dB in the frequency range of 8.2-12.4 GHz at a density of 0.0036 g cm^3 and a thickness of 1.5 mm. The Si_3N_4 NWs are not only a template for obtaining gad structure, but also a silicon source for forming SiC nanocrystals after high-temperature heat treatment.

Pure graphene films and inorganic/graphene composite films often have impedance mismatches and insufficient toughness, while polymer-based composites can effectively utilize the advantages of graphene and polymer based on the controllable amount of graphene. The combination of the two can increase impedance matching, reduce electromagnetic wave reflection and prevent secondary pollution to a certain extent. However, due to the nanoscale graphene sheets' agglomeration effect and weak adhesion between graphene and matrix materials, mechanical brittleness often increases while the EMI properties of the material deteriorate. To address these issues, strategies such as uniform dispersion, separation and orientation structures have been adopted. Currently, the primary preparation methods for blending composite films and coatings include the following steps: selecting the appropriate solvent to prepare the two as a uniformly dispersed solution, then evenly mixing them using ultrasonic or mechanical blending. Finally, the solvent is completely volatilized by coating, extraction and filtration to produce the final sample. Alternating self-assembly using the interaction between the two is also another method used for preparing these composites.

Jiajie Liang et al.[165] produced a graphene/epoxy resin composite film through dissolution treatment, exhibiting a low filtration rate threshold and strong EMI performance, demonstrating the feasibility of blending. Additionally, flexibility is a significant development trend for EMI materials. Ahmed A. al-Ghamdi et al.[166] utilized traditional rubber roller grinding technology to prepare graphene/acrylonitrile-butadiene rubber nanocomposite films, acting as a reference for flexible electromagnetic shielding material preparation. Biao Zhao[167] selected a flexible polyether-block-amide polymer combined with graphene to create a film material with a negative pressure effect achieved by adjusting the graphene content. Wenge Zheng et al.[168] fabricated a flexible graphene/TPU composite membrane with a sandwich structure through brushing methods. Ultimately, the film had graphene loading at approximately 20 wt% and a total coating thickness of about 50 nm. The film not only displayed excellent mechanical flexibility and strong strength of around 60 MPa but also demonstrated an ideal EMI SE of roughly 15–26 dB, coupled with a qualified frequency bandwidth of SE totaling 20 dB, as wide as approximately 49.1 GHz.

The preparation of composite films through blending methods requires comprehensive considerations of various factors, including the selection of blending solvents, regulation of blending parameters, compatibility of composite materials and modification, all with the goal of achieving high-performance and multifunctional electromagnetic shielding films. One way to achieve this is by utilizing the interaction between components to prepare graphene composite films, where the interface between components is more tightly connected. The generation of multiple interfaces is also beneficial for interface polarization, which increases the dissipation of EM waves. Various kinds of interactions, such as electrostatic interaction, ionic interaction, hydrophilic and hydrophobic interaction, conjugate inter-action, covalent bond and hydrogen bond, can be employed. Sheng-Tsung Hsiao et al.[169] prepared graphene nanosheet/WPU composite film materials by adsorbing cationic surfactant on the graphene surface using sulfonate-containing WPU as the polymer matrix. This improved the compatibility between materials due to the good interface interaction caused by elec-trostatic attraction, which led to even dispersion of graphene and a good electrical network. Cristina Valles et al.[170] prepared a layer-by-layer com-posite film by alternately coating negatively rGO and positively charged polyelectrolytes. This method resulted in a good orientation of graphene and excellent EMI performance, mainly through EMWA (EMI SE 4830 dB/mm). This study proposes the idea of using the least absorbing material to achieve the penetration threshold and perfect coverage, based on the theory of penetration threshold and coverage. This provides a useful reference for the preparation of low-cost, strong-performance, and thin-thickness-coated electromagnetic shielding materials. Ching Ping Wong et al.[171] skill-fully utilized the potential difference between Zn/Zn^{2+} and GO/rGO to form an rGO film on the surface of Zn foil through a gelation process at an acidic condition. They formed a layer-by-layer stacking structure by coating the film with PDMS, resulting in the best performance of SE reaching 53 dB at X-band for a sample of 1.547 mm in thickness. This method maintains various loss mechanisms of porous graphene membranes mentioned above while further increasing the impedance matching performance and solving the defects of its toughness.

Furthermore, another method for preparing polymer/graphene films involves loading graphene nanoparticles onto fabric through a suitable coating process. This method can also achieve excellent properties such as light weight, flexibility, stretchability, and bending resistance[172]. MD Zahidul Islam et al.[173] prepared graphene coating by circulating cotton fabric in a graphene dispersion without an adhesion agent, resulting in good conductivity and electromagnetic SE, as well as certain tensile strength and elongation at break. In addition, researchers have also used in-situ polymerization of conductive polymer nanoparticles on graphene nanocrystals. J. Yan et al.[174] successfully polymerized an ordered high-density PANI nanorod array in situ on an amino fossil graphene sheet (AFG). To better achieve graphene electromagnetic shielding composite films in the future, further development and exploration are necessary not only in the preparation methods, composite materials and material ratio but also in understanding the relationships between materials and properties. Xiaodong Xia et al.[175] developed an effective medium theory that considers various factors such as interface effect, percolation threshold, electron tunneling, MWS polarization effect, frequency dependence electronic jumping and Debye dielectric relaxation. This theory can calculate the nanocomposites' conductivity, dielectric constant and magnetic permeability and provide a reference for subsequent systems through the analysis of plane wave Maxwell equations.

6.2.3 MXene-based EMI shielding films/fabrics

Developing 2D MXene-based film with outstanding flexibility can satisfy demands of smart wearable MA materials. Meanwhile, flexible MXene-based films with other functions can solve more practical problems. Two common strategies can be used to fabricate flexible 2D film: (i) using textile-based materials as substrate, and (ii) assembly of nanostructured materials by vacuum filtration. During the formation of MXene, MXene is prone to lamellar self-stacking, resulting in the reduction of the active sites on its surface. And its specific properties are affected such as reducing its loss of EM waves, hindering ion transmission and limiting the effective load of other functional materials.

6.2.3.1 Pure MXene film/fabric

Koo's group reported that a 45-μm $Ti_3C_2T_x$ film displayed EMI SE of 92 dB (a 2.5-μm film showed > 50 dB), which is the highest among synthetic materials of comparable thickness produced to date[176]. The outstanding electrical conductivity of $Ti_3C_2T_x$ films and multiple internal reflections led to this excellent performance. After that, they systematically studied the EMI shielding of $Ti_3C_2T_x$ MXene-assembled films over a broad range of film thicknesses, monolayer by monolayer[177]. Theoretical research showed that multiple reflections, the surface reflection and bulk absorption become significant in the shielding mechanism below skin depth. The 24-layer film of 55 nm thickness showed an EMI SE of 20 dB, revealing an extraordinarily large absolute SE (3.89×10^6 dB/cm²/g). Meanwhile, they prepared Ti_3CNT_x and $Ti_3C_2T_x$ MXene free-standing films of different thicknesses by vacuum-assisted filtration and investigated their EMI shielding performance under different annealing temperatures[178]. It is found that Ti_3CNT_x

film provided a higher EMI SE compared with more conductive $Ti_3C_2T_x$ or metal foils of the same thickness. This excellent EMI shielding performance of Ti_3CNT_x was achieved by thermal annealing, owing to anomalously high absorption of EM waves in its layered, metamaterial-like structure. Han et al. systematically studied the shielding properties of 16 different MXene films[179]. All MXene films with micrometer thickness displayed excellent EMI shielding performance (> 20 dB). Among them, $Ti_3C_2T_x$ film displayed the best EMI shielding performance. For example, $Ti_3C_2T_x$ film with a thickness of only ~ 40 nm showed the EMI shielding performance of 21 dB.

6.2.3.2 Porous MXene-based film/fabric

Constructing a porous structure can effectively solve the self-stacking problem of MXene sheets, and MXene porous films are believed to show better performance than other film materials. Moreover, absorption-dominant lightweight porous film exhibits more attractiveness than traditional reflection-dominant shielding materials and can minimize secondary pollution from reflected EMW. With the rapid advancement of flexible device hardware and portable electronic devices, ultrathin EMI shielding films with both high flexibility and brilliant mechanical are more required.

Xu et al. adopted a simple freeze-drying strategy to synthesize porous f-Ti_2CT_x/PVA foams and films[180]. Fig. 6.13(a) shows the fabrication strategy for f-Ti_2CT_x/PVA foams and films and f-Ti_2CT_x nanosheets were obtained using LiF and HCl to etch the Ti_2AlC MAX phase. Then, the PVA solution was mixed with the f-Ti_2CT_x solution to obtain the precursor pure sol (Fig. 6.13b). The f-Ti_2CT_x/PVA foam was placed on the dandelion without falling off, proving that it exhibits the characteristics of light density, and the measured density is only 10.9 mg cm^{-3}. In addition, f-Ti_2CT_x/PVA-1 also has an extremely high porosity (99.3%), which stands out in the category of ultralight aerogels. Fig. 6.13(c) depicts the strategy by which f-Ti_2CT_x/PVA composite thin films can be fabricated by compressing the f-Ti_2CT_x/PVA composite. The 2D profile of the f-Ti_2CT_x/PVA foam-1 (Fig. 6.13d) shows that the RL_{min} is -18.7 dB and the thickness is 3.9 mm at 8.2 GHz. The average shielding efficiency (SE_A) of f-Ti_2CT_x/PVA foam-1 is 26 dB, SE_T is 28 dB, and SE_R is 2 dB, as shown in Fig. 6.13(e). As reflection, transmission, and absorption occur when the EMW reaches the surface of the prepared sample, the underlying mechanism for the absorption-dominated

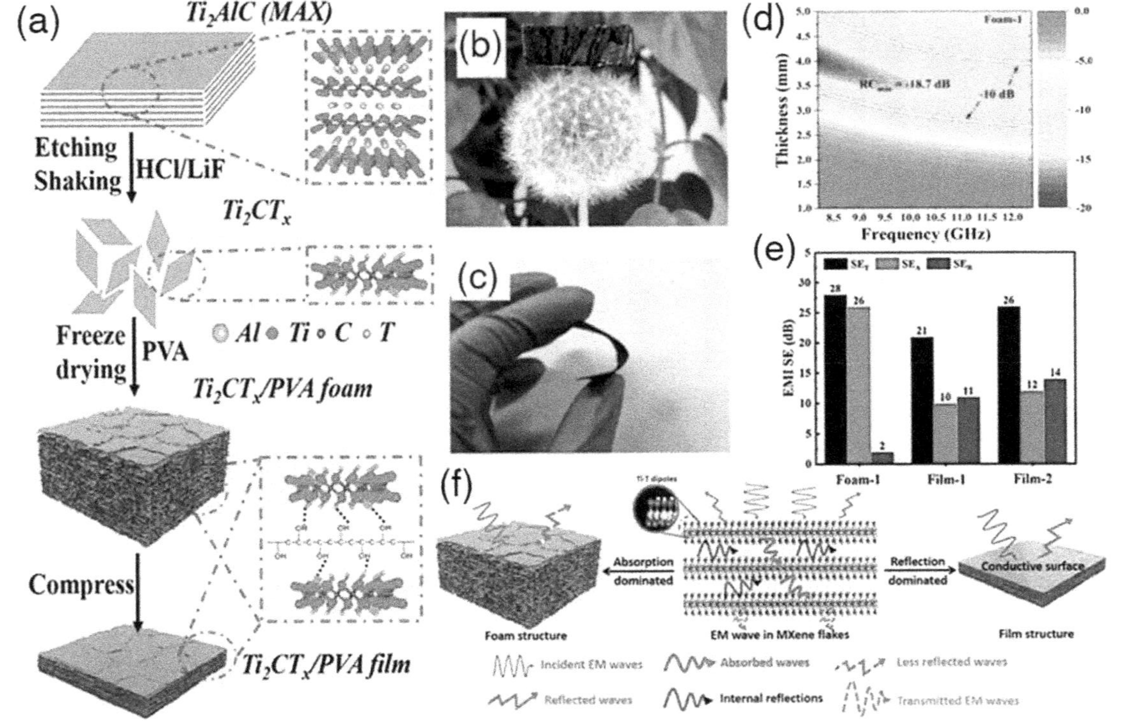

FIG. 6.13   (a) Schematic diagram of the synthesis of f-Ti$_2$CT$_x$/PVA syntactic foams and films. (b) A sheet off-Ti$_2$CT$_x$/PVA foam standing on a dandelion, showing its ultralow density. (c) The picture of flexible f-Ti$_2$CT$_x$/PVA thin film. (d) 2D profile and reflection coefficient curves versus frequency. (e) EMI SE$_T$, SE$_A$, and SE$_R$ of Ti$_2$CT$_x$/PVA hybrid materials. (f) Schematic illustration of the EMI shielding mechanism of MXene/PVA composites with foam or film structure. (Reproduced with permission from Ref.[180] Copyright 2019, American Chemical Society.)

EMI shielding properties of MXene/PVA composites can be observed in Fig. 6.13(f). Whether absorption or reflection contributes the most to the shielding properties depends largely on the type of product being manufactured. For the MXene/PVA composite with a film-like structure, the conductivity is 8.0×10^{-4} S m^{-1}, and EM wave reflection is the primary shielding mechanism, while for foamed MXene/PVA composites, the structure of the product leads to better impedance matching, and EM wave reflection is relatively low. When the EM wave enters the inside of the foam, due to the porous structure of the foam and the layered structure of the MXene flakes, the EM wave is reflected inside, which is beneficial to the dissipation of the EMW. Dipoles and interface polarization can dissipate as much of the multi-reflected EM wave as possible, resulting in absorption-dominated EMI shielding performance.

6.2.3.3 Flexible MXene-based film/fabric

With the rapid advancement of flexible device hardware and portable electronic devices, the requirements for ultrathin EMI shielding films with both high flexibility and brilliant mechanical performance are increasing. Focusing on solving the problem of EMW pollution, Zhang et al. prepared for the first time an electrically insulating sandwich structure film based on calcium ion-cross-linked sodium alginate (SA) montmorillonite (MMT) and $Ti_3C_2T_x$ MXene using a vacuum-assisted stepwise filtration process[181]. This strategy of designing an EMI shielding network inside the material also satisfies the other two demands, both as a remarkable fire barrier and as exceptional protection for electronic equipment in the event of an accidental fire. The interlayer film can maintain prominent EMI shielding properties (50.01 dB), significant mechanical properties (84.4 MPa), and distinguished fire prevention performance compared with other $Ti_3C_2T_x$ layers. In particular, comparing the EMI shielding effect of the hybrid film with the interlayer film, the EMI shielding effect of the interlayer film is more significant, and the interlayer film also performs well in the long-term thermal aging test at 80 °C.

Faced with the increasingly critical threat of EM radiation, highly conductive $Ti_3C_2T_x$ MXene-based EMI shielding materials show remarkable application potential, but their single loss mechanism limits their application potential[4,182]. Liang et al. assembled $Ti_3C_2T_x$ MXenes and conductive CNTs into brick-like NiCo/MX CNT films with both magnetic and elastic properties while achieving high electrical conductivity and brilliant EMA attenuation capability[183]. Notably, the fabricated brick-like NiCo/MX-CNT films comprehensively demonstrate the necessity of compositional and beneficial structural design. The EMI SE is 99.99999991% (90.7 dB) and the NiCo/MX-CNT composite film with a thickness of only 53 mm can reflect and absorb the SE of EMW at the same time, which is one of the composite films with the best SE, compared with pure CNT films (71 dB) and pure MXene films (61 dB), both exhibiting more prominent advantages. By adjusting the film thickness from 9 to 116 mm, the composite film can achieve 46–105 dB of EMI shielding performance modulation. More importantly, the dense brick-like-layered structure endows the composite membrane with remarkable flexibility, bendable foldability and stable mechanical properties, which greatly improve the practical application potential of the composite membrane in complex application environments.

The widespread popularity of portable and wearable electronic products requires researchers to develop high-performance and flexible EMI shielding materials to deal with the increasingly severe EMW pollution problem[184]. On this basis, it is compatible with outstanding EMI shielding ability and excellent mechanical flexibility. Luo et al. adopted a convenient vacuum-assisted filtration strategy to construct a cross-linked MXene network in natural rubber (NR) matrix and then fabricated flexible and highly conductive $Ti_3C_2T_x$ MXenes NR nanocomposite films (Fig. 6.14a)[185]. The electrostatic repulsion generated by the negative charges of MXene and NR latex enables the MXene flakes to be uniformly distributed at the interface of the NR particles to create an interconnected network, leading to efficient electron transport at lower MXene content. When the MXene content is 6.71 vol%, the conductivity of the nanocomposite reaches 1400 S m⁻¹ and the EMI shielding performance reaches -53.6 dB (Fig. 6.14b). The 3D MXene network can significantly enhance the NR matrix, and its tensile strength and modulus are significantly increased by 700% and 150% compared with pure NR, respectively. The MXene/NR nanocomposite films possess stable EMI shielding ability and tensile properties under cyclic deformation and are expected to be a significant component of next-generation flexible and foldable electronic devices. To analyze the EMI shielding mechanism, Fig. 6.14c schematically illustrates the possible EMW attenuation mechanism. It has been reported that polymer nanocomposites possess large surfaces and interfaces with multiple reflections, and their shielding mechanisms against EMI are more complex than homogeneous conductive

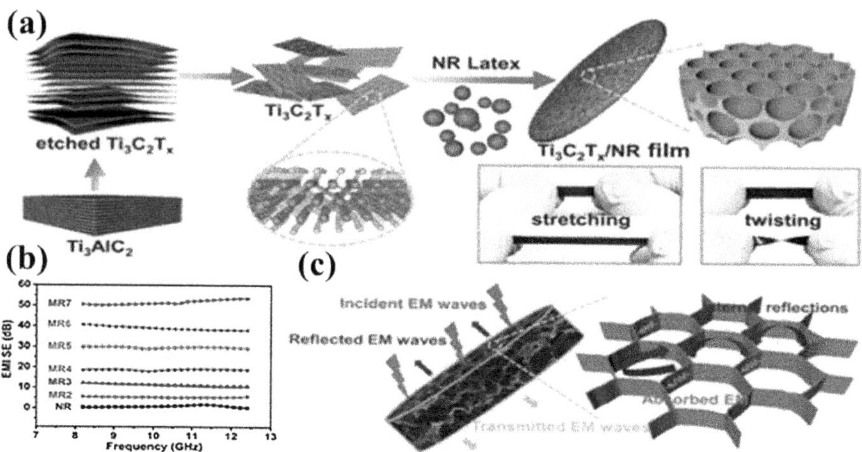

FIG. 6.14 (a) Synthesis schematic of the $Ti_3C_2T_x$/NR film. (b) EMI SE of MXene/rubber composites under different MXene loadings. (c) Schematic of nanocomposite film's potential for shielding EMI properties. (Reproduced with permission from Ref.[185] Copyright 2019, Elsevier Ltd.)

materials. For MXene/NR nanocomposites, the porous crosslinked network of MXenes facilitates the penetration of incident EMW through the internal porous structure instead of direct reflection. Through multiple scattering and interface polarization, the incident wave is effectively attenuated and dissipated on the surface and interface of the porous structure, resulting in an uncommonly improved EMA performance. The adequate end groups and a large number of charge carriers on the MXene sheet can interact with the incident EMW and slake the incident EMW by converting it into heat.

Transparent conductive films with excellent EMI shielding performance show a great potential in the field of optoelectronic instruments. More specific properties such as robust, stable, flexible and tolerated at harsh environments are required for transparent conductive film, but it is still a challenge. Zhou et al. presented a flexible transparent conductive film based on 2D MXene nanosheets and 1D AgNWs hybrid structure with strong interfacial adhesive, superior EMI shielding performance and electro-photo-thermal response[186]. The MXene/AgNWx-PVA film exhibited a low electrical resistance of 18.3 Ω sq^{-1} and an optical transmittance of 52.3% and hence possesses an excellent EMI shielding efficiency of 32 dB in the X-band at a thickness of 10 μm. The unique attributes of our multifunctional films demonstrated their great potential as high-performance transparent EMI shielding materials for emerging optoelectronic devices used in harsh environments. Jin et al. developed transparent, conductive and flexible MXene grid/silver nanowire hierarchical films with both high optical transmittances and high EMI shielding performances[187]. The MXene grid exhibits an average EMI shielding performance of 21.7 dB, which is very rare on MXenes and other graphene transparent conductive films with high optical transmittance. Furthermore, by introducing 1D AgNWs, the MXene grid/AgNW hierarchical structures are constructed and optimized, exhibiting high optical transmittance, high electrical conductivity, excellent EMI shielding efficiency and superior mechanical stability.

6.3 FOAM/AEROGEL/SPONGE MATERIALS FOR EMI SHIELDING

6.3.1 Construction strategies of foam, aerogel and sponge materials

In order to achieve foams, aerogels and sponges with excellent EMI shielding performance, two different construction strategies are employed: self-assembly and template methods. These strategies aim to enhance conductivity, improve EM wave absorption and regulate the pore structure of the materials. The self-assembly method, which is the most commonly used, typically involves three stages: (a) Formation of a stable precursor/dispersion through processes such as hydrogen bonding or gelation[188]. (b) Construction of the porous network structure using techniques like freeze-drying[189], supercritical drying[190] or ambient pressure drying[153]. (c) Improvement of material performance through processes such as high-temperature heat treatment.

6.3.1.1 Fabrication of homogeneous precursors

Achieving uniform dispersion of matrix and fillers within the system is crucial in order to construct a homogeneous network structure for aerogel and sponge composites. The use of stable precursors is often facilitated by various mechanisms such as hydrogen bonding, electrostatic interaction, gelation process and ion interaction. (i) Hydrogen bonding interaction: Materials containing hydrophilic groups can interact through hydrogen bonds. (ii) Electrostatic interaction: Electrostatic attraction occurs between oppositely charged groups of different polymers, while repulsive forces arise between identically charged groups (e.g., carboxyl groups). (iii) Gelation process: Gelation refers to the connection of colloidal particles in a sol or solution, or high molecular polymers under specific temperature conditions, resulting in the formation of a spatial network structure. The voids within this structure are filled with liquid as a dispersion medium. During the gelation process, the dispersion system gradually becomes viscous, eventually losing fluidity and transforming into an elastic colloid that can withstand greater pressure. This process helps maintain the stability of the structure[191]. (iv) Ion interaction: The interaction between functional groups on the polymer chain and metal ions (e.g., Ca^{2+}, Zn^{2+}) promotes the formation of chelates within the precursor. This facilitates ionic interaction with reversible physical crosslinking[192], which is advantageous for the subsequent drying process of porous aerogels and sponges.

Certain additives, such as TEMPO-oxidized cellulose nanofibrils (CNFs)[193] and SA as dispersants[19], lignin as an adhesive[194], PVP[195] as a protective agent, and phenolic resin[18] and poly(vinyl alcohol)[196] as binders, have been chosen to stabilize the dispersion of easily agglomerated nanomaterials. For instance, GO sheets can be dispersed in an aqueous solution by mixing and sonication treatment (Fig. 6.15a), utilizing cellulose fibers with active groups on their surfaces that exhibit strong hydrogen bonds and electrostatic repulsion effects[197].

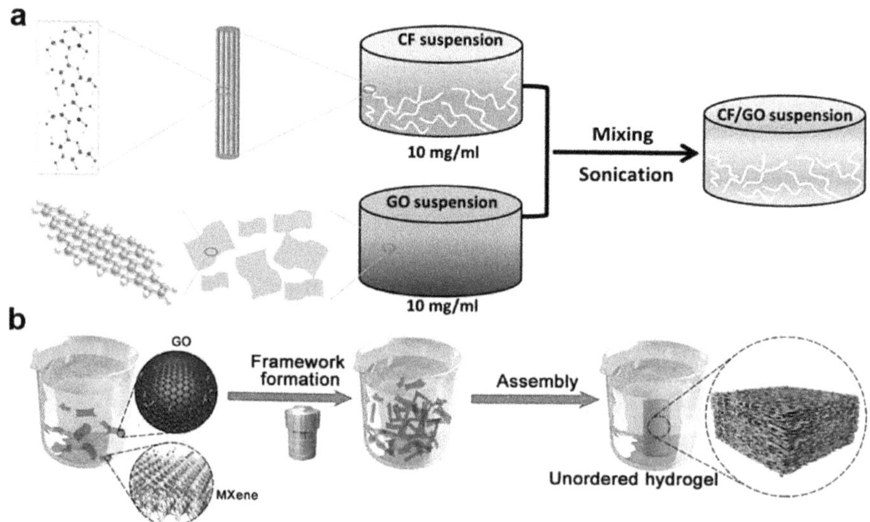

FIG. 6.15 (a) The fabrication process of the cellulose/graphene oxide precursor dispersion. (Reprinted with permission from Ref.[197] Copyright 2017, Elsevier Ltd.) (b) The fabrication for the $Ti_3C_2T_x$/reduced graphene oxide hydrogel. (Reproduced with permission from Ref.[198] Copyright 2018, American Chemical Society.)

However, although GO is effectively dispersed, its electrical conductivity remains poor, which hampers its effectiveness in shielding EM waves during application. Therefore, GO is often chemically reduced using L-ascorbic acid through hydrothermal reaction, resulting in interconnected reduced GO (rGO) sheets and the removal of oxygen-containing groups from GO[198]. This process enables the construction of a stable 3D hydrogel framework with high electrical conductivity. Additionally, during this process, $Ti_3C_2T_x$ sheets can be embedded into the frame structure of rGO (Fig. 6.15b), forming an unordered $Ti_3C_2T_x$/rGO hybrid hydrogel precursor that exhibits improved electrical performance due to synergistic effects between the two materials[199]. Apart from $Ti_3C_2T_x$, other materials like copper nanowires and CNTs can also achieve similar effects, providing a foundation for subsequent functional applications in aerogel and sponge composites.

Cellulose, which is the most abundant natural biomass polymer on Earth, has gained considerable attention as a green matrix for aerogel skeleton structures through a sol-gel process. This is due to its biodegradability, biocompatibility and chemical stability[200,201]. To illustrate, the cellulose chains are first dissolved in an inexpensive and highly effective alkali (LiOH or NaOH)/urea aqueous system at low temperatures to form a homogeneous cellulose solution. Subsequently, a cellulose hydrogel precursor is obtained through gelation, regeneration, and solvent exchange, resulting in robust regenerated hydrogen-bonding networks. During the gelation process, nanofillers such as CNTs, GO and Fe_3O_4 can also be incorporated into the cellulose matrix. This incorporation increases the viscosity of the system and transforms it into an elastic colloid that can withstand higher pressure. As a result, a hybrid hydrogel is formed, enhancing the stability of the structure.

In general, chemical hydrogels have superior frame strength and structural stability compared to physical dispersions, leading to aerogels and sponges with excellent mechanical properties. However, the gelation process for hydrogels typically requires a longer timeframe, often taking hours or even days. Therefore, the choice of precursor preparation primarily depends on factors such as reactivity, dispersibility and compatibility between the fillers and matrix.

6.3.1.2 Formation of the porous network

The drying process is a crucial step in removing solvents from aerogels and sponges, resulting in a porous network structure with open pores that prevent the propagation of EM waves. Three commonly used drying methods are ambient pressure drying, supercritical drying and freeze-drying.

Ambient pressure drying is a simple, cost-effective method that involves evaporating solvents in the air to obtain porous structures[153,202]. However, the capillary pressure at the gas-liquid interface can compress the pores, leading to structural collapse and cracking of aerogels and sponges. To overcome this issue, supercritical drying is used as a traditional technique by substituting the precursor's solvent (e.g., water) with supercritical fluids to form ultra-light and porous aerogels and sponges. Carbon dioxide (CO_2) is a commonly used supercritical fluid with readily achievable critical temperature (~40 °C) and pressure (~10 MPa)[203]. Since water is not miscible with CO_2, aqueous precursor gels often require immersion in an intermediate solvent (e.g., ethanol or acetone) before exchanging with CO_2[204]. Ethanol can also be utilized as a supercritical fluid for fabricating supercritical-dried aerogels and sponges[205]. However, its high critical temperature (~260 °C) poses certain security risks under ~9 MPa. In general, supercritical drying is effective in maintaining a stable and intact structure with uniformly small pores in aerogels and sponges. However, its high equipment cost and complex process restrict its widespread usage. Another approach, freeze-drying, can also prevent structural collapse of the aerogel and sponge framework caused by capillary forces. Freeze-drying has become a widely used method for preparing porous aerogel and sponge composites due to its simplicity, flexibility and controllability. However, challenges remain, such as high energy consumption, long processing time, low temperature and pressure requirements, and the relatively large pores composed mainly of ice crystals. These limitations hinder large-scale production and practical applications, and researchers need to address these challenges in the future.

6.3.1.3 Methods to improve related performance

In general, certain specific modification treatments, such as regulating the structure's orientation, high-temperature heat treatment or polymer impregnation, are commonly employed to enhance the performance of aerogel and sponge composites. These enhancements include improvements in electrical conductivity and mechanical strength, which in turn contribute to enhancing the EMI properties of the material, as well as its durability and practicality.

Structure orientation regulation has been adopted in recent years to facilitate the adjustment of internal structure and maximize material performance in the preparation of ordered aerogels and sponges[196,206]. Unlike conventional irregular freeze-drying, directional freeze-drying applies a temperature gradient in one (unidirectional freeze-drying) or two directions (bidirectional freeze-drying), resulting in anisotropic pore structures[207,208]. The anisotropy in structure leads to distinct absorption and loss effects on electromagnetic (EM) waves, allowing the aerogel and sponge material to selectively shield EM signals based on specific requirements.

For instance, Han et al. utilized bidirectional freeze-drying to design the porous structure of three types of MXene aerogels, achieving a long-range aligned lamellar architecture for EMI shielding[207]. These MXene aerogels demonstrated excellent structural stability and compressibility, maintaining electrical resistance stability over 20 compression cycles at 50% strain. Moreover, MXene aerogels exhibited superior EMI SE compared to films, primarily due to their improved impedance match with the free space enabled by the macroporous structures of the aerogel. Specifically, $Ti_3C_2T_x$, Ti_2CT_x and Ti_3CNT_x aerogels achieved EMI SE values of 70.5 dB, 69.2 dB and 54.1 dB, respectively. The presence of a large number of surface terminations and defects in MXene materials induced intrinsic dipole polarizations, leading to EM wave dissipation through relaxation in the alternating EM field.

High-temperature heat treatment can be employed to enhance the electrical properties of materials by converting original aerogels and sponges into carbon aerogels and sponges through processes such as pyrolysis of organic components, thermal reduction, or graphitization of primary carbon nanomaterials at elevated temperatures[209]. In the high-temperature reduction process, oxygen-containing functional groups can be removed, resulting in the reduction of GO to rGO[210]. Unlike chemical reduction, thermal reduction of graphene leads to a higher degree of reduction and the formation of sp2 graphite lattices with minimal side polar groups, facilitating π-π stacking between adjacent graphene sheets (Fig. 6.16a)[211]. Zhao et al. demonstrated a significant decrease in the ID/IG ratio from 1.06 (GO aerogel) to 0.47 (chemically reduced graphene aerogel, GAC) and 0.23 (800 °C thermally reduced graphene aerogel, GAT), indicating that thermal reduction treatment effectively repairs numerous structural defects. Consequently, GAT exhibited superior electrical conductivity compared to GO aerogels or GAC, resulting in enhanced EMI SE (Fig. 6.16b).

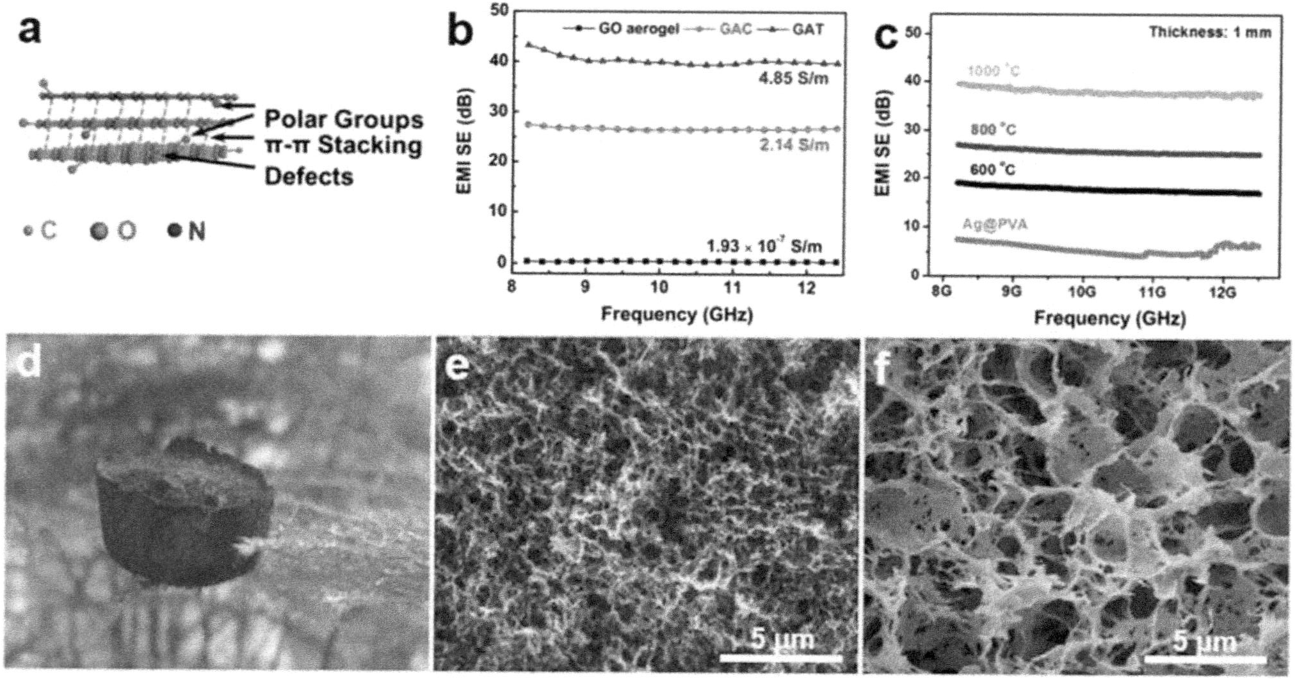

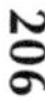

FIG. 6.16   (a) The molecular structure of the graphene sheets inside the thermally reduced graphene aerogel. (b) Electrical conductivity and EMI SE of the GAT, GAC and GO aerogel. (Reprinted with permission from Ref.[211] Copyright 2017, Elsevier Inc.) (c) EMI SE of Ag/PVA sponges by annealing with different temperatures. (d) The ultra-light Co/C@CNF aerogel. SEM images of the (e) carbon aerogel and (f) cellulose aerogel. (Reproduced with permission from Ref.[213] Copyright 2019, Elsevier Ltd.)

Furthermore, different heat treatment temperatures play a crucial role in determining the electrical conductivity of the material[18,197]. Wan et al. annealed Ag/poly(vinyl alcohol) (PVA) sponges at increasing temperatures (600 °C, 800 °C and 1000 °C) and observed notable improvements in carbonization degree, electrical conductivity (from 47.5 S m^{-1} to 363.1 S m^{-1}) and EMI SE reaching 37.9 dB with a thickness of 1 mm (Fig. 6.16c)[212]. Additionally, heat treatment induces volume shrinkage and density reduction in aerogels and sponges (Fig. 6.16d). Carbon aerogels also exhibit smaller pore size and denser structures (Fig. 6.16e) compared to their uncarbonized counterparts (Fig. 6.16f) due to partial component pyrolysis, as well as improved hydrophobic properties and thermal stability[213].

Polymer impregnation is a common method used to enhance the mechanical strength and structural stability of aerogels and sponges by filling them with certain polymers, such as polydimethylsiloxane (PDMS), resorcinol-formaldehyde (RF) and epoxy. One example is the vacuum-assisted impregnation process employed by Wu et al. to create a porous PDMS-coated MXene/SA composite[19]. The addition of PDMS had a minimal impact on the complete and aligned porous structures of the composite similar to that of the MXene/SA aerogel. Even after undergoing multiple compression-release cycles, the structure and electrical conductivity of the composite remained largely unchanged, indicating that PDMS impregnation could protect the porous architecture and enhance its structural stability and durability.

Moreover, polymer impregnation plays a crucial role in improving EMI shielding performance. On one hand, the porous architecture created by polymer impregnation effectively reduces the impedance mismatch between the composite and air interfaces, thereby weakening surface reflection and allowing most EM waves to penetrate into the interior of the shielding composite. On the other hand, the numerous filler/matrix interfaces within the composite promote the attenuation of incident EMW through multiple reflections, further enhancing EMI shielding capabilities[11].

6.3.1.4 Template method

The aforementioned methods, which involve building the network structure first and then drying to obtain aerogels and sponges, often require complex processes and are time-consuming. As a result, some commercial sponges, such as melamine and PU, are directly used as the skeleton matrices for functional fillers. This is achieved through a simple dipping

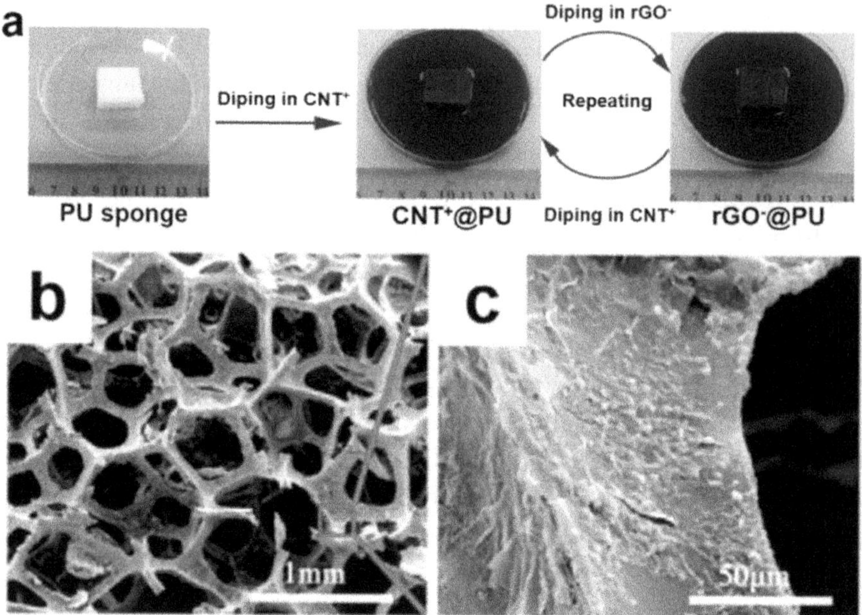

FIG. 6.17   (a) The dipping process of the PU sponge, and (b, c) the SEM images of the rGO/CNT@PU sponge. (Reprinted with permission from Ref.[214] Copyright 2019, American Chemical Society.)

process. For example, Chen's group soaked a yellow commercial PU sponge in a CNT and rGO aqueous solution using repetitive squeezing to create the rGO/CNT@PU sponge (Fig. 6.17a)[214]. The PU sponge serves as a 3D skeleton for the rGO and CNTs (Fig. 6.17b), resulting in a rougher surface coated with carbon layers (Fig. 6.17c). By adjusting the number of dipping cycles, the amount of rGO/CNTs can be controlled to improve the electrical conductivity of the composite sponge. To enhance the adhesion between the MF sponge and silver nanowires (AgNWs), PDA is introduced as a surface modifier in the dipping process[208]. This ensures reliable bonding between the nanofillers and matrix.

Additionally, Li et al. successfully obtained a SiC-coated carbonized loofah sponge by soaking it in a phenol-formaldehyde resin/SiC/ethanol slurry and subjecting it to carbonization treatment. The carbonization process not only transforms the loofah sponge into a good electrical conductor due to its carbon-rich nature, but also improves its mechanical properties and EMI shielding effects through the incorporation of SiC. Similarly, Liang et al. constructed an MXene aerogel/wood-derived porous carbon (WPC) composite by carbonizing wood and impregnating it with MXene[215]. The WPC skeleton provides stability to the composite and, in combination with the highly conductive MXene, extends the transmission path of EM waves. It absorbs and dissipates incident waves in the form of heat and electric energy, resulting in superior EMI shielding performance. In summary, while the direct-dipping method using commercial sponges helps prevent stacking and agglomeration of nanomaterials through the porous framework, further exploration is needed to ensure uniform filler penetration into the matrix and compatibility of the interfaces.

The synthesis of carbon sponges can be achieved using the chemical vapor deposition (CVD) template method[216]. In this method, 1,2-dichlorobenzene is used as the carbon source, while ferrocene acts as a catalyst to hinder the aligned growth of CNTs. As a result, during the CVD process, CNT sponges with a spontaneously interconnected structure are formed[217]. Unlike most CNTs that grow in an aligned manner, the CNT sponge is characterized by a 3D macro-porous network frame, which is the result of physical stacking and entanglement between the nanotubes. This structure creates numerous open pores and small gaps.

In another approach, nickel foam is employed as a 3D template for CVD growth of graphene sponge. By depositing graphene onto the template and subsequently etching it, a free-standing graphene sponge is obtained. The skeleton structure provided by the template ensures a continuously porous network in the graphene sponge, allowing it to serve as a precursor carrier for functional modifications. However, the mechanical strength and stability of the CVD-prepared sponge may be limited, requiring reinforcement through polymer impregnation. For example, epoxy can be impregnated into a highly porous CNT sponge via CVD, followed by curing to obtain a CNT/epoxy composite[192]. The pre-existing 3D conductive framework of the CNT sponge embedded within the reinforced epoxy matrix enables the preparation of a composite material with high electrical and mechanical properties, even with a low CNT loading.

6.3.2 Carbon nanotube-based materials

The CNTs exhibit lightweight properties, excellent mechanical strength, electrical conductivity, thermal conductivity and stability[188]. Therefore, it is significant to combine CNTs with various polymers to create flexible and self-supporting conductive polymer composite aerogels and sponges for highly efficient EMI shielding. However, there are challenges to overcome.

On one hand, CNTs tend to entangle and agglomerate due to their large specific surface area, high aspect ratio and strong Van der Waals forces between adjacent tubes. On the other hand, the smooth surfaces of CNTs lack active functional groups, making them hydrophobic and chemically stable, which hinders their dispersion within polymers[218]. The poor dispersibility of CNTs directly affects the formation of effective conductive networks within the polymer matrix, consequently impacting the electrical conductivity and EMI shielding performance of the composite material. This limitation greatly restricts the functional applications of CNTs as reinforcing fillers in composite aerogels and sponges. Therefore, achieving good dispersion of CNTs within the matrix is a prerequisite for harnessing their optimal performance.

Currently, there are three main approaches to improve the dispersibility of CNTs: mechanical dispersion methods (e.g., high-shear mixing or ultrasonic mixing), chemical modification techniques (e.g., acid oxidation), and surfactant modification methods (e.g., using cetyltrimethyl ammonium bromide (CTAB) or poly(vinylpyrrolidone) (PVP)). Physical methods such as mechanical stirring or ultrasonic treatment can disrupt CNT aggregates through strong mechanical forces or high-energy impacts. However, these methods may cause interruptions in the CNT structure, leading to reduced aspect ratios and mechanical properties. Furthermore, CNTs can still settle and aggregate over time, making it challenging to maintain a stable dispersion state for extended periods. These physical methods are typically used to assist dispersion during the preparation process of CNT aerogels and sponges[219]. Zhao et al. demonstrated that SWCNTs can be effectively dispersed in an aqueous solution by leveraging the good dispersibility of GO with the assistance of high-pressure homogenization and ultrasonication[11]. Achieving an optimal dispersion of SWCNTs was possible by using a mass ratio of 1:3 for SWCNTs and GO. The presence of GO facilitated the even distribution of SWCNTs, allowing them to tightly adhere to the surfaces of GO nanosheets due to Van der Waals, π–π, and hydrophobic interactions between SWCNTs and GO. Additionally, a homogeneous MWCNT/$Ti_3C_2T_x$ solution could be obtained after a 2-hour-assisted ultrasonic treatment[14]. This solution retained the high electrical conductivity of MWCNTs and minimized its impact on the EMI shielding performance of the resulting hybrid aerogel.

Huangfu et al.[198] introduced carboxylated MWCNTs, which were modified with H_2SO_4/HNO_3, into the rGO/polyaniline system. The addition of carboxyl groups significantly increased the number of oxygen-containing active groups on the CNT surface. These surface functional groups created steric hindrance, which helped prevent the agglomeration of CNTs to some extent. As a result, the dispersibility of CNTs in polar aqueous solutions was improved. The modified CNTs, acting as secondary conductive fillers, further enhanced the EMI SE of the composite aerogel. This was due to their high electrical conductivity, good dispersion and corrosion resistance. However, the acid modification process could damage the overall structure of CNTs and have certain effects on their mechanical and electrical properties.

Surfactant modification is another method used to improve the dispersibility of CNTs in polar solvents when combined with polymers. Surfactants consist of hydrophilic groups that coat the CNT surface and lipophilic groups that extend into the solvents. This configuration, along with electrostatic repulsion and steric hindrance, allows CNTs to be stably dispersed in the mixtures. Huang et al.[220] designed and prepared CTAB-modified CNT/cellulose aerogels with a scaffold structure by controlling the concentration of cellulose in a sodium hydroxide/urea solution. The resulting cylindrical cellulose composite aerogel exhibited a uniform outer surface and a highly porous inner structure. The surfactant-modified CNTs formed well-defined conductive networks, thanks to their excellent electrical properties derived from long-range π-conjugation. Even with a low content of CNTs (0.5 vol%), the composites achieved an ultra-low density, high porosity and an EMI SE of 20.8 dB and specific EMI SSE/t of 875.8 dB cm^2/g in the microwave frequency range of 8.2–12.4 GHz. Similarly, Zhou et al.[209] used PVP to promote the uniform dispersion of CNTs into a cellulose matrix. They then developed a CNT/cellulose carbon aerogel through freeze-drying, pyrolysis and KOH activation treatment. The resulting carbon aerogel exhibited a well-designed hierarchical porous structure and achieved an ultrahigh EMI SE of approximately 96.4 dB with a low density of around 30.5 mg/cm^3 (Fig. 6.18a). Two mechanisms contributed to the effective EMI shielding performance of the CNT/cellulose aerogel: the high porosity of the composite aerogel allowed easy penetration of incident EM microwaves with less reflectivity, and the hierarchically porous structure trapped the microwaves among the CNTs of high electrical conductivity and cell walls with strong absorption and dissipation (Fig. 6.18b). Additionally, the natural cellulose matrix provided excellent characteristics such as biocompatibility, biodegradability and thermal/chemical stability, which further enhanced the potential applications of the composite aerogel.

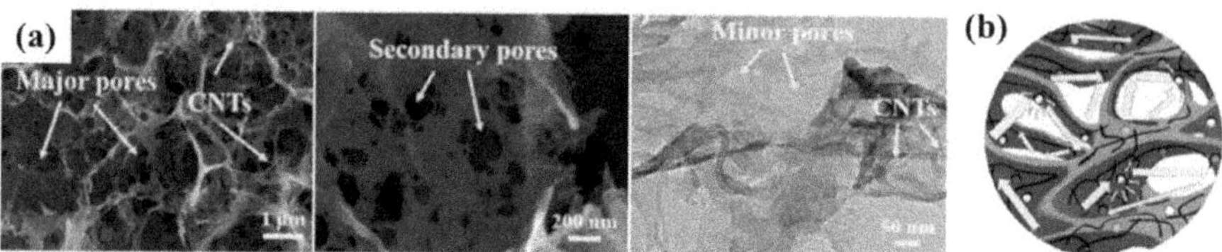

FIG. 6.18   (a) The hierarchically porous structures of the cellulose/CNT aerogel. (b) Schematic illustration for multiple scattering of EM waves inside the composite aerogel. (Reprinted with permission from Ref.[209] Copyright 2020, American Chemical Society.)

Furthermore, the dispersion of CNTs in dimethyl sulfoxide (DMSO) was achieved by mixing them with ANFs to form an ANF/CNT aerogel film. This film was then coated using a blade, followed by gelation, solvent exchange and freeze-drying processes (Fig. 6.19a)[221]. The resulting composite aerogel film exhibited remarkable mechanical properties, allowing it to be bent and twisted easily (Fig. 6.19b). This was due to the favorable compatibility and synergistic effect between the filler (CNTs) and the matrix. The porous conductive network formed by the CNTs with a large aspect ratio endowed the aerogel with an exceptional EMI SSE/t of approximately 33,528.3 dB cm^2/g, despite its thin thickness of 568 μm. According to the transmission line theory, when EMW encounter the surface of the aerogel film, there is less immediate reflection due to the impedance mismatch. The remaining waves penetrate the aerogel film and interact with the CNTs, which have a high electron density, generating electric current that leads to conductance loss and polarization relaxation loss. Moreover, the porous conductive network within the aerogel film facilitates the absorption loss of EMW (Fig. 6.19c).

To enhance the mechanical strength and reduce the CNT content, the flexible CNT sponge can be compressed and impregnated with epoxy[219]. As the compaction ratio increases from 0% to 70%, the electrical conductivity of the composite gradually improves, resulting in enhanced EMI shielding performance. This is attributed to the increased contact between CNTs, allowing for more efficient pathways for electron transport. Additionally, CNTs with their high aspect ratio and large specific surface area can host special materials such as Fe and Au, forming unique core-shell nanostructures[217]. These structures further enhance the magnetic and electrical properties of the resulting sponges, increasing the dielectric loss and magnetic loss.

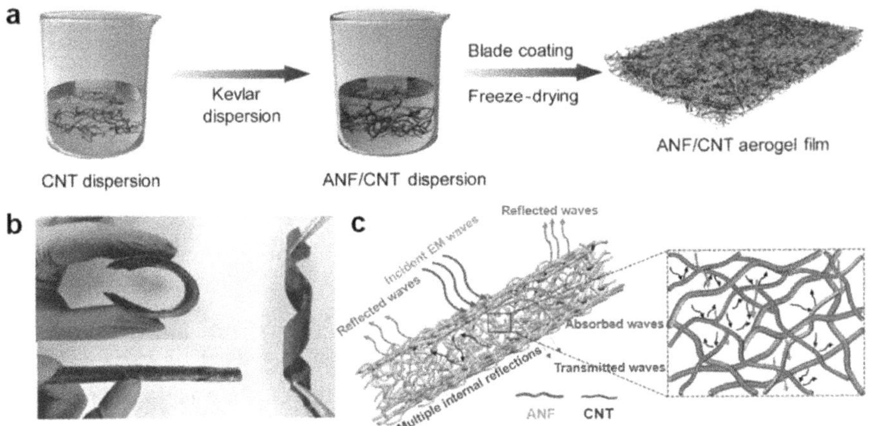

FIG. 6.19 (a) Schematic illustration of the fabrication and structural composition of the ANF/CNT aerogel film. (b) Photographs of the composite aerogel film under bending, curling and twisting. (c) Proposed EMI shielding mechanism of the aerogel film. (Reprinted with permission from Ref.[221] Copyright 2020, American Chemical Society.)

6.3.3 Graphene-based materials

For a long time, researchers have been studying carbon-based materials for EMI shielding, such as CNTs, carbon black, graphite, carbon fibers and silicon carbide fibers. However, these materials have certain drawbacks including high density and susceptibility to oxidation. Therefore, new EMI shielding materials are sought after with the characteristics of being lightweight, wide-ranging, and strong. As quasi-1D carbon nanomaterials, CNTs have been considered as alternative candidates for creating novel EMI shielding materials[214]. However, their development in the field of EMIS has been limited due to their high capacitance and low permeability.

In comparison to existing carbon materials, graphene exhibits higher electrical conductivity, a larger specific surface area and greater chemical stability. It is regarded as an optimal material for EMIS. One of graphene's advantages is its 3D network structure, which forms a continuous conductive skeleton in fabricated materials, resulting in higher conductivity. Additionally, the porous structure of graphene increases the surface for electromagnetic wave reflection, thereby further enhancing the EMIS performance of materials. Furthermore, graphene's excellent mechanical strength and chemical stability expand the practical applications of EMIS materials[222].

The assembly of graphene sheets into graphene foam (GF) not only enables the utilization of graphene's characteristics but also enhances the EMI shielding of the material through structural control. Various methods, including hydrothermal reaction, template synthesis, self-assembly, sol-gel synthesis and others, have been reported for fabricating an interconnected network structure of GF. Moreover, the formation of porous structures can reduce the overall density of graphene-based materials and meet the requirements for lightweight applications such as satellites, space stations, unmanned aerial vehicles, commercial aircraft and more.

Graphene serves as the primary raw material for creating the GF structure. The stacking and folding of graphene sheets increase the scattering and absorption (SE_A) of EMW, while the continuous interconnected network between the graphene sheets forms a larger conductive network, thereby enhancing the EMI shielding efficiency of the designed materials[223]. Additionally, the numerous porous structures inside the foam allow EMW to be trapped within the pores. Furthermore, it is worth noting that lightweight and porous GF can be effectively adjusted and controlled to improve the EMIS performance through simple physical pressing techniques.

6.3.3.1 Pure graphene aerogel, foam and sponge

The self-assembly method, also known as the solvothermal method, involves the formation of GF with macropores and mesopores structures by assembling graphene sheets through physical crosslinking points formed under π-π interactions. The preparation process involves treating a sealed container containing a GO suspension in a high-temperature environment for several hours, followed by freeze-drying and annealing to obtain rGO foams. The self-assembly method was initially proposed by Shi et al. in 2010 for preparing GF[224]. Their team utilized GO as a precursor and stacked layers of GO to successfully create interconnected 3D structures of graphene exhibiting electrical conductivity of 5×10^{-3} S/cm. However, the prepared GF suffered from looseness and inadequate mechanical strength, mainly due to the collapse of interconnected network structures or the loss of hydrogen bonds during drying and dehydration in the preparation process. Additionally, the structure of GF is influenced by the dispersion of the precursor. Wu et al. reported the preparation of GF with ultralow density and high compression ratio using a solvent-thermal method[225]. They employed ethanol as the dispersed phase in the solvent-thermal reaction, which resulted in a dispersion solution placed in a reactor at 180 °C for solvent-thermal reduction. This process yielded an intermediate phase solid with specific strength. Subsequent freeze-drying, annealing and other processes produced GF with high elasticity. By modifying the dispersed phase, the mechanical strength of GF increased, and a compressible 3D GF structure with an almost zero Poisson ratio was achieved. This compressible 3D graphene structure allows for controllable shielding of EMW, which is valuable for intelligent materials applications. The EMI shielding efficiency can be appropriately adjusted according to changes in the environment. Shen et al. explored the enhancement of GO reduction and the production of ultrathin GF with high EMI SE (~25.2 dB, 0.3 mm)[154]. They introduced reducing agents during GO hydrothermal reduction, resulting in GF with higher reduction degree achieved through hydrothermal and chemical reduction. Furthermore, the GF obtained through this method exhibited a significant network structure, increasing the energy loss of EMW within the material.

6.3.3.2 Graphene-based composite aerogel, foam and sponge

Although there have been improvements in the strength of 3D graphene by modifying the preparation process, it still falls short of the strength requirements in certain applications. As a result, researchers have been exploring the combination of graphene with other materials to enhance the mechanical strength of GFs.

For example, Chen et al. attempted to enhance the 3D network structure of GFs by incorporating phenolic resin and its pyrolysis derivatives[18]. This resulted in a 67% increase in bending strength and a 20.2% increase in bending modulus of the aerogel. The addition of phenolic resin and its pyrolysis derivative amorphous carbon led to the development of high-strength and lightweight electromagnetic shielding materials that can meet the requirements of various application scenarios, such as shielding rooms, supercomputers, base stations and other fields. The EMI shielding performance of these composite materials (with EMI SE of 25 dB for a 1 mm thickness and 33 dB for a 2 mm thickness) was comparable to or even surpassed that of GFs prepared using the CVD method (with about 20 dB for a 1 mm thickness and less than 28 dB for a 2 mm thickness in the X-band range). Zhao et al. reported the successful synthesis of porous $Ti_3C_2T_x$/GF with high conductivity through a combination of hydrothermal and directional-freezing methods[199]. This method reduced the high-temperature reduction process of rGO and simplified the preparation process of rGO foam. The materials prepared through directional freezing exhibited anisotropic structure characteristics, which contributed to the EMI shielding efficiency of the material. The researchers proposed that during reduction, GO sheets, which have a strong self-gelling ability, form a 3D network structure. Under the influence of polar interactions, $Ti_3C_2T_x$ is adsorbed to the outer surface of the graphene skeleton, resulting in the continuous construction of a $Ti_3C_2T_x$/rGO shell and core structure, as shown in Fig. 6.20(a). Remarkably, the $Ti_3C_2T_x$ hybrid graphene composite foam exhibited an impressive conductivity of 1085 S/m due to its well-connected structure and tight hole alignment. The EMIS efficiency of the material at a low loading (0.74 vol %) reached 56.4 dB for a 2 mm thickness, surpassing the qualified level, as shown in Fig. 6.20(b).

In addition to hybrid conductive particles, researchers are also exploring the introduction of magnetic particles into the GF structure through self-assembly methods to achieve a wider shielding bandwidth. For instance, Prasad et al. reported the two-step hydrothermal synthesis of MoS_2-rGO composites with magnetic particles $CoFe_2O_4$ on the surface[226]. The incorporation of magnetic particles and MoS_2 layers resulted in the formation of a multiple interface structure and moderate impedance matching, which improved the material's magnetic loss and shield effectiveness. To further investigate the influence of magnetic particle hybridization on the EMIS performance of graphene composite foam, Sushmita et al. prepared ferromagnetic and paramagnetic hybrid particle composite graphene structures using the self-assembly method[227] (Figs. 6.20c–d). They found that in the ferromagnetic hybrid composite GF, magnetic particles were the primary factor affecting the EMIS performance, while in the paramagnetic hybrid composite GF, dielectric loss played a significant role. Furthermore, they demonstrated that materials with only high conductivity or permeability may not exhibit better shielding performance. Therefore, it is crucial to select nanoparticles with stronger dissipation to improve the EMI shielding performance of materials.

6.3.4 MXene-based materials

The 3D macrostructures fabricated from 2D MXenes exhibit exceptionally impressive electrochemical and mechanical properties. In light of the inadequate interfacial interactions involving MXenes, Pu et al. developed a 3D layered-columnar PINF/$Ti_3C_2T_x$ MXene composite aerogel (PINF/MA) material employing a synergistic assembly approach. PINF/MA possesses

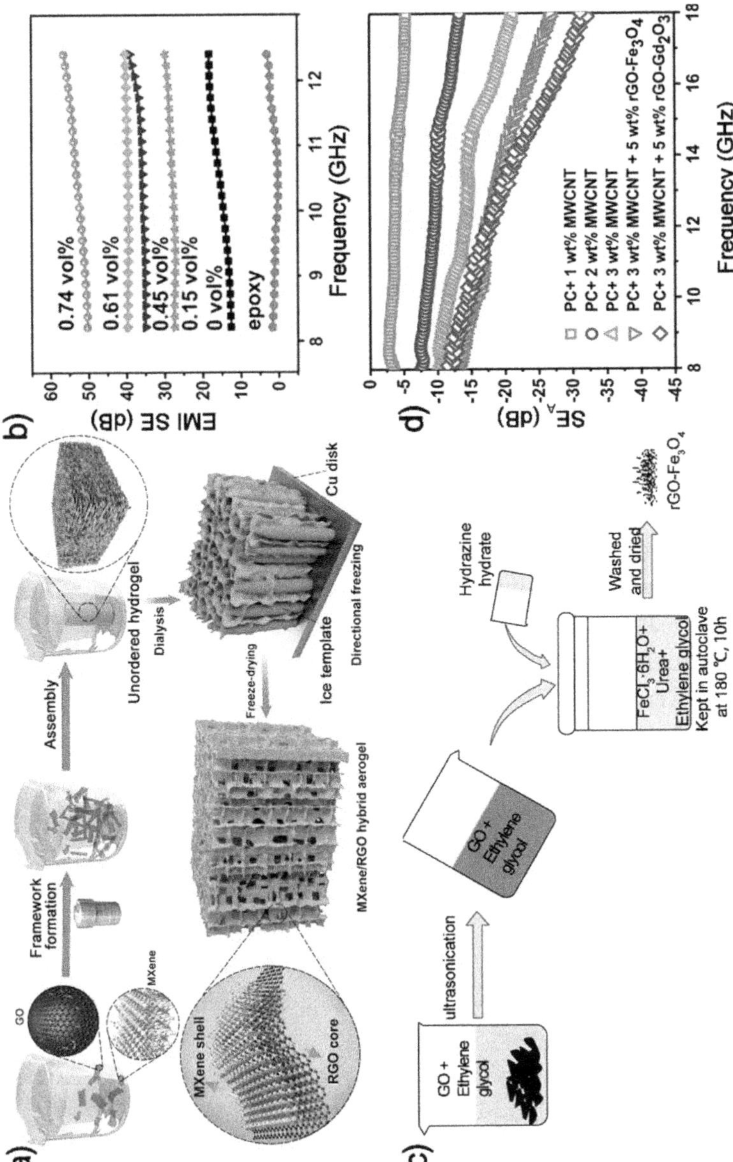

FIG. 6.20   (a) Schematic illustrating the fabrication process of a $Ti_3C_2T_x$ MXene/rGO hybrid aerogel by GO-assisted hydrothermal assembly, directional freezing, and freeze-drying and (b) EMI-shielding performances of epoxy/MGA nanocomposites. (Panels a and b reprinted with permission from Ref.[199] Copyright 2018, American Chemical Society.) (c) Synthesis procedure of rGO-Fe_3O_4 and (d) SE_A vs. frequency. (Panels (c) and (d) reprinted with permission from Ref.[227] Copyright 2019, American Chemical Society.)

a unique structure and demonstrates remarkable mechanical properties, piezoresistive sensing properties, compression and rebound stability properties, human motion detection capabilities, as well as EMA properties[228]. The preparation process of the PINF/MA composite aerogel is illustrated in Fig. 6.21(a). Initially, the aluminum component in Ti_3AlC_2 was eliminated using a mixed solution of HCl/LiF, exposing surface groups including F, O, and OH on the $Ti_3C_2T_x$ flakes. Subsequently, the hydrophilic PAANFs in their as-spun state prior to imidization were selected as the fundamental building blocks for MXene aerogel synthesis. A suitable concentration of PAA spinning solution was prepared by polycondensation reaction of ODA and PMDA monomers. PAANF films were shear dispersed through electrospinning and high-speed homogenization treatment. Finally, MXene flakes were assembled with PAANF in water under gentle magnetic stirring, resulting in homogeneous PAANF/MXene materials that were freeze-dried and thermally imidized to obtain PINF/MA. PINFs act as crosslinking agents facilitating the 3D shaping of MXene flakes. As depicted in the inset of Figure 7a, the PINF/MA aerogel can effortlessly stand on a dandelion, demonstrating its lightweight characteristic. Moreover, PINF/MA materials also exhibit exceptional thermal insulation and favorable hydrophobicity to ensure stability and adaptability in harsh environmental conditions (Figs. 6.21b–c).

Lu et al. conducted a study on the fabrication of composite aerogels by combining MXene ($Ti_3C_2T_x$) and ANFs through freeze-drying[229]. The ANFs possess characteristics such as oxidation resistance, compressibility, and EMI shielding properties. The EMI shielding performance of the MXene/ANFs composite aerogels can be adjusted by varying the content and thickness of the MXene ($Ti_3C_2T_x$). The MXene/ANFs composite aerogels exhibit a distinct porous structure and are lightweight. In terms of EMI shielding, the composite aerogels demonstrate remarkable performance with SE values of 13.12, 30.74, and 56.8 dB at 12.4 GHz under loadings of 7, 14 and 21 wt%. The EMI shielding mechanism of the MXene/ANFs composite aerogels involves functional groups present in the MXene nanosheets, such as F, O and OH, which provide polarization centers leading to dielectric relaxation and extensive dielectric loss.

Liang et al. aimed to design novel electromagnetic absorption (EMA) materials by combining microstructure design and a multi-component strategy. They synthesized magnetic Ni nanochain-anchored $Ti_3C_2T_x$ MXene/rGO aerogels using a directional freezing method and hydrazine

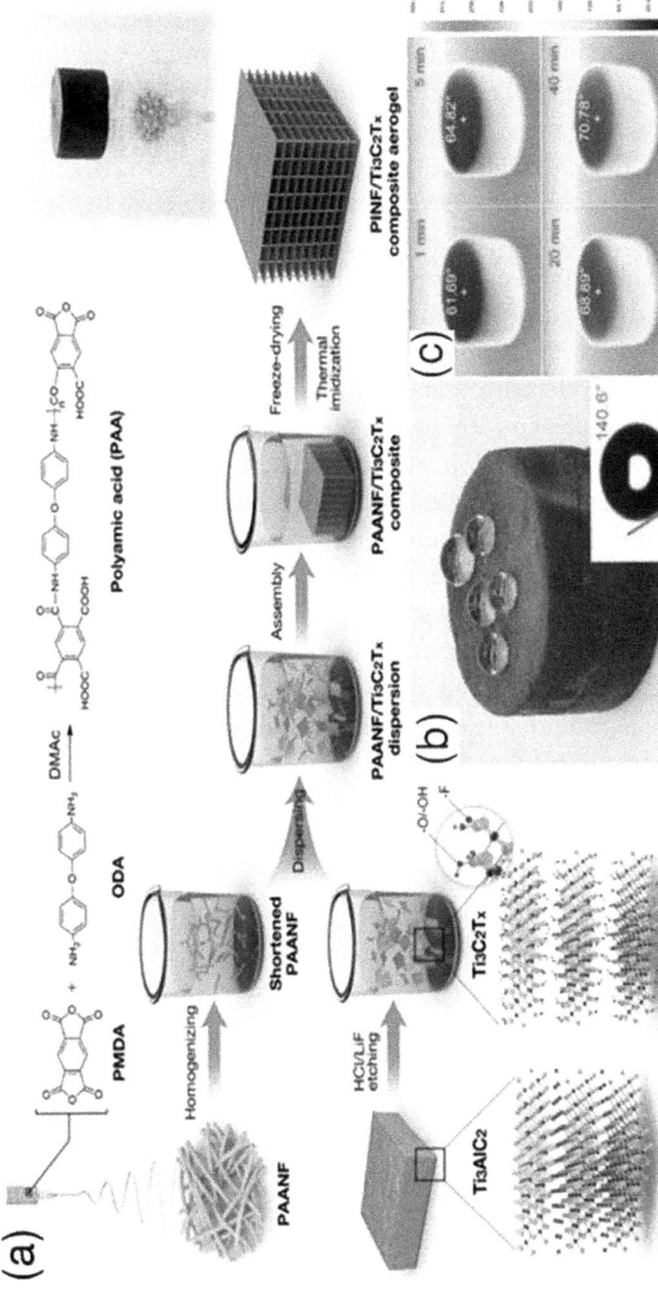

FIG. 6.21 (a) Schematic diagram of the synthesis of PINF/MA composite aerogels. The photo in the corner proves that this composite aerogel is lightweight. (b) Photographs of water droplets on the surface of the PINF/MA composite aerogel and the corresponding water contact angle. (c) Infrared thermal images of PINF/MA at different times (heating stage 300 °C). (Reproduced with permission from Ref.[228] Copyright 2021, American Chemical Society.)

gas-phase reduction method[230]. The fabrication process of the Ni/MXene/ GO (NiMRH) aerogel. The prepared Ni nanochains, GO and $Ti_3C_2T_x$ MXene nanosheets can be directly assembled into the NiMRH aerogel in an aqueous solution. Off-axis electron holography demonstrates the magnetic coupling phenomenon, where magnetic currents from adjacent Ni nanochains interact to form a dense 3D magnetic coupling network on the NiMRH aerogel. The EMA performance of the NiMRH aerogel, showing a minimum reflection loss (RL_{min}) of -75.2 dB at a thickness of 2.15 mm and a wide EAB of 5.4 GHz. The EMA mechanism of the NiMRH aerogels is attributed to the good impedance matching provided by the porous structure, allowing the aerogel to absorb more EMW while minimizing surface reflection. Within the cell space, the incident EMW is captured, scattered and attenuated by the 3D electrical/magnetic coupling network. Additionally, EMA is gradually reduced due to dielectric losses (dipole polarization, heterogeneous interface polarization, conduction losses) and magnetic losses (magnetic coupling effects, magnetic resonance, eddy current losses, etc.). Combinations of various 2D nanomaterials can be effectively utilized to develop innovative EMA materials. Li et al. fabricated a composite aerogel microsphere material by incorporating GO and $Ti_3C_2T_x$ MXene through a rapid freezing and electrospinning method[231]. This unique absorber capitalizes on the newly formed heterointerface and abundant surface groups, resulting in remarkable EMA performance of the $Ti_3C_2T_x$ MXene@ GO hybrid aerogel microspheres (M@GAMS) (Figures 7m, n).

To achieve thinness, wide absorption bandwidth and lightweight properties in EMA materials, researchers have made significant efforts to overcome various challenges. Li et al. employed a combination of wet chemical, self-assembly and sacrificial template methods to fabricate a unique rGO/ $Ti_3C_2T_x$ syntactic foam, allowing for the regulation of EMA properties[232]. In this approach, polymethyl methacrylate (PMMA) spheres and $Ti_3C_2T_x$ flakes were dispersed in water, mixed and centrifuged to form $Ti_3C_2T_x$/PMMA spheres through hydrogen bonding and van der Waals forces. Similarly, GO/$Ti_3C_2T_x$/PMMA spheres were synthesized using the same method. SEM images and corresponding schematic diagrams show that RGO nanoflakes act as stability bridges within the foam structure. Compared to other reported foam-based materials, the RGO/$Ti_3C_2T_x$ foam exhibits superior EMA performance, covering almost the entire X-band in terms of EMA. The EMA mechanism of the RGO/$Ti_3C_2T_x$ syntactic foam. It primarily relies on increased scattering of EMW due to the microporous core-shell structure, enhanced polarization loss resulting from abundant heterointerfaces, and conduction loss caused by various defects. All of these factors play crucial roles in improving the EMA performance.

The structural design principle is not limited to high-performance EMA materials but also extends to high-performance EMI shielding materials. Ma et al. successfully developed a lightweight, self-healing, and tunable EMI shielding sponge composed of shielding capsules using a concise dip-coating method[233]. In this method, the melamine sponge's uniform pores are covered by a formed MXene film, connecting the porous framework to the shielding capsule structure (Fig. 6.22). This unique structural design significantly enhances the interaction between EMW and the MXene film, resulting in the composite sponge achieving an EMI shielding effect of 90.49 dB in the X-band. Additionally, the incorporation of a PU interlayer contributes to the self-healing function of the hybrid sponge. Consequently, even after repeated cutting and repairing, the EMI SE of the hybrid sponge remains at 72.89 dB, demonstrating the superior application potential of

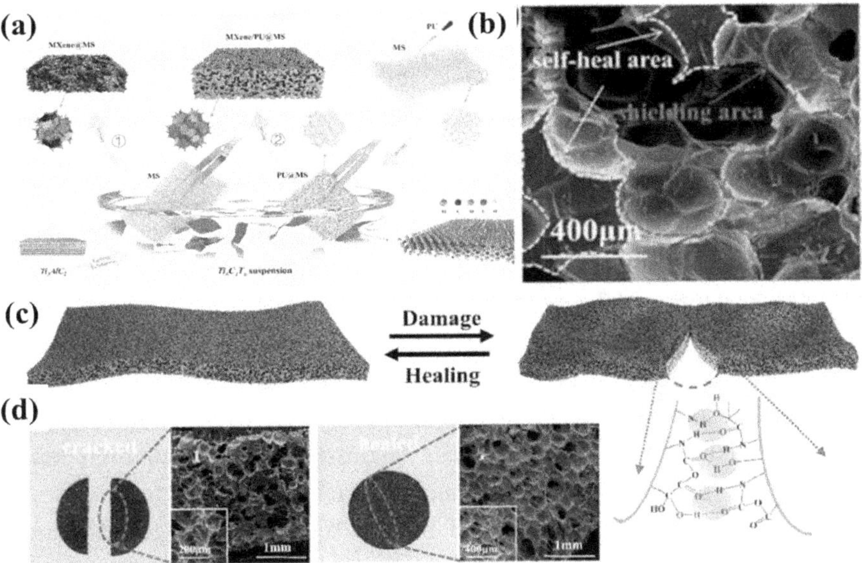

FIG. 6.22   (a) Schematic diagram of the preparation process of MXene-supported MS (1) and MXene/PU@MS (2). (b) Schematic illustration of the dynamic fracture and self-healing mechanism of MXene/PU@MS induced by interfacial hydrogen bonding. (c–d) Photographs and SEM images of damaged MXene/PU@MS and healed MXene/PU@MS samples. (Reproduced with permission from Ref.[233] Copyright 2021, Elsevier Ltd.)

the MXene/PU@MS composite sponge in wearable devices. What's even more intriguing is that the sponge foam material can be integrated with terahertz (THz) absorption technology, opening up possibilities in the fields of EMI shielding, radar stealth and the emerging 6G communication. To address the challenges of complex manufacturing processes and narrow-band THz absorbing materials, Shui et al. proposed a solution to obtain $Ti_3C_2T_x$ MXene sponge foam (MSF) through a dip-coating method. They successfully created a lightweight, broadband and hydrophobic THz absorber[234]. By immersing a PU sponge with pore sizes larger than 300 μm into a $Ti_3C_2T_x$ colloidal solution and applying extrusion for 5 minutes, the $Ti_3C_2T_x$ quickly covered the sponge skeleton. After the drying process, the $Ti_3C_2T_x$ flakes completely filled the entire sponge. It shows three different filling states: filling, adhering to the framework and forming a thin film on the pores. Each state exhibits distinct THz absorption properties. Furthermore, the lightweight and flexibility of the MSF are evident in Figure 8l and 8m. Hence, this innovative approach of combining large-aperture porous structures with 2D nanosheets lays the foundation for the development of high-performance multifunctional terahertz absorbers.

6.3.5 Metal nanowire-based materials

In contrast to conventional bulk metals, metal nanomaterials (such as 0D nanoparticles and 1D nanowires or nanorods) possess unique characteristics, including smaller size, higher strength, larger specific surface area, and better suitability for EMI shielding. These nanomaterials can be broadly categorized into two groups: conductive metal nanomaterials (e.g., silver nanoparticles (AgNPs), silver nanowires (AgNWs), and copper nanowires (CuNWs) and magnetic metal nanomaterials (e.g., Fe_3O_4 and nickel nanoparticles (NiNPs))[202,235]. Conductive metal nanomaterials primarily contribute to the dielectric loss of EMW, while magnetic metal nanomaterials affect the magnetic loss. To mitigate the issues of high density and agglomeration, these metal nanomaterials are often incorporated as nanofillers in polymer matrices to fabricate 3D porous aerogel and sponge composites.

In recent years, nanosilver has emerged as a preferred conductive metal filler for lightweight and porous EMI shielding materials due to its high conductivity and corrosion resistance. To enhance the EMI shielding performance, nanosilver particles (AgNPs) have been successfully attached to commercial melamine sponges or shaped cellulose sponges through simple dip-coating or electroless plating methods. This attachment allows for multiple scattering of EMW within the sponge, effectively improving the EMI shielding capabilities. However, these strategies often come with drawbacks such as high demand for filler content, high density determined by the commercial sponge itself, limited filler adhesion and permeability to the sponge, and increased aggregation of nanoparticles with an increase in dipping or plating time. In comparison to AgNPs, silver nanowires (AgNWs) exhibit similar high conductivity but possess a higher aspect ratio. This characteristic enables AgNWs to form an effective conductive network using less filler content, thereby lowering the percolation threshold of conductive fillers in composite aerogels and sponges. Additionally, AgNWs can serve as connectors to bridge other components[236]. Wan et al. pioneered the development of an ultralightweight metal composite sponge for efficient EMI shielding by freeze-drying and annealing a crosslinked AgNW/PVA sponge[212]. The resulting carbon-wrapped AgNW (Ag@C) hybrid sponge, with an ultralow density of 3.83 mg/cm^3, exhibited an extremely high specific surface area to total mass (SSE/t) ratio of approximately 61,169.3 dB cm^2/g. This high SSE/t ratio was mainly attributed to the geometrical absorption/scattering effect caused by the 3D interconnected porous structure of the sponge, which provided a large surface area and interface area. The presence of electrical conductivity differences between the silver core and carbon shell promoted dielectric relaxation, tunneling effect, inherent electric dipole polarization and interfacial polarization. These effects synergistically led to higher losses, effectively attenuating EM waves and improving the overall shielding effect. Furthermore, the Ag@C sponge exhibited super-hydrophobicity, strong corrosion resistance, outstanding mechanical strength and structural stability, making it suitable for applications under harsh conditions.

However, the weak interactions, easy entanglements and large contact resistance among high-aspect-ratio AgNWs posed challenges in assembling them into 3D EMI shielding aerogels and sponges. To address this issue, our group employed TEMPO-treated CNFs to aid in the dispersion and network bonding of AgNWs[237]. Hydrogen bonding interactions formed between the carbonyl groups of PVP on the AgNW surface and the hydroxyl groups of CNFs facilitated the dispersion and bonding process. As a result, a wood-inspired CNF/AgNW sponge with anisotropic macrostructures was created using unidirectional freeze-drying[235]. Due to the introduction of high-charge-density AgNWs and oriented structures, the EMI SE of the sponge in the transverse direction was significantly better than that in the longitudinal direction, resulting in anisotropy in the EMI shielding performance. CuNWs, which have excellent electrical conductivity of approximately 5.7×10^7 S m^{-1} and are easy to prepare, are also utilized as conductive metal nanofillers for EMI shielding aerogels and sponges. Moreover, they are more cost-effective compared to AgNWs. However, bare CuNWs are prone to oxidation when exposed to air, resulting in a rapid decline in performance. To address this issue, Wu et al. introduced a protective layer of PVP to effectively inhibit CuNW surface oxidation. They then created a core-shell aerogel wrapped with graphene (CuNW@G) through thermal annealing[195]. The graphene outer layer not only prevented CuNW oxidation, maintaining high electrical conductivity and EMI performance for an extended period, but also significantly enhanced the mechanical strength, modulus and elasticity of the aerogel. This approach offers new insights into preventing oxidation in metal nanowire-based aerogels and sponges.

Although electrical conductivity is crucial for the EMI shielding performance of aerogels and sponges, it should not be excessively high according to formulas (3) and (4). Excessive conductivity can cause impedance mismatch at the interfaces between the shield and air, leading to increased SER and secondary reflections. As a result, magnetic metal nanomaterials such as iron, cobalt, nickel and their oxides have been doped with highly conductive fillers into the polymer matrix to create multi-element hybrid aerogels and sponges in recent years. Examples include cellulose/polypyrrole/Fe_2O_3[13], graphene/cobalt ferrite (CFO)/zinc oxide (ZnO)[238], graphene/Fe_3O_4/epoxy[239], cellulose/graphene/Fe_3O_4[240] and Co/C@CNFs[17]. On one hand, incorporating magnetic metal nanomaterials in combination with the porous network structure of the aerogel and sponge reduces overall conductivity, improves impedance mismatch and lowers the skin effect, allowing more EM waves to enter the shield. On the other hand, in addition to the dielectric loss resulting from dipole polarization and interfacial polarization of the conductive fillers and multiple reflections between porous interfaces, when EM waves encounter the surfaces of highly magnetized magnetic materials, various relaxation mechanisms such as magnetization relaxation, magnetic domain movement, spin resonance, natural resonance and eddy currents synergistically occur. This leads to high magnetic losses of EM waves and an absorption-dominant shielding mechanism with enhanced power absorption of up to 87.8%[238].

Overall, the combined effects of the porous structure, conductive and magnetic components contribute to the composite aerogel and sponge's highly efficient EMI shielding for EM waves.

6.4 3D SELF-SUPPORTING MATERIALS FOR EMI SHIELDING

6.4.1 Fabrication of wood-based composites

6.4.1.1 Direct internal synthesis method

Direct internal synthesis is a widely employed technique for synthesizing absorbents within wood. The impregnation process, in particular, is a significant approach used to synthesize metal or compound particles directly inside wood as an absorbent. The key to this method is maximizing the amount of absorbent within the wood. Initially, researchers impregnated wood with pre-prepared particles to create magnetic wood. For example, H. Oka et al. impregnated cedar sapwood with a water-based magnetic fluid under specific pressure to prepare impregnated-type magnetic wood. However, this method has limitations regarding particle size and wood structure. It becomes difficult for particles to penetrate the wood's interior when they are too large or when the wood's pore channels are too small. Therefore, a desirable solution to overcome this issue is the direct synthesis of nano-absorbents inside the wood. In recent years, the in-situ co-precipitation method has emerged as a popular technique for synthesizing nanoparticles inside wood due to its cost-effectiveness, precise control, production of ultrafine powders and low energy consumption. Wood-based composites such as wood/Fe_3O_4, wood/$MnFe_2O_4$, wood/ZnO, wood/$FeNi_3$ and others have been successfully synthesized using the in-situ co-precipitation method.

Co-precipitation is generally a straightforward method for synthesizing nanoparticles. However, when it comes to fabricating the necessary absorbents inside wood, the corresponding ions need to penetrate the unique structure of wood effectively. This poses a challenge, and extensive research efforts have been made to address this problem. Jana S et al. developed a microwave-assisted in situ synthesis method to synthesize superparamagnetic iron oxide nanoparticles (SPIONs) inside wood[241]. They first pretreated wood cubes in a NaOH solution to enhance porosity before particle synthesis. Then, the wood cubes were immersed in an iron salt solution and heated using a microwave for a specific duration. This approach facilitated the diffusion of ions within the wood, resulting in a uniform distribution of synthesized particles throughout the complex wood cell wall structure, which provides advantages for microwave attenuation.

6.4.1.2 Surface deposition method

Reflection applications of EMI shielding and antistatic materials made with natural wood are hampered by deteriorated magnetism and electrical conductivity. Therefore, a great number of research studies have been carried out to improve the electrical conductivity and magnetism of wood. The surface deposition method is considered to be a feasible approach to achieve these goals. Compared to the internal synthesis method, the porous structure inside wood has little impact on the surface deposition method. Meanwhile, the deposition layer of the wood surface is easy to control, which can conveniently adjust the electromagnetic parameters. Typical deposition methods for synthesizing inorganic particles on the material surface include electroless plating[242], hydrothermal synthesis[243], the wet chemical method[244]. Among them, electroless plating has become the preferred method for depositing nanometer-sized conductive or magnetic particles on various insulating substrates because of the advantages of uniform coating, low cost, facile operation, energy-saving and so forth. Specifically, many researchers have prepared EMI shielding materials with remarkable electrical conductivity and magnetism by depositing metal particles on the surface of wood via the electroless plating method such as Cu, Ni, Ni/Cu, etc. When specific inorganic nanoparticles are chemically deposited on the wood surface, the EMI shielding performance could be greatly improved. Compared to coating directly on the wood surface, the electroless plating method can form a uniform and continuous coating on the wood surface. Moreover, the wood surface is rich in hydroxyl groups, which is an effective matrix for the nucleation and growth of inorganic nanomaterials. Thus, the coating can firmly exist on the surface of wood through chemical bonding.

As we all know, wood's non-metallic nature results in its lack of auto-catalytic abilities. Therefore, it is crucial to activate the wood surface before the deposition of inorganic nanomaterials in electroless plating. The activation process plays a significant role in determining the quality of electroless plating. Common chemical activators used in this process include colloidal palladium, $PdCl_2$, $NaBH_4$ and others. Researchers, such as Wang et al.,[245] have utilized electroless plating technology to deposit nickel particles on the wood surface. To achieve strong chemical bonding, the wood is pretreated with aminopropyltrihydroxysilane (APTHS), which acts as a coupling agent to enhance adhesion between the activator and wood substrate. A monolayer of APTHS self-assembles on the wood surface, facilitating the combination of the activator and wood surface. Subsequently, the activated wood is immersed in a solution containing nickel ions to form the metal layer. High magnification SEM image reveals a uniform and continuous metal layer, indicating successful electroless plating that forms a tightly bonded and high-quality metal layer. This approach achieves strong EMI shielding properties of over 60 dB at frequencies ranging from 10 MHz to 1.5 GHz. Similarly, Song et al. have developed a multifunctional wood-based composite with superhydrophobicity and EMI shielding properties[246]. Upon activating and electroless plating, the composite transforms from its original appearance to a gray color (Figs. 6.23a–c). Fig. 6.23(c) demonstrates the sample's superhydrophobicity with a water contact angle of 160 degrees, and the roll-off angle is less than 5 degrees after fluoralkylsilane (FAS) modification. The superhydrophobic wood surface can easily remove solid particles if polluted when water is dropped on it, resulting in a self-cleaning surface (Fig. 6.23d). Additionally, the EMI shielding properties are evaluated using an electromagnetic field tester. Copper-plated wood veneers are assembled into a box containing a mobile phone. By making a call to the mobile phone inside the box using another phone (Fig. 6.23e), the electromagnetic field tester shows that the intensity value of the electromagnetic field inside the box is only 0.358 mT, significantly lower than that of other samples. This experiment demonstrates that copper-plated wood effectively reduces electromagnetic radiation. Moreover, these multifunctional features (superhydrophobicity and self-cleaning) contribute to the practical application of wood-based composites.

6.4.1.3 Hot-pressing method

The aforementioned methods involve using wood veneers, wood blocks and other large wooden materials to construct materials with EMI shielding and MA properties. Although these methods have yielded wood-based composites with excellent EMI shielding and MA properties, they still have certain limitations that restrict their applications. For example, the internal synthesis method is influenced by the depth of penetration of the absorbent into the wood, while coatings prepared through surface deposition can easily be damaged or fall off in harsh external environments. These

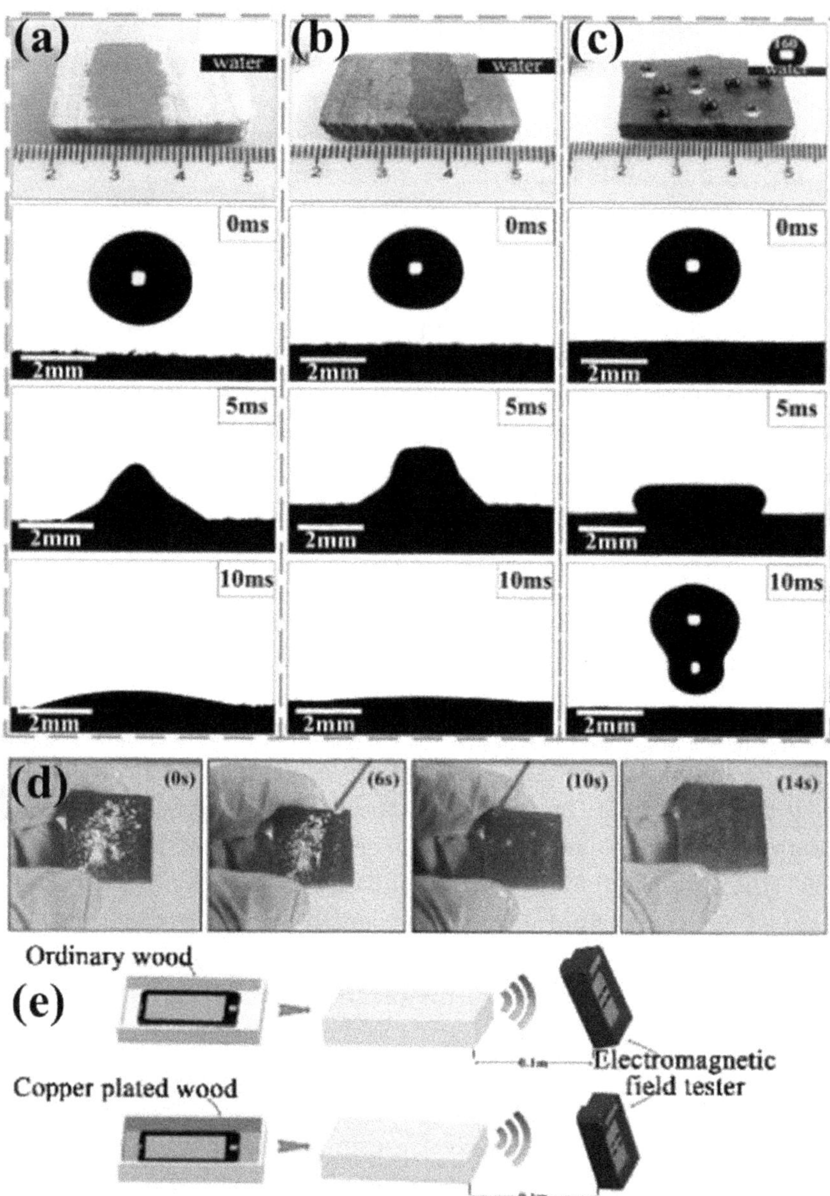

FIG. 6.23 (a) The natural wood surface. (b) The wood surface by electroless copper plating for 10 min. (c) The wood surface with electroless copper plating for 10 min and FAS modification. (d) Self-cleaning test after powder-contaminations. (e) The process of testing electromagnetic radiation intensity. (Reproduced with permission from Ref.[246] Copyright 2018, Elsevier Ltd.)

drawbacks have a negative impact on the electromagnetic performance of these materials. To address these issues, using smaller wooden components as raw materials appears to be an ideal solution.

The hot-pressing method is a powerful technique for producing wood-based composites for EMI shielding and MA, especially using smaller wooden components such as wood fiber, wood flour, and wood cellulose[247]. Typically, these smaller wooden components are pretreated or directly combined with absorbents and adhesives. Subsequently, boards or films are obtained through hot-pressing[248]. This "bottom-up" synthesis approach provides convenience in regulating the electromagnetic parameters of wood-based composites. For instance, in a study by Sun et al., the hot-pressing method was used to design a fiberboard with Fe_3O_4[249]. The fibers were pretreated by mechanical grinding and then hot-pressed to form a fiberboard. The synthesis process did not require the use of adhesives due to the formation of hydrogen bonds and covalent bonds between the fiber components at high temperature and pressure, effectively bonding the fibers together[250]. The fiberboards exhibited different RL values depending on the addition of Fe_3O_4, with the nano-Fe_3O_4/fiber-4 (NFF4) composite achieving a minimum RL value of 31.9 dB at 17.44 GHz, covering a frequency range of 5.6 GHz from 12.4 GHz to 18 GHz.

Furthermore, to adjust the MA properties of wood-based composites, researchers such as Lou et al. prepared multilayer boards using the hot-pressing method[251]. The immersion time of the fibers in an iron salt solution greatly influenced the MA performance. Longer immersion times resulted in significantly enhanced MA ability. The uniform distribution of wood fibers and magnetic particles in the fiberboard contributed to strong dielectric loss and excellent magnetic loss, thereby enhancing the MA properties. By adjusting the number of layers in the magnetic fiberboard, the RL value increased from 14.14 dB to 60.16 dB, with the corresponding number of layers ranging from three to seven. This adjustment was achieved by introducing multiple interfaces between the treated fiberboard and natural wood veneer in the direction of microwave penetration. These interfaces facilitated multiple reflections and dissipation of microwaves, thus improving the electromagnetic performance of the material[252,253]. This work presents a novel concept for wood-based MA materials with tunable electromagnetic parameters.

In summary, the fabrication methods mentioned above have both advantages and disadvantages. The direct internal synthesis method is limited by the thickness of the wood, as thicker wood makes it difficult for other materials to penetrate the interior, resulting in poor electromagnetic performance. However, this method utilizes the natural porous structure of wood, which facilitates the formation of an inner conductive network and microwave absorption rather than reflection. On the other hand, the surface deposition method is feasible and easily accessible. It is efficient and has a high yield. Additionally, the surface coating can protect the wood from corrosion, significantly improving the service life of wood-based composites. This method also provides a dense and uniform coating, ensuring excellent EMI shielding performance. However, it is not environmentally friendly, and the surface coating may easily come off, hindering its further development. Furthermore, the hot-pressing method allows for easy adjustment of the morphological structure and electromagnetic parameters of composites due to the "bottom-up" strategy. This method can be used to design wood-based materials with specific structures for MA and EMI shielding. However, the drawbacks of this method include high energy consumption and the destruction of the natural wood structure. In conclusion, the preparation strategies greatly impact the morphologies, structures and components of wood-based composites, which in turn determine their physical performances such as conductivity and magnetization, thereby affecting their electromagnetic properties. Therefore, the appropriate preparation method should be selected based on specific requirements.

6.4.2 Wood biomass-derived carbon materials for EMI shielding

Wood, with its natural cellular structure, is an important renewable resource that can be utilized as an economical and abundant carbon source for carbon-based EMI shields. In addition to cellulose, lignin is another significant biopolymer found in wood biomass, making it a subject of great interest. Lignin has advantages such as high yield, low cost and high carbon yield, which make it a promising material for carbon-based EMI shielding. However, research on utilizing lignin for this purpose is still in the early stages. In the following sections, we will explore the applications of both wood and lignin-derived carbon. They can serve three main functions: (1) Forming a conductive network: Wood and lignin-derived carbon materials can be used to create a conductive network, facilitating the effective transmission of electromagnetic signals. (2) Loading functional materials: Wood and lignin-derived carbon can act as carriers for other functional materials, enhancing their EMI shielding properties and expanding their applications. (3) Building a porous structure: The inherent porous structure of wood and lignin-derived carbon allows for the creation of materials with specific pore sizes and distributions. This porous structure is beneficial for improving EMI shielding performance by trapping and absorbing EMW. Overall, wood and lignin-derived carbon offer promising avenues for the development of carbon-based EMI shielding materials, and their functions can be categorized into forming a conductive network, loading functional materials and building a porous structure.

6.4.2.1 Serving as conductive networks

The physical and chemical properties of lignin, being a heterogeneous biopolymer, are complex[254]. Additionally, its high graphitization difficulty poses a challenge in obtaining highly conductive materials through lignin carbonization. Numerous efforts have been made to address this issue. Thomas et al. proposed that the source material and carbonization temperature of lignin significantly impact the quality and morphology of the resulting lignin-derived carbon[255]. Among four types of lignin (kraft lignin, soda lignin, lignoboost lignin and hydrolysis lignin), soda lignin exhibited the highest electrical conductivity, reaching 335 S m^{-1} under a high temperature of 1400 °C. This level of conductivity meets the standards required for EMI shielding materials and supercapacitor electrodes. Additionally, Zeng et al. demonstrated that annealing a single-fold lignin-derived carbon aerogel at 900 °C can yield a maximum conductivity of 25.6 S m^{-1} [256]. This indicates that lignin-derived carbon can establish an electronic conduction network, albeit with limitations imposed by factors such as carbonization temperature, lignin source and material structure.

Similarly, high-temperature pyrolysis is necessary to obtain highly conductive wood biomass-derived carbon. Prior delignification is also crucial to prevent the adverse effects of lignin on the graphitization of carbonized wood[257]. Zhao et al. reported the development of a multifunctional anisotropic carbon scaffold derived from natural wood for EMI shielding and thermal management[258] (Fig. 6.24). Through delignification and carbonization at 1000 °C, the longitudinal wood-derived carbon exhibited a significantly enhanced electrical conductivity of 17.5 S m^{-1}, resulting in an EMI SE of up to 55 dB. Furthermore, Feng and Shen fabricated a composite material using wood-derived carbon scaffold and epoxy[259]. Since epoxy is electrically insulating, carbonized wood requires a sufficiently high conductivity to ensure effective EMI shielding. By employing a high

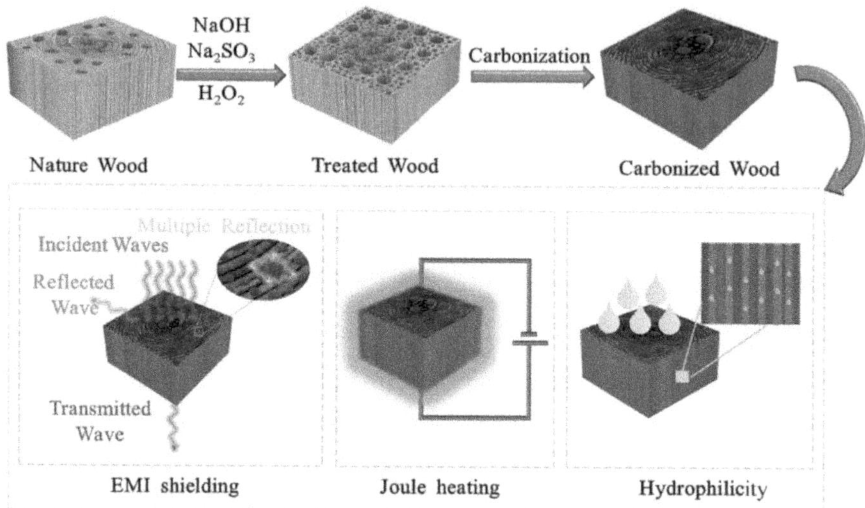

FIG. 6.24 Schematic for the preparation process of wood-derived carbon monoliths with multifunctionality. (Reproduced with permission from Ref.[258] Copyright 2021, American Chemical Society.)

carbonization temperature of 1200 °C, they obtained wood-derived carbon with enhanced graphitization. The resulting carbonized wood/epoxy composite demonstrated a decent conductivity of 12.5 S m^{-1}. Ultimately, this composite showcased an average EMI SE of 27.8 dB at frequencies ranging from 8 to 12 GHz.

In composite materials, carbonized wood can play a crucial role in establishing conductive networks. Ma et al. conducted a study where they carbonized a ZIF-67/wood composite to develop a 3D porous carbon skeleton/magnetic composite known as Co/C@WC[260]. The carbonized wood provided a continuous carbon structure that facilitated the rapid transmission of electrons. Because of the high-temperature carbonization of the wood and the graphitization catalyzed by Co particles, the maximum conductivity of Co/C@WC reached an impressive 3247 S m^{-1}. This resulted in an EMI SE of 43.2 dB and a specific SE per unit weight (SSE/t) of 557.1 dB cm^2 g^{-1}, primarily relying on a reflection-dominated shielding mechanism. Similarly, Zheng et al. fabricated Ni/porous carbon (Ni/PC) composites using wood as the starting material[261]. The pure wood-derived porous carbon exhibited a conductivity of 9.30 S m^{-1}, while the introduction of Ni catalysts during the carbonization process significantly increased the conductivity to 16.36 S m^{-1} in Ni/PC composites. The authors found that Ni catalyzed the graphitization of carbonized wood and accelerated electron transport. Consequently, the Ni/PC composites achieved an EMI SE of up to 50.8 dB, further confirming the conductive nature of carbonized wood in composite materials.

Furthermore, when compounded with insulating materials, carbonized wood can contribute significantly to the overall conductivity. Li et al. embedded γ-Fe$_2$O$_3$ nanoparticles into a wood-derived porous carbon matrix to form a γ-Fe$_2$O$_3$/porous carbon composite[261]. The porous carbon matrix alone exhibited a conductivity of 5.35 S m^{-1}, providing a solid foundation for EMI shielding. With the addition of magnetic γ-Fe$_2$O$_3$ particles, the optimized conductivity of the γ-Fe$_2$O$_3$/porous carbon composite reached 1.73 S m^{-1}. This, combined with the magnetic loss induced by the presence of γ-Fe$_2$O$_3$, resulted in an impressive SE of 44.80 dB.

Researchers have found that incorporating functional fillers can effectively enhance the EMI shielding performance of carbon-based EMI shields. Wood-derived carbon, with its aligned pore structure, high specific surface area and large internal space, serves as an ideal framework for facilitating the growth of various materials. In one study, Liu et al. used wood-derived carbon as a support to prepare NbC/pyrolytic carbon foams[262]. By immersing wood pieces in an NbCl$_5$/ethanol solution and subjecting them to high-temperature annealing, they successfully loaded NbC nanoparticles onto the carbonized wood. The carbonized wood provided support for NbC nanoparticles and acted as a continuous carbon source during the annealing process, promoting the formation of highly conductive NbC on the cell walls and enhancing electron conduction. As a result, the NbC/pyrolytic carbon foam exhibited an excellent EMI SE of 75.7 dB.

Similarly, Ren et al. created Ni/CNTs@carbonized wood foam by depositing Ni^{2+} on carbonized wood and employing a CVD process[263]. The porous carbonized wood served as a building block unit, allowing easy impregnation of a nickel chloride solution and providing a favorable environment for the growth of CNTs. This composite foam demonstrated enhanced EMI shielding ability due to polarization loss caused by the unbalanced electronegativity among Ni nanoparticles, CNTs and carbonized wood, as well as magnetic loss induced by Ni nanoparticles. The highest EMI SE achieved was 48.2 dB. Moreover, supported by carbonized wood, functional fillers could be uniformly dispersed, preventing agglomeration.

In another study, Li et al. developed AgNWs/N-doped graphene/wood-derived carbon monolith (WCM@N-G@AgNWs)[264]. As shown in Figs.

6.25(b)–(c), N-doped graphene sheets were grafted onto a wood-derived carbon framework (WCM@N-G), creating numerous interfaces due to their separation from each other, resembling a "moss on tile" structure (Fig. 6.25a). Subsequently, AgNWs were loaded onto the carbon framework, forming a highly conductive network. The resulting WCM@N-G@AgNWs exhibited an EMI SE of 60 dB and a specific SE per unit weight (SSE/t) of 465.1 dB cm^3 g^{-1} (Figs. 6.25d–e).

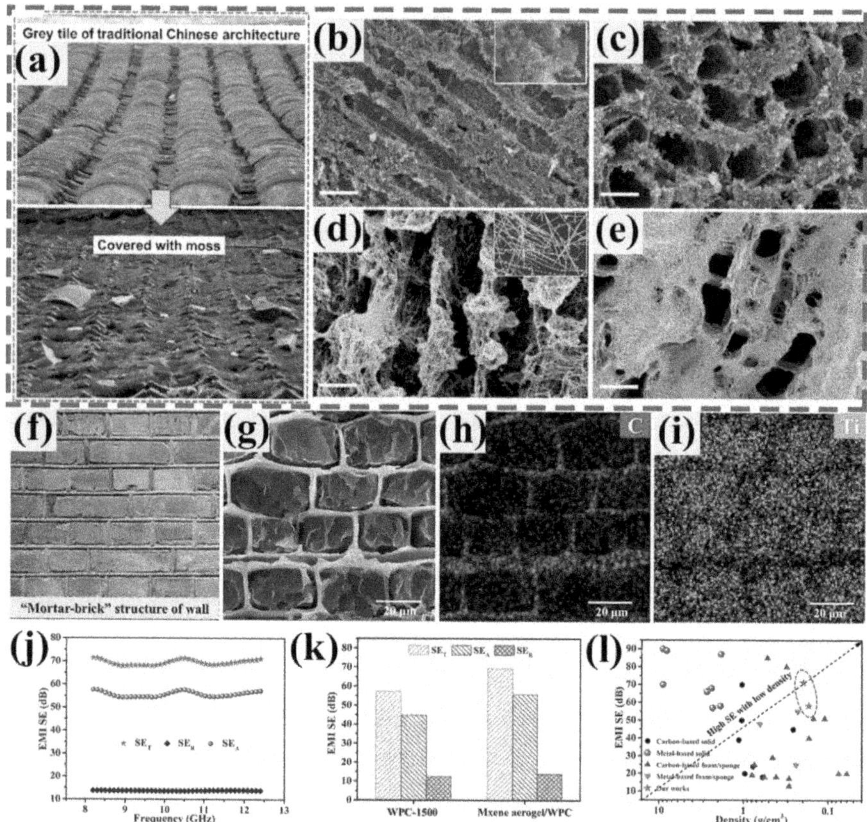

FIG. 6.25   (a) Schematic of proposed "moss on tile" structure. Cross-section SEM images of (b, c) WCM@NG and (d, e) WCM@N-G@AgNWs. (Reprinted with permission from Ref.[264] Copyright 2017, American Chemical Society.) (f) Schematic, (g) SEM image and corresponding (h, i) EDS mapping of MXene aerogel/wood-derived porous carbon composite. (j) EMI SE of MXene aerogel/wood-derived porous carbon composites. (k) Comparation of EMI SE of wood-derived porous carbon between MXene aerogel/wood-derived porous carbon composites. (l) Comparison of EMI SE values vs. density. (Reproduced with permission from Ref.[215] Copyright Elsevier 2020.)

Furthermore, by leveraging the adhesion properties of polymers, researchers can simplify the in-situ growth process. Polymer solutions containing functional materials can be coated onto wood-derived carbon frameworks. For example, Liu et al. encapsulated polyethylene glycol (PEG) with magnetic Fe3O4 nanoparticles in wood-derived porous carbon, achieving satisfactory EMI shielding and heat storage capabilities[265]. Jia et al. coated a trans-1, 4-polyisoprene (TPI)-MXene layer on wood-derived carbon foam[266]. This composite carbon foam exhibited an impressive EMI SE of 44.7 dB, along with excellent thermally and electrically stimulated shape-memory behaviors.

In addition to loading nanomaterials, carbonized wood can also support monolithic materials in an impressive manner. Gu and his team proposed a new approach to encapsulate aerogels within the carbonized wood framework, which has a cellular structure[215]. The researchers first prepared a dispersion of MXene in water and then obtained carbonized wood by subjecting natural wood to pyrolysis. Subsequently, the carbonized wood was immersed in the MXene dispersion and freeze-dried. The resulting product was then annealed at 200 °C for 2 hours to create MXene aerogel/wood-derived porous carbon composites with a wall-like "mortar/brick" structure (Fig. 6.25f). The corresponding SEM images and energy-dispersive X-ray spectroscopy (EDS) analysis were presented in Figs. 6.25(g)–(i). Within the wood-derived porous carbon, the MXene aerogels formed as the "bricks," while the porous carbon skeleton acted as the "mortar" that protected the fragile MXene aerogel through hydrogen bonding. This innovative composite material exhibited excellent EMI SE of 71.3 dB at a low density of 0.197 g cm^{-3} (Figs. 6.25j–l). This novel approach provides inspiration for the development of ultra-light and high-performance EMI shields, while expanding the range of functional materials that can be incorporated.

6.4.2.2 Other wood-based EMI shielding composite materials

By introducing different additives and materials to establish diversified composition/multi-level structure for a variety of loss mechanisms (such as conduction loss, dipole polarization loss, absorption loss, multiple RL, interface polarization loss, etc.), combined electromagnetic synergy can improve the electromagnetic shielding performance of composites[265,267,268]. Pu et al.[267] inserted carbon nanotube (CNT)/MXene composite nanosheets into natural wood with lignin removed by vacuum impregnation, and then wrapped the above wood-based composites with PDMS to prepare hydrophobic and multifunctional wood derived composites (Fig. 6.26). The composite has a high compressive strength of 1.53 MPa and a maximum breaking strain of 74.1%. It also shows excellent EMI shielding performance (29.3 dB). The improvement of EMI shielding performance is closely related to the reflection, absorption, and multiple internal reflection of EMW. Because the high porosity of the composite reduces the impedance mismatch, the incident EMW easily enters the channel with low reflectivity. In addition, the incident EMW is reflected and attenuated many times along the narrow channel "trap."

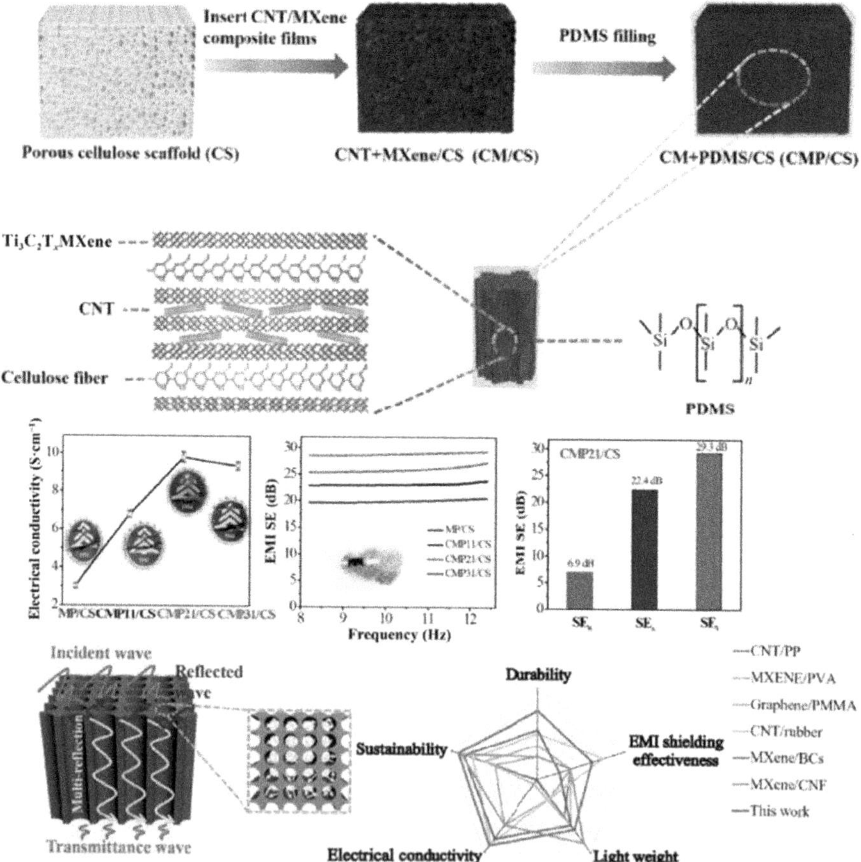

FIG. 6.26 Schematic diagrams of other wood-derived composites. Synthesis and EMI shielding of elastic, multifunctional, and wood-derived composites. (Reproduced with permission from Ref.[267] Copyright 2021, Elsevier Ltd.)

While at the same time, the non-uniform interface channel increases the scattering inside the composite. Besides, $Ti_3C_2T_x$ MXene causes impedance mismatch, as the integration of 2D planar MXene and 1D CNT greatly enhances interface polarization or additional attenuation mechanism. Therefore, electromagnetic energy is gradually absorbed and dissipated by CNT/MXene composite nanosheets deposited on wood, making it difficult for EMW to escape from the porous composite before being absorbed. Therefore, the composite has excellent EMI shielding properties and meets the requirements of commercial applications. Adjusting electromagnetic parameters by combining with magnetic particles to improve impedance matching is an important strategy to obtain lightweight and efficient carbon-based EMW absorbing materials. Qu et al.[265] used biological wood-derived porous carbon as the support material to encapsulate PEG, and further prepared a series of WPC/PEG/ Fe, O: phase change composites. with excellent shape stability and EMI shielding properties through a simple vacuum impregnation method. Due to the free electrons on the conductive WPC-10-7/surface, when the EMW is incident on the WPC-0 surface, part of the EMW will be reflected back immediately. Moreover, the introduction of magnetic Fe_3O_4 allows more EMW to enter the phase change composite, because of good impedance matching between magnetic materials and EMW which can effectively. reduce the direct reflection on the surface. In addition, more eddy current and hysteresis loss of the incident EMW can be realized through the resonance and offset of the magnetic dipole. At the same time, WPC-10 has a conductive continuous 1D wall connection, which makes it difficult for the incident EMW to escape before dissipation or absorption in the form of heat energy. Finally, the EMI SE of the prepared material can be as high as 55.08 dB between 8.2 and 12.4 GHz.

CONCLUSION

Benefiting from the tough crosslinking between polymers derived from widely used physical and chemical cross-linking methods, the hydrogels possess mechanical ultra-flexibility and stretchability beyond the traditional porous materials. Hydrogel-based EMI shields with various inorganic conductive fillers or conductive polymer networks demonstrated excellent EMI shielding performance. In addition, due to the conductivity, flexibility and biocompatibility, the multifunctionalities of hydrogel materials also contributed to a promising application prospect. Conceivably, hydrogels with pore structure design will be desirable next-generation EMI shielding materials.

Although the researchers have made decent progress in exploring the EMI shielding performance, there are still several issues that are not involved. First, the controllable morphology of micrometer-sized pores such as aligned channels and anisotropic pore structure, which have been proven to efficiently improve the EMI shielding performance of porous materials, is not considered for the current hydrogel-based EMI shields. In addition, apart from the EMI shielding, the attempts of hydrogels in multifunctional applications are barely carried out, which does not fully utilize the superiorities in mechanical properties and biocompatibility of the hydrogels. Thus, we suggest that illuminating and further understanding the relationships between controllable interior porous structures, components and EMI shielding performance should be the priority in hydrogel-based EMI shields. It is believed that if hydrogel-based shielding materials are combined with the design of controllable pore structures, they will be more promising for high-performance EMI shielding materials. In the meanwhile, the multifunctionalities such as photothermal therapy, antisepsis coating, and adhesiveness should be paid more attention to extend the application potential of hydrogel-based EMI shields.

However, as science and technology continue to advance, the demands for electromagnetic shielding materials are becoming increasingly stringent. To better meet these evolving requirements, numerous challenges still need to be addressed. Firstly, achieving the integration of structure and function design holds great significance. While electromagnetic shielding materials should satisfy the performance requirements of being "thin, light, strong, and wide," incorporating multiple functions such as flexibility, corrosion resistance, environmental stability and intelligence can expand their practical applications. Currently, there have been some studies on flexibility, thermal conductivity, transparency, intelligent response, etc. However, most of these studies remain limited to isolated investigations, with only a few comprehensive and systematic research endeavors. Therefore, there is ample room for further extending the integrated design that combines structure and performance.

Secondly, in order to better guide the development of materials that meet the specified requirements, it is essential to clarify and refine the system theory that integrates material, structure and performance. Although studies have explored the effects of factors such as the size, integrity and annealing temperature of GO on the properties of the final prepared films, there is limited research on crucial aspects such as the dosage and orientation of graphene, as well as the dispersion within the composite component and structure. Thus, methodological approaches for addressing these areas more comprehensively are largely unreported. Thirdly, the ultimate goal of material design and preparation is to facilitate practical applications. Therefore, it is imperative to explore low-cost, easy-to-operate and controllable mass production methods.

An ideal fabric material for EMI shielding should possess characteristics such as flexibility, stretchability, light weightness, durability and comfort. The widespread use of electronic devices has led to an increase in unwanted EMW, which can pose potential risks to both humans and nearby electronic equipment. Consequently, the development of fabric as an EMI shielding material has become a crucial discovery for personal protection. Given the wide range of natural and synthetic fabrics available, there is ample opportunity to explore various materials and their intrinsic properties to effectively transform them into long-lasting EMI shields through appropriate modifications and scientific processes. However, the use of synthetic polymers as conductive fillers in textile matrices is generally non-biodegradable and expensive, making the final product unaffordable for many. Therefore, research should focus on utilizing natural chemicals, polymers and reinforcing materials that provide the required strength and shielding parameters while remaining cost-effective and environmentally friendly. Improving the fabric's reusability is also important, necessitating the development of materials that can withstand the stresses and strains of different environments. Laundering the fabric often reduces its effectiveness, as reported in some studies. Therefore, it is crucial to promote the creation of self-cleaning and smart textiles. Certain studies have shown the possibility of converting electromagnetic wave energy into heat or electrical energy, particularly in capacitors. Implementing such technology on a large scale could not only minimize conventional electricity usage but also revolutionize energy generation. This opens up vast opportunities for manufacturing textiles with EMI shielding properties and efficient energy storage applications, applicable in various settings such as space stations, radiation shielding facilities, hospitals, power plants and workstations in polar regions. Considering the fundamental need for fabrics in our daily lives, transforming them into multifunctional materials would allow individuals to carry them for various purposes. The exploration of fabric as an

EMI shielding material with diverse applications holds significant potential for advancements in multiple fields.

Regarding EM shielding fabric, current theoretical methods aim to simulate metal properties using yarns, such as metal plates, perforated metal plates or metal grid structures through various combinations. However, these methods are suitable for single-layer or double-layer fabric composites, and they require conductive intersection points in the fabric grid, which has certain limitations. Specifically, these methods result in significant errors when applied to 2D fabrics with high buckling or multi-layer fabrics. Fabricating EM shielding fabrics can be achieved through surface metallization, metal coating or weaving fabrics with conductive fibers. However, without structure optimization, these processes lead to low efficiency and high production costs. Additionally, there is a lack of research on EM shielding mechanisms, analysis methods for EM properties based on fabric structures, and optimal design approaches for shielding fabrics. In terms of absorbing fabrics, current theoretical research mainly focuses on calculating the EM parameters using equivalent methods to achieve high-loss angle fiber materials. However, these equivalent methods fail to meet the requirements of wide frequencies and ideal EM parameters. Experimental studies on absorbing fabrics primarily concentrate on fiber materials, where the desired complex dielectric constant and permeability can be obtained through surface treatment or fiber modification. Nevertheless, these experimental procedures are costly and complicated. Moreover, there are limited reports on fabric structure-based theories, despite the fact that different fabric structures have varying effects on absorbing performance. Thus, future research should focus on developing design methods for fabric structures that meet absorbing performance requirements. On the other hand, theoretical methods for wave-transparent fabrics have matured, with key factors being the dielectric constant and loss angle tangent. Current research primarily revolves around reducing the dielectric constant and loss angle tangent while meeting mechanical and ablative properties. However, there is a scarcity of reports on the influence of fabric structure on wave-transparent properties. Consequently, the development and optimization of wave-transparent fabric structure designs can enhance the diversity of wave-transparent fabrics.

The future development trend for EMW shielding materials is centered on achieving low reflection and high absorption. It is widely recognized that EMI shielding is primarily determined by the electrical conductivity of the material. A high conductivity level typically leads to strong EMI shielding, mainly through reflection. Nevertheless, excessive reflection can generate secondary EM pollution. Therefore, there is a pressing need to identify EM shielding materials with a cellular structure that exhibits low reflection and high absorption capabilities. In particular, sandwich-type shielding materials should also be taken into account during their development. For the next generation of sandwich shielding composites, the first layer must consist of materials that are transparent to EM waves to enable microwaves to propagate as far as possible. The second layer primarily utilizes microwave-absorbing materials to effectively dissipate the EM energy. The third layer should be constructed using highly conductive components, such as Ag nanowires, CNTs, graphene, and MXene, to reflect any residual EM waves back to the second absorber layer. By following this approach, it is possible to obtain sandwich shielding materials that are predominantly absorption-dominant.

The incorporation of a cellular structure in composites can enhance the effectiveness of EMI shielding. This approach not only significantly reduces the density of the materials but also enables their application in sectors where weight reduction is crucial, such as aircraft. Controlled porosity within the cellular structure allows for the selective placement and dispersion of conductive nanoparticles in the cell walls. This reduces the average separation between nanoparticles, optimizing the formation of an efficient conductive network even at lower concentrations of conductive fillers. Furthermore, the introduction of a microcellular structure promotes multiple reflections as the primary EMI shielding mechanism. The larger interfaces between the cells and matrix, combined with the layered structure of graphene and MXene and the high aspect ratio of CNTs and metallic nanowires, enhance the reflection and scattering of incident EM waves. These waves become trapped within the cellular structure until they dissipate as heat, resulting in a significant increase in EMI SE. As a result, these materials find expanded applications in high-demand electronics and aerospace industries.

The majority of the reported materials have primarily undergone laboratory-level testing. However, it is crucial to focus on the synthesis mechanism in order to scale up and apply these materials in commercial settings. In the future, researchers will likely turn their attention towards lightweight multilayered nanostructures and 3D architectures, such as foams, hydrogels, and aerogels, to achieve absorption-dominant EMI shielding at reduced thicknesses. However, effective methods for controlling the formation of lightweight materials and utilizing existing pores and interfaces appropriately still need to be determined. With the rapid advancements and breakthroughs in interdisciplinary science and technology, it becomes possible to integrate all the necessary requirements into future EMI shielding materials. Therefore, the following research directions are anticipated to be explored in the near future.

(1) Developing and implementing lightweight shields, specifically composite foams, to mitigate EMI in sensitive electronic devices and systems presents practical challenges in various applications such as portable electronic devices, satellites, aircraft, telecommunications and more. The choice of conductive foams as the most efficient EMI shielding materials typically relies on designing and optimizing cost-effective EMI shielding solutions. The selection of composite foams must be tailored through EMI shielding design optimization. The optimal type of conductive foam is determined by considering the specific shield design requirements and the environmental factors.

(2) The adjustable fabrication of conductive foams and the optimization of their structures and constituents are crucial for developing lightweight and flexible EMI shielding materials. Additionally, a deep understanding of the properties resulting from the composition and structure is vital for optimizing the SE of conductive foams. This serves as the primary and most significant step in the development of high-performance shielding materials. The properties of composites can be effectively controlled by incorporating metallics, CNTs, graphene, MXene and other constituents, as well as utilizing configurations such as porous 2D multi-layer films and 3D honeycomb structures.

(3) MXenes, a new category of 2D nanomaterials, have demonstrated remarkable efficiency in EMI shielding. Their combination of conductivity and magnetic properties makes them an excellent choice for EMI shielding, benefiting from the synergistic effects of these two properties. MXene-based composite foams have significant potential to replace carbon-based foams, offering superior performance. To achieve maximum internal reflection and total absorption, it is crucial to have appropriate pore size and structure in these composite foams.

(4) The discovery of new EMI shielding foams has introduced borophene, sheets of boron atoms, as a promising alternative to carbon, metal and magnetic conducting filler-based materials. Borophene, particularly in its 2D form, could outperform graphene as an electrical conductor. Calculations suggest that corrugated borophene exhibits enhanced conductivity along its ridges compared to other directions. These properties make borophene well-suited for designing novel EMI shielding foams for future electronic devices. Additionally, liquid metallic or liquid ion composite foams are also considered as promising EMI shielding materials.

(5) Smart EMI shielding foams need to withstand external stimuli in real-world applications. Therefore, it is crucial to focus on the EMI shielding properties of materials under various external factors, including temperature, strain, humidity, electric field, magnetic field and sunlight irradiation. Furthermore, future development should aim at creating multifunctional EMI shielding materials that offer quick heat dissipation, hydrophobicity, anticorrosion properties and flexibility.

To tackle the issue of serious electromagnetic pollution, it is expected to develop EMI shielding and MA materials with strong absorption capacity, wide absorption bandwidths, light-weight, and low-cost. 3D Wood-based composites have shown tremendous potential as excellent materials for EMI shielding and MA (magnetic absorption) applications. These composites offer advantages such as high yield, sustainability and the ability to tune their electromagnetic properties.

Despite recent significant progress, comprehensive information regarding wood-based composites for EMI shielding and MA is still lacking, hindering their further development. Unsatisfactory interfacial compatibility between wood and other materials significantly impacts the stability of their electromagnetic performance. Additionally, current methods of synthesizing wood-based composites are complicated and time-consuming. Moreover, intrinsic defects of wood, such as anisotropic heterogeneous features, moist expansion, dry shrinkage, easy corrosion, etc., negatively impact the properties of wood-based composites. Thus, it is crucial to develop effective strategies to overcome these obstacles and enhance the properties of wood-based composites.

The interface plays a crucial role in shaping the microstructure of composites. It serves as a connection between the composite and the wood component, significantly influencing the overall properties of the composites. Wood contains numerous polar functional groups (such as -OH and -COOH) that can serve as anchor points for enhancing the compatibility between wood and other materials, thereby improving the stability of wood composites. Therefore, it is crucial to leverage the inherent properties of wood to enhance the interface compatibility with other materials.

Furthermore, to efficiently achieve the desired wood-based EMI shielding and MA composites, the development of micro-nano synthesis methods is of utmost importance. Wood typically exhibits microstructure sizes ranging from micrometers to nanometers. Therefore, the materials combined with wood should have a small enough size to evenly infiltrate the interior of wood. Utilizing micro-nano synthesis methods can be an effective approach to obtaining desirable wood-based EMI shielding and MA composites. Additionally, it is essential to focus on the research and development of multifunctional wood-based EMI shielding and MA composites, rather than solely relying on composites with single properties to meet practical application requirements. In the future, wood-based EMI shielding and MA composites should not only possess excellent electromagnetic properties but also exhibit other desirable characteristics such as weather resistance, corrosion resistance and fire resistance. Overall, sustainable wood-based composites for EMI shielding and MA applications have gained significant attention due to their excellent properties. However, there is still ample room for innovation and improvement in the research field of wood-based composites. Although the journey ahead may present challenges, it holds great potential for advancement.

REFERENCES

1. Yang, Y., Han, M., Liu, W., et al. (2022). Hydrogel-based composites beyond the porous architectures for electromagnetic interference shielding. Nano Research *15*: 9614–30.
2. Sano, K., Ishida, Y., and Aida, T. (2018). Synthesis of anisotropic hydrogels and their applications. Angewandte Chemie International Edition *57*: 2532–43.
3. Zeng, Z., Wang, C., Wu, T., et al. (2020). Nanocellulose assisted preparation of ambient dried, large-scale and mechanically robust carbon nanotube foams for electromagnetic interference shielding. Journal of Materials Chemistry A *8*: 17969–79.
4. Zhan, Z., Song, Q., Zhou, Z., et al. (2019). Ultrastrong and conductive MXene/cellulose nanofiber films enhanced by hierarchical

nano-architecture and interfacial interaction for flexible electromagnetic interference shielding. Journal of Materials Chemistry C 7: 9820–29.

5. Hamedi, H., Moradi, S., Hudson, S.M., et al. (2018). Chitosan based hydrogels and their applications for drug delivery in wound dressings: a review. Carbohydr Polym *199*: 445–60.

6. Distler, T., and Boccaccini, A.R. (2020). 3D printing of electrically conductive hydrogels for tissue engineering and biosensors – a review. Acta Biomater *101*: 1–13.

7. Abbasi, H., Antunes, M., and Velasco, J.I. (2019). Recent advances in carbon-based polymer nanocomposites for electromagnetic interference shielding. Progress in Materials Science *103*: 319–73.

8. Iqbal, A., Sambyal, P., and Koo, C.M. (2020). 2D MXenes for electromagnetic shielding: a review. Advanced Functional Materials *30*: 2000883.

9. Singh, A.K., Shishkin, A., Koppel, T., et al. (2018). A review of porous lightweight composite materials for electromagnetic interference shielding. Composites Part B: Engineering *149*: 188–97.

10. Wang, C., Murugadoss, V., Kong, J., et al. (2018). Overview of carbon nanostructures and nanocomposites for electromagnetic wave shielding. Carbon *140*: 696–733.

11. Zhao, S., Yan, Y., Gao, A., et al. (2018). Flexible polydimethylsilane nanocomposites enhanced with a three-dimensional graphene/carbon nanotube bicontinuous framework for high-performance electromagnetic interference shielding. ACS Applied Materials & Interfaces *10*: 26723–32.

12. Chen, Z., Xu, C., Ma, C., et al. (2013). Lightweight and flexible graphene foam composites for high-performance electromagnetic interference shielding. Advanced Materials *25*: 1296–300.

13. Wan, C., and Li, J. (2017). Synthesis and electromagnetic interference shielding of cellulose-derived carbon aerogels functionalized with α-Fe2O3 and polypyrrole. Carbohydrate Polymers *161*: 158–65.

14. Sambyal, P., Iqbal, A., Hong, J., et al. (2019). Ultralight and mechanically robust Ti3C2Tx hybrid aerogel reinforced by carbon nanotubes for electromagnetic interference shielding. ACS Applied Materials & Interfaces *11*: 38046–54.

15. Zhang, Y., Zhou, W., Chen, H., et al. (2021). Facile preparation of CNTs microspheres as improved carbon absorbers for high-efficiency electromagnetic wave absorption. Ceramics International *47*: 10013–18.

16. Chen, Y., Pang, L., Li, Y., et al. (2020). Ultra-thin and highly flexible cellulose nanofiber/silver nanowire conductive paper for effective electromagnetic interference shielding. Composites Part A: Applied Science and Manufacturing *135*: 105960.

17. Fei, Y., Liang, M., Yan, L., et al. (2020). Co/C@cellulose nanofiber aerogel derived from metal-organic frameworks for highly efficient electromagnetic interference shielding. Chemical Engineering Journal *392*: 124815.

18. Chen, Y., Zhang, H.-B., Wang, M., et al. (2017). Phenolic resin-enhanced three-dimensional graphene aerogels and their epoxy nanocomposites with high mechanical and electromagnetic interference shielding performances. Composites Science and Technology *152*: 254–62.

19. Wu, X., Han, B., Zhang, H.-B., et al. (2020). Compressible, durable and conductive polydimethylsiloxane-coated MXene foams for high-performance electromagnetic interference shielding. Chemical Engineering Journal *381*: 122622.

20. Cao, W.-T., Chen, F.-F., Zhu, Y.-J., et al. (2018). Binary strengthening and toughening of MXene/cellulose nanofiber composite paper with nacre-inspired structure and superior electromagnetic interference shielding properties. ACS Nano *12*: 4583–93.

21. Wang, Q.W., Zhang, H.B., Liu, J., et al. (2018). Multifunctional and water-resistant MXene-decorated polyester textiles with outstanding electromagnetic interference shielding and Joule heating performances. Advanced Functional Materials *29*: 1806819.

22. Raagulan, K., Kim, B.M., and Chai, K.Y. (2020). Recent advancement of Electromagnetic Interference (EMI) shielding of two dimensional (2D) MXene and graphene aerogel composites. Nanomaterials *10*: 702.

23. Liang, C., Hamidinejad, M., Ma, L., et al. (2020). Lightweight and flexible graphene/SiC-nanowires/ poly(vinylidene fluoride) composites for electromagnetic interference shielding and thermal management. Carbon *156*: 58–66.

24. Wang, M.-L., Zhang, S., Zhou, Z.-H., et al. (2021). Facile heteroatom doping of biomass-derived carbon aerogels with hierarchically porous architecture and hybrid conductive network: towards high electromagnetic interference shielding effectiveness and high absorption coefficient. Composites Part B: Engineering *224*: 109175.

25. Kumar, R., Singh, R.K., Tiwari, V.S., et al. (2017). Enhanced magnetic performance of iron oxide nanoparticles anchored pristine/ N-doped multi-walled carbon nanotubes by microwave-assisted approach. Journal of Alloys and Compounds *695*: 1793–801.

26. Sun, H., Che, R., You, X., et al. (2014). Cross-stacking aligned carbon-nanotube films to tune microwave absorption frequencies and increase absorption intensities. Advanced Materials *26*: 8120–25.

27. Li, Z., Lin, Z., Han, M., et al. (2021). Flexible electrospun carbon nanofibers/silicone composite films for electromagnetic interference

shielding, electrothermal and photothermal applications. Chemical Engineering Journal *420*: 129826.

28. Kumar, R., Macedo, W.C., Singh, R.K., et al. (2019). Nitrogen–sulfur co-doped reduced graphene oxide-nickel oxide nanoparticle composites for electromagnetic interference shielding. ACS Applied Nano Materials *2*: 4626–36.

29. Kumar, R., Sahoo, S., Joanni, E., et al. (2021). Microwave as a tool for synthesis of carbon-based electrodes for energy storage. ACS Applied Materials & Interfaces *14*: 20306–25.

30. Diaz, L.A., Lister, T.E., Rae, C., et al. (2018). Anion exchange membrane electrolyzers as alternative for upgrading of biomass-derived molecules. ACS Sustainable Chemistry & Engineering *6*: 8458–67.

31. Shen, X., Zhang, C., Han, B., et al. (2022). Catalytic self-transfer hydrogenolysis of lignin with endogenous hydrogen: road to the carbon-neutral future. Chemical Society Reviews *51*: 1608–28.

32. Xia, Q., Chen, Z., Shao, Y., et al. (2016). Direct hydrodeoxygenation of raw woody biomass into liquid alkanes. Nature Communications *7*: 11162.

33. Lou, Z., Yuan, C., Zhang, Y., et al. (2019). Synthesis of porous carbon matrix with inlaid Fe3C/Fe3O4 micro-particles as an effective electromagnetic wave absorber from natural wood shavings. Journal of Alloys and Compounds *775*: 800–09.

34. Hu, P., Dong, S., Li, X., et al. (2019). A low-cost strategy to synthesize MnO nanorods anchored on 3D biomass-derived carbon with superior microwave absorption properties. Journal of Materials Chemistry C *7*: 9219–28.

35. Xiong, Y., Xu, L., Yang, C., et al. (2020). Implanting FeCo/C nanocages with tunable electromagnetic parameters in anisotropic wood carbon aerogels for efficient microwave absorption. Journal of Materials Chemistry A *8*: 18863–71.

36. Li, G., Li, C., Li, G., et al. (2022). Development of conductive hydrogels for fabricating flexible strain sensors. Small *18*: e2101518.

37. Zhang, H., Tang, N., Yu, X., et al. (2022). Natural glycyrrhizic acid-tailored hydrogel with in-situ gradient reduction of AgNPs layer as high-performance, multi-functional, sustainable flexible sensors. Chemical Engineering Journal *430*: 132779.

38. Mao, J., Zhao, C., Li, Y., et al. (2020). Highly stretchable, self-healing, and strain-sensitive based on double-crosslinked nanocomposite hydrogel. Composites Communications *17*: 22–27.

39. Lee, K.H., Zhang, Y.Z., Kim, H., et al. (2021). Muscle fatigue sensor based on Ti(3) C(2) T(x) MXene hydrogel. Small Methods *5*: e2100819.

40. Zhao, M., Zhang, W., Wang, D., et al. (2021). A packaged and reusable hydrogel strain sensor with conformal adhesion to skin for human motions monitoring. Advanced Materials Interfaces 9: 2101786.

41. Han, X., Xiao, G., Wang, Y., et al. (2020). Design and fabrication of conductive polymer hydrogels and their applications in flexible supercapacitors. Journal of Materials Chemistry A 8: 23059–95.

42. Lu, B., Yuk, H., Lin, S., et al. (2019). Pure PEDOT:PSS hydrogels. Nature Communications 10: 1043.

43. Hu, S., Zhou, L., Tu, L., et al. (2019). Elastomeric conductive hybrid hydrogels with continuous conductive networks. J Mater Chem B 7: 2389–97.

44. Gan, D., Han, L., Wang, M., et al. (2018). Conductive and tough hydrogels based on biopolymer molecular templates for controlling in situ formation of polypyrrole nanorods. ACS Applied Material Interfaces 10: 36218–28.

45. Zhou, L., Fan, L., Yi, X., et al. (2018). Soft conducting polymer hydrogels cross-linked and doped by tannic acid for spinal cord injury repair. ACS Nano 12: 10957–67.

46. Sun, X., Yao, F., and Li, J. (2020). Nanocomposite hydrogel-based strain and pressure sensors: a review. Journal of Materials Chemistry A 8: 18605–23.

47. Zhang, Y.Z., El-Demellawi, J.K., Jiang, Q., et al. (2020). MXene hydrogels: fundamentals and applications. Chemical Society Review 49: 7229–51.

48. Chen, C., Wang, Y., Meng, T., et al. (2019). Electrically conductive poly-acrylamide/carbon nanotube hydrogel: reinforcing effect from cellulose nanofibers. Cellulose 26: 8843–51.

49. Wang, Q., Pan, X., Wang, X., et al. (2021). Fabrication strategies and application fields of novel 2D Ti3C2Tx (MXene) composite hydrogels: a mini-review. Ceramics International 47: 4398–403.

50. Ai, J., Li, J., Li, K., et al. (2021). Highly flexible, self-healable and conductive poly(vinyl alcohol)/Ti3C2Tx MXene film and it's application in capacitive deionization. Chemical Engineering Journal 408: 127256.

51. Deng, Y., Shang, T., Wu, Z., et al. (2019). Fast gelation of Ti(3) C(2) T(x) MXene initiated by metal ions. Adv Mater 31: e1902432.

52. Shi, X., Wang, H., Xie, X., et al. (2019). Bioinspired ultrasensitive and stretchable MXene-based strain sensor via nacre-mimetic microscale "Brick-and-Mortar" architecture. ACS Nano 13: 649–59.

53. Rong, Q., Lei, W., and Liu, M. (2018). Conductive hydrogels as smart materials for flexible electronic devices. Chemistry 24: 16930–43.

54. Liu, H., Li, Q., Zhang, S., et al. (2018). Electrically conductive polymer composites for smart flexible strain sensors: a critical review. Journal of Materials Chemistry C 6: 12121–41.

55. Lin, F., Wang, Z., Shen, Y., et al. (2019). Natural skin-inspired versatile cellulose biomimetic hydrogels. Journal of Materials Chemistry A 7: 26442–55.

56. Liao, M., Liao, H., Ye, J., et al. (2019). Polyvinyl alcohol-stabilized liquid metal hydrogel for wearable transient epidermal sensors. ACS Applied Materials & Interfaces 11: 47358–64.

57. Le Bideau, J., Viau, L., and Vioux, A. (2011). Ionogels, ionic liquid based hybrid materials. Chemical Society Reviews 40: 907–25.

58. Yang, C., and Suo, Z. (2018). Hydrogel ionotronics. Nature Reviews Materials 3: 125–42.

59. Sun, H., Zhao, Y., Wang, C., et al. (2020). Ultra-stretchable, durable and conductive hydrogel with hybrid double network as high performance strain sensor and stretchable triboelectric nanogenerator. Nano Energy 76: 105035.

60. Liu, J., Wang, H., Ou, R., et al. (2021). Anti-bacterial silk-based hydrogels for multifunctional electrical skin with mechanical-thermal dual sensitive integration. Chemical Engineering Journal 426: 130722.

61. Wang, Q., Zhang, Q., Wang, G., et al. (2022). Muscle-inspired anisotropic hydrogel strain sensors. ACS Applied Materials & Interfaces 14: 1921–28.

62. Ganguly, S., and Margel, S. (2020). Review: remotely controlled magneto-regulation of therapeutics from magnetoelastic gel matrices. Biotechnology Advances 44: 107611.

63. Pan, F., Yu, L., Xiang, Z., et al. (2021). Improved synergistic effect for achieving ultrathin microwave absorber of 1D Co nanochains/2D carbide MXene nanocomposite. Carbon 172: 506–15.

64. Zeng, Z., Wu, T., Han, D., et al. (2020). Ultralight, flexible, and biomimetic nanocellulose/silver nanowire aerogels for electromagnetic interference shielding. ACS Nano 14: 2927–38.

65. Zeng, Z., Chen, M., Pei, Y., et al. (2017). Ultralight and flexible polyurethane/silver nanowire nanocomposites with unidirectional pores for highly effective electromagnetic shielding. ACS Applied Materials & Interfaces 9: 32211–19.

66. Yang, R., Gui, X., Yao, L., et al. (2021). Ultrathin, lightweight, and flexible CNT Buckypaper enhanced using MXenes for electromagnetic interference shielding. Nano-Micro Letters 13: 66.

67. Zeng, Z., Jin, H., Chen, M., et al. (2016). Lightweight and anisotropic porous MWCNT/WPU composites for ultrahigh performance electromagnetic interference shielding. Advanced Functional Materials 26: 303–10.

68. Zhou, Q., Lyu, J., Wang, G., et al. (2021). Mechanically strong and multifunctional hybrid hydrogels with ultrahigh electrical conductivity. Advanced Functional Materials *31*: 2104536.

69. Zeng, Z., Jin, H., Chen, M., et al. (2017). Microstructure design of lightweight, flexible, and high electromagnetic shielding porous multiwalled carbon nanotube/polymer composites. Small *13*: 1701388.

70. Yang, W., Shao, B., Liu, T., et al. (2018). Robust and mechanically and electrically self-healing hydrogel for efficient electromagnetic interference shielding. ACS Applied Materials & Interfaces *10*: 8245–57.

71. Yu, Y., Yi, P., Xu, W., et al. (2022). Environmentally tough and stretchable MXene organohydrogel with exceptionally enhanced electromagnetic interference shielding performances. Nanomicro Lett *14*: 77.

72. Yang, Y., Wu, N., Li, B., et al. (2022). Biomimetic porous MXene sediment-based hydrogel for high-performance and multifunctional electromagnetic interference shielding. ACS Nano *16*: 15042–52.

73. Sarkar, B., Li, X., Quenneville, E., et al. (2021). Lightweight and flexible conducting polymer sponges and hydrogels for electromagnetic interference shielding. Journal of Materials Chemistry C *9*: 16558–65.

74. Feig, V.R., Tran, H., Lee, M., et al. (2018). Mechanically tunable conductive interpenetrating network hydrogels that mimic the elastic moduli of biological tissue. Nature Communication. *9*: 2740.

75. Qian, W., Fu, H., Sun, Y., et al. (2022). Scalable assembly of high-quality graphene films via electrostatic-repulsion aligning. Advanced Materials e2206101.

76. Liu, J., McKeon, L., Garcia, J., et al. (2022). Additive manufacturing of Ti3 C2 -MXene-functionalized conductive polymer hydrogels for electromagnetic-interference shielding. Advanced Materials *34*: e2106253.

77. Liu, J., Lin, S., Huang, K., et al. (2020). A large-area AgNW-modified textile with high-performance electromagnetic interference shielding. NPJ Flexible Electronics *4*: 10.

78. Hu, Y., Zhuo, H., Zhang, Y., et al. (2021). Graphene oxide encapsulating liquid metal to toughen hydrogel. Advanced Functional Materials *31*: 2106761.

79. Bai, Y., Qin, F., and Lu, Y. (2020). Multifunctional electromagnetic interference shielding ternary alloy (Ni-W-P) decorated fabric with wide-operating-range Joule heating performances. ACS Applied Materials & Interfaces *12*: 48016–26.

80. Pan, T., Zhang, Y., Wang, C., et al. (2020). Mulberry-like polyaniline-based flexible composite fabrics with effective electromagnetic shielding capability. Composites Science and Technology *188*: 107991.

81. Gao, J., Luo, J., Wang, L., et al. (2019). Flexible, superhydrophobic and highly conductive composite based on non-woven polypropylene fabric for electromagnetic interference shielding. Chemical Engineering Journal *364*: 493–502.

82. Jia, L.-C., Xu, L., Ren, F., et al. (2019). Stretchable and durable conductive fabric for ultrahigh performance electromagnetic interference shielding. Carbon *144*: 101–08.

83. Lan, C., Jia, H., Qiu, M., et al. (2021). Ultrathin MXene/polymer coatings with an alternating structure on fabrics for enhanced electromagnetic interference shielding and fire-resistant protective performances. ACS Applied Materials & Interfaces *13*: 38761–72.

84. Ren, W., Zhu, H., Yang, Y., et al. (2020). Flexible and robust silver coated non-woven fabric reinforced waterborne polyurethane films for ultra-efficient electromagnetic shielding. Composites Part B: Engineering *184*: 107745.

85. Mahmoodi, M., Arjmand, M., Sundararaj, U., et al. (2012). The electrical conductivity and electromagnetic interference shielding of injection molded multi-walled carbon nanotube/polystyrene composites. Carbon *50*: 1455–64.

86. Guo, H., Chen, Y., Li, Y., et al. (2021). Electrospun fibrous materials and their applications for electromagnetic interference shielding: a review. Composites Part A: Applied Science and Manufacturing *143*: 106309.

87. Gao, D., Guo, S., Zhou, Y., et al. (2022). Hydrophobic, flexible electromagnetic interference shielding films derived from hydrolysate of waste leather scraps. Journal of Colloid and Interface Science *613*: 396–405.

88. Sun, M., Li, P., Qin, H., et al. (2022). Liquid metal/CNTs hydrogel-based transparent strain sensor for wireless health monitoring of aquatic animals. Chemical Engineering Journal. *454:* Part 3, 140459.

89. Guan, H., and Chung, D.D.L. (2019). Effect of the planar coil and linear arrangements of continuous carbon fiber tow on the electromagnetic interference shielding effectiveness, with comparison of carbon fibers with and without nickel coating. Carbon *152*: 898–908.

90. Kursun Bahadir, S., Mitilineos, S.A., Symeonidis, S., et al. (2019). Electromagnetic shielding and reflection loss of conductive yarn incorporated woven fabrics at the S and X radar bands. Journal of Electronic Materials *49*: 1579–87.

91. Lee, S., Park, J., Kim, M.C., et al. (2021). Polyvinylidene fluoride Core–Shell nanofiber membranes with highly conductive shells for electromagnetic interference shielding. ACS Applied Materials & Interfaces *13*: 25428–37.

92. Kim, M., Kim, S., Seong, Y.C., et al. (2021). Multiwalled carbon nano-tube buckypaper/polyacrylonitrile nanofiber composite membranes for

electromagnetic interference shielding. ACS Applied Nano Materials *4*: 729–38.

93. Kim, D.G., Choi, J.H., Choi, D.-K., et al. (2018). Highly bendable and durable transparent electromagnetic interference shielding film prepared by wet sintering of silver nanowires. ACS Applied Materials & Interfaces *10*: 29730–40.

94. Xiao, Y.-Y., He, Y.-J., Wang, R.-Q., et al. (2022). Mussel-inspired strategy to construct 3D silver nanoparticle network in flexible phase change composites with excellent thermal energy management and electromagnetic interference shielding capabilities. Composites Part B: Engineering *239*: 109962.

95. Wu, L.-P., Li, Y.-Z., Wang, B.-J., et al. (2018). Electroless Ag-plated sponges by tunable deposition onto cellulose-derived templates for ultra-high electromagnetic interference shielding. Materials & Design *159*: 47–56.

96. Kandasaamy, P.V., and Rameshkumar, M. (2020). Electromagnetic interference shielding effectiveness of sol-gel coating on Cu-plated fabrics. The Journal of The Textile Institute *112*: 855–63.

97. Ameri, Z., Soleimani, E., and Shafyei, A. (2022). Preparation and identification of a biocompatible polymer composite: shielding against the interference of electromagnetic waves. Synthetic Metals *283*: 116983.

98. Mao, Y., Zhang, S., Wang, W., et al. (2018). Electroless silver plated flexible graphite felt prepared by dopamine functionalization and applied for electromagnetic interference shielding. Colloids and Surfaces A: Physicochemical and Engineering Aspects *558*: 538–47.

99. Zeng, Z., Jiang, F., Yue, Y., et al. (2020). Flexible and ultrathin waterproof cellular membranes based on high-conjunction metal-wrapped polymer nanofibers for electromagnetic interference shielding. Advanced Materials *32*: e1908496.

100. Zhu, L., Bi, S., Zhao, H., et al. (2018). Cu–Ni–Gd coating with improved corrosion resistance on linen fabric by electroless plating for electromagnetic interference shielding. Journal of Materials Science: Materials in Electronics *29*: 16348–58.

101. Chen, Y., Wei, W., Zhu, Y., et al. (2019). Noncovalent functionalization of carbon nanotubes via co-deposition of tannic acid and polyethyleneimine for reinforcement and conductivity improvement in epoxy composite. Composites Science and Technology *170*: 25–33.

102. Wang, J., Zhu, J., Tsehaye, M.T., et al. (2017). High flux electroneutral loose nanofiltration membranes based on rapid deposition of polydopamine/polyethyleneimine. Journal of Materials Chemistry A *5*: 14847–57.

103. Wang, W., Li, W., Gao, C., et al. (2015). A novel preparation of silver-plated polyacrylonitrile fibers functionalized with antibacterial and electromagnetic shielding properties. Applied Surface Science *342*: 120–26.

104. Jiu, J., Nogi, M., Sugahara, T., et al. (2012). Strongly adhesive and flexible transparent silver nanowire conductive films fabricated with a high-intensity pulsed light technique. Journal of Materials Chemistry *22*: 23561.

105. Wang, G., Hao, L., Zhang, X., et al. (2022). Flexible and transparent silver nanowires/biopolymer film for high-efficient electromagnetic interference shielding. Journal of Colloid and Interface Science *607*: 89–99.

106. Yu, J., Gu, W., Zhao, H., et al. (2021). Lightweight, flexible and freestanding PVA/PEDOT: PSS/Ag NWs film for high-performance electromagnetic interference shielding. Science China Materials *64*: 1723–32.

107. Wang, G., Zhao, Y., Yang, F., et al. (2022). Multifunctional integrated transparent film for efficient electromagnetic protection. Nanomicro Letters *14*: 65.

108. Arjmand, M., Moud, A.A., Li, Y., et al. (2015). Outstanding electromagnetic interference shielding of silver nanowires: comparison with carbon nanotubes. RSC Advances *5*: 56590–98.

109. Wang, M.-Q., Yan, J., Du, S.-G., et al. (2014). Microstructure and electromagnetic interference shielding effectiveness of electroless Ni-P alloy coating on PVC plastic. Fibers and Polymers *15*: 1175–81.

110. An, Z., Zhang, X., and Li, H. (2015). A preliminary study of the preparation and characterization of shielding fabric coated by electrical deposition of amorphous Ni–Fe–P alloy. Journal of Alloys and Compounds *621*: 99–103.

111. Shen, Y., Zhang, H.F., Wang, L.M., et al. (2014). Fabrication of electromagnetic shielding polyester fabrics with carboxymethyl chitosan-palladium complexes activation. Fibers and Polymers *15*: 1414–21.

112. Bi, S., Zhao, H., Hou, L., et al. (2017). Comparative study of electroless Co-Ni-P plating on Tencel fabric by Co0-based and Ni0-based activation for electromagnetic interference shielding. Applied Surface Science *419*: 465–75.

113. Bi, S., Zhao, H., Hou, L., et al. (2018). Enhanced electromagnetic interference shielding effects of cobalt-nickel polyalloy coated fabrics with assistance of rare earth elements. Fibers and Polymers *19*: 1084–93.

114. Liu, X., Ye, Z., Zhang, L., et al. (2021). Highly flexible electromagnetic interference shielding films based on ultrathin Ni/Ag composites on paper substrates. Journal of Materials Science *56*: 5570–80.

115. Liu, Q., Yi, C., Chen, J., et al. (2021). Flexible, breathable, and highly environmental-stable Ni/PPy/PET conductive fabrics for efficient

electromagnetic interference shielding and wearable textile antennas. Composites Part B: Engineering *215*: 108752.

116. Sun, G., Ren, K., Wang, Y., et al. (2020). Fabrication of highly conducting nickel-coated graphite composite particles with low Ni content for excellent electromagnetic properties. Journal of Alloys and Compounds *834*: 155142.

117. Zhou, J., Sun, Y., Du, X., et al. (2010). Dual-modality in vivo imaging using rare-earth nanocrystals with near-infrared to near-infrared (NIR-to-NIR) upconversion luminescence and magnetic resonance properties. Biomaterials *31*: 3287–95.

118. Xu, Y., Lin, Z., Rajavel, K., et al. (2021). tailorable, lightweight and superelastic liquid metal monoliths for multifunctional electromagnetic interference shielding. Nanomicro Letters *14*: 29.

119. Sharma, G.K., and James, N.R. (2021). Progress in electrospun polymer composite fibers for microwave absorption and electromagnetic interference shielding. ACS Applied Electronic Materials *3*: 4657–80.

120. Shi, M., Wu, S., Han, Z.-D., et al. (2020). Utilization of electroless plating to prepare Cu-coated cotton cloth electrode for flexible Li-ion batteries. Rare Metals *40*: 400–08.

121. Tugirumubano, A., Vijay, S.J., Go, S.H., et al. (2019). Characterization of electromagnetic interference shielding composed of carbon fibers reinforced plastics and metal wire mesh based composites. Journal of Materials Research and Technology *8*: 167–72.

122. Jiao, Y., Wan, C., Zhang, W., et al. (2019). Carbon fibers encapsulated with nano-copper: a core-shell structured composite for antibacterial and electromagnetic interference shielding applications. Nanomaterials *9*: 460.

123. Lee, J., Liu, Y., Liu, Y., et al. (2017). Ultrahigh electromagnetic interference shielding performance of lightweight, flexible, and highly conductive copper-clad carbon fiber nonwoven fabrics. Journal of Materials Chemistry C *5*: 7853–61.

124. Liang, L.-L., Liu, Z., Xie, L.-J., et al. (2021). Bamboo-like N-doped carbon tubes encapsulated CoNi nanospheres towards efficient and anticorrosive microwave absorbents. Carbon *171*: 142–53.

125. Zheng, J., Yu, Z., Ji, G., et al. (2014). Reduction synthesis of FexOy@SiO2 Core–Shell nanostructure with enhanced microwave-absorption properties. Journal of Alloys and Compounds *602*: 8–15.

126. Liu, C., and Kang, Z. (2019). Facile fabrication of conductive silver films on carbon fiber fabrics via two components spray deposition technique for electromagnetic interference shielding. Applied Surface Science *487*: 1245–52.

127. Lee, S.H., Kim, J.Y., Koo, C.M., et al. (2017). Effects of processing methods on the electrical conductivity, electromagnetic parameters, and EMI shielding effectiveness of polypropylene/nickel-coated carbon fiber composites. Macromolecular Research 25: 936–43.

128. Wang, R., He, F., Wan, Y., et al. (2012). Preparation and characterization of a kind of magnetic carbon fibers used as electromagnetic shielding materials. Journal of Alloys and Compounds 514: 35–39.

129. Yim, Y.-J., Lee, J.J., Tugirumubano, A., et al. (2021). Electromagnetic interference shielding behavior of magnetic carbon fibers prepared by electroless FeCoNi-plating. Materials 14: 3774.

130. Zhu, L., Zeng, S., Teng, Z., et al. (2019). Significantly enhanced electromagnetic interference shielding in Al2O3 ceramic composites incorporated with highly aligned non-woven carbon fibers. Ceramics International 45: 12672–76.

131. Cui, J., Xu, J., Li, J., et al. (2020). A crosslinkable graphene oxide in waterborne polyurethane anticorrosive coatings: experiments and simulation. Composites Part B: Engineering 188: 107889.

132. Xu, Y., Yang, Y., Yan, D.-X., et al. (2018). Gradient structure design of flexible waterborne polyurethane conductive films for ultraefficient electromagnetic shielding with low reflection characteristic. ACS Applied Materials & Interfaces 10: 19143–52.

133. Zhang, Y., Pan, T., and Yang, Z. (2020). Flexible polyethylene terephthalate/polyaniline composite paper with bending durability and effective electromagnetic shielding performance. Chemical Engineering Journal 389: 124433.

134. Wang, H., Ren, H., Jing, C., et al. (2021). Two birds with one stone: graphene oxide@sulfonated polyaniline nanocomposites towards high-performance electromagnetic wave absorption and corrosion protection. Composites Science and Technology 204: 108630.

135. Tang, W., Lu, L., Xing, D., et al. (2018). A carbon-fabric/polycarbonate sandwiched film with high tensile and EMI shielding comprehensive properties: an experimental study. Composites Part B: Engineering 152: 8–16.

136. Lu, H., Li, Z., Qi, X., et al. (2021). Flexible, electrothermal-driven controllable carbon fiber/poly(ethylene-co-vinyl acetate) shape memory composites for electromagnetic shielding. Composites Science and Technology 207: 108697.

137. Park, J., Hu, X., Torfeh, M., et al. (2020). Exceptional electromagnetic shielding efficiency of silver coated carbon fiber fabrics via a roll-to-roll spray coating process. Journal of Materials Chemistry C 8: 11070–78.

138. Zhu, S., Shi, R., Qu, M., et al. (2021). Simultaneously improved mechanical and electromagnetic interference shielding properties of carbon fiber

fabrics/epoxy composites via interface engineering. Composites Science and Technology *207*: 108696.

139. Wang, H., Li, N., Wang, W., et al. (2019). Bead nano-necklace spheres on 3D carbon nanotube scaffolds for high-performance electromagnetic-interference shielding. Chemical Engineering Journal *360*: 1241–46.

140. Liu, X., Yin, X., Kong, L., et al. (2014). Fabrication and electromagnetic interference shielding effectiveness of carbon nanotube reinforced carbon fiber/pyrolytic carbon composites. Carbon *68*: 501–10.

141. Mei, H., Han, D., Xiao, S., et al. (2016). Improvement of the electromagnetic shielding properties of C/SiC composites by electrophoretic deposition of carbon nanotube on carbon fibers. Carbon *109*: 149–53.

142. Pothupitiya Gamage, S., Yang, K., Braveenth, R., et al. (2017). MWCNT coated free-standing carbon fiber fabric for enhanced performance in EMI shielding with a higher absolute EMI SE. Materials *10*: 1350.

143. Ramírez-Herrera, C.A., Gonzalez, H., Torre, F.d.l., et al. (2019). Electrical properties and electromagnetic interference shielding effectiveness of interlayered systems composed by carbon nanotube filled carbon nanofiber mats and polymer composites. Nanomaterials *9*: 238.

144. Chen, Y., Li, J., Li, T., et al. (2021). Recent advances in graphene-based films for electromagnetic interference shielding: review and future prospects. Carbon *180*: 163–84.

145. Chen, J., Wu, J., Ge, H., et al. (2016). Reduced graphene oxide deposited carbon fiber reinforced polymer composites for electromagnetic interference shielding. Composites Part A: Applied Science and Manufacturing *82*: 141–50.

146. Wu, J., Ye, Z., Ge, H., et al. (2017). Modified carbon fiber/magnetic graphene/epoxy composites with synergistic effect for electromagnetic interference shielding over broad frequency band. Journal of Colloid and Interface Science *506*: 217–26.

147. Lin, N., Chen, H., Mei, X., et al. (2021). A carbon composite film with three-dimensional reticular structure for electromagnetic interference shielding and electro-photo-thermal conversion. Materials *14*: 2423.

148. Kim, T.-S., Shin, D.-W., Kim, S.-G., et al. (2017). Large area nanometer thickness graphite freestanding film without transfer process. Chemical Physics Letters *690*: 101–04.

149. Vallés, C., David Núñez, J., Benito, A.M., et al. (2012). Flexible conductive graphene paper obtained by direct and gentle annealing of graphene oxide paper. Carbon *50*: 835–44.

150. Rozada, R., Paredes, J.I., Villar-Rodil, S., et al. (2013). Towards full repair of defects in reduced graphene oxide films by two-step graphitization. Nano Research *6*: 216–33.

151. Liu, Y., Li, J., Ge, X., et al. (2020). Macroscale superlubricity achieved on the hydrophobic graphene coating with glycerol. ACS Applied Materials & Interfaces *12*: 18859–69.

152. Wei, Q., Pei, S., Qian, X., et al. (2020). Superhigh electromagnetic interference shielding of ultrathin aligned pristine graphene nanosheets film. Advanced Materials *32*: 1907411.

153. Xi, J., Li, Y., Zhou, E., et al. (2018). Graphene aerogel films with expansion enhancement effect of high-performance electromagnetic interference shielding. Carbon *135*: 44–51.

154. Shen, B., Li, Y., Yi, D., et al. (2016). Microcellular graphene foam for improved broadband electromagnetic interference shielding. Carbon *102*: 154–60.

155. Lai, D., Chen, X., and Wang, Y. (2020). Controllable fabrication of elastomeric and porous graphene films with superior foldable behavior and excellent electromagnetic interference shielding performance. Carbon *158*: 728–37.

156. Li, C.-B., Li, Y.-J., Zhao, Q., et al. (2020). Electromagnetic interference shielding of graphene aerogel with layered microstructure fabricated via mechanical compression. ACS Applied Materials & Interfaces *12*: 30686–94.

157. Peng, L., Xu, Z., Liu, Z., et al. (2017). Ultrahigh thermal conductive yet superflexible graphene films. Advanced Materials *29*: 1700589.

158. Fan, C., Wu, B., Song, R., et al. (2019). Electromagnetic shielding and multi-beam radiation with high conductivity multilayer graphene film. Carbon *155*: 506–13.

159. Liu, L.X., Chen, W., Zhang, H.B., et al. (2019). Flexible and multifunctional silk textiles with biomimetic leaf-like MXene/silver nanowire nanostructures for electromagnetic interference shielding, humidity monitoring, and self-derived hydrophobicity. Advanced Functional Materials *29*(44), 1905197.

160. Zhou, T., Wu, C., Wang, Y., et al. (2020). Super-tough MXene-functionalized graphene sheets. Nature Communications *11*: 2077.

161. Zhou, E., Xi, J., Guo, Y., et al. (2018). Synergistic effect of graphene and carbon nanotube for high-performance electromagnetic interference shielding films. Carbon *133*: 316–22.

162. Ma, X., Li, Y., Shen, B., et al. (2018). Carbon composite networks with ultrathin skin layers of graphene film for exceptional electromagnetic interference shielding. ACS Applied Materials & Interfaces *10*: 38255–63.

163. Fu, H., Yang, Z., Zhang, Y., et al. (2020). SWCNT-modulated folding-resistant sandwich-structured graphene film for high-performance electromagnetic interference shielding. Carbon *162*: 490–96.

164. Yin, X., Li, H., Han, L., et al. (2020). Lightweight and flexible 3D graphene microtubes membrane for high-efficiency electromagnetic-interference shielding. Chemical Engineering Journal 387: 124025.

165. Liang, J., Wang, Y., Huang, Y., et al. (2009). Electromagnetic interference shielding of graphene/epoxy composites. Carbon 47: 922–25.

166. Al-Ghamdi, A.A., Al-Ghamdi, A.A., Al-Turki, Y., et al. (2016). Electromagnetic shielding properties of graphene/acrylonitrile butadiene rubber nanocomposites for portable and flexible electronic devices. Composites Part B: Engineering 88: 212–19.

167. Zhao, B., Zhang, X., Deng, J., et al. (2020). Flexible PEBAX/graphene electromagnetic shielding composite films with a negative pressure effect of resistance for pressure sensors applications. RSC Advances 10: 1535–43.

168. Shen, B., Li, Y., Yi, D., et al. (2017). Strong flexible polymer/graphene composite films with 3D saw-tooth folding for enhanced and tunable electromagnetic shielding. Carbon 113: 55–62.

169. Hsiao, S.-T., Ma, C.-C.M., Tien, H.-W., et al. (2013). Using a non-covalent modification to prepare a high electromagnetic interference shielding performance graphene nanosheet/water-borne polyurethane composite. Carbon 60: 57–66.

170. Vallés, C., Zhang, X., Cao, J., et al. (2019). Graphene/polyelectrolyte layer-by-layer coatings for electromagnetic interference shielding. ACS Applied Nano Materials 2: 5272–81.

171. Han, D., Zhao, Y.-H., Bai, S.-L., et al. (2016). High shielding effectiveness of multilayer graphene oxide aerogel film/polymer composites. RSC Advances 6: 92168–74.

172. Samanta, A., and Bordes, R. (2020). Conductive textiles prepared by spray coating of water-based graphene dispersions. RSC Advances 10: 2396–403.

173. Islam, M.D.Z., Dong, Y., Khoso, N.A., et al. (2019). Continuous dyeing of graphene on cotton fabric: Binder-free approach for electromagnetic shielding. Applied Surface Science 496: 143636.

174. Yan, J., Huang, Y., Wei, C., et al. (2017). Covalently bonded polyaniline/graphene composites as high-performance electromagnetic (EM) wave absorption materials. Composites Part A: Applied Science and Manufacturing 99: 121–28.

175. Xia, X., Mazzeo, A.D., Zhong, Z., et al. (2017). An X-band theory of electromagnetic interference shielding for graphene-polymer nanocomposites. Journal of Applied Physics 122: 025104.

176. Shahzad, F., Alhabeb, M., Hatter, C.B., et al. (2016). Electromagnetic interference shielding with 2D transition metal carbides (MXenes). Science 353: 1137–40.

177. Yun, T., Kim, H., Iqbal, A., et al. (2020). Electromagnetic shielding of monolayer MXene assemblies. Advanced Materials *32*: e1906769.
178. Iqbal, A., Shahzad, F., Hantanasirisakul, K., et al. (2020). Anomalous absorption of electromagnetic waves by 2D transition metal carbonitride Ti3CNTx (MXene). Science *369*: 446–50.
179. Han, M., Shuck, C.E., Rakhmanov, R., et al. (2020). Beyond Ti3C2Tx: MXenes for electromagnetic interference shielding. ACS Nano *14*: 5008–16.
180. Xu, H., Yin, X., Li, X., et al. (2019). Lightweight Ti2CTx MXene/poly(vinyl alcohol) composite foams for electromagnetic wave shielding with absorption-dominated feature. ACS Applied Materials & Interfaces *11*: 10198–207.
181. Zhang, Y., Cheng, W., Tian, W., et al. (2020). Nacre-inspired tunable electromagnetic interference shielding sandwich films with superior mechanical and fire-resistant protective performance. ACS Applied Materials & Interfaces *12*: 6371–82.
182. Zhao, H., Yue, Y., Zhang, Y., et al. (2016). Ternary artificial nacre reinforced by ultrathin amorphous alumina with exceptional mechanical properties. Advanced Materials *28*: 2037–42.
183. Liang, L., Yao, C., Yan, X., et al. (2021). High-efficiency electromagnetic interference shielding capability of magnetic Ti3C2Tx MXene/CNT composite film. Journal of Materials Chemistry A *9*: 24560–70.
184. Lin, Z., Liu, J., Peng, W., et al. (2020). Highly stable 3D Ti3C2Tx MXene-based foam architectures toward high-performance terahertz radiation shielding. ACS Nano *14*: 2109–17.
185. Luo, J.-Q., Zhao, S., Zhang, H.-B., et al. (2019). Flexible, stretchable and electrically conductive MXene/natural rubber nanocomposite films for efficient electromagnetic interference shielding. Composites Science and Technology *182*: 107754.
186. Zhou, B., Su, M., Yang, D., et al. (2020). Flexible MXene/silver nanowire-based transparent conductive film with electromagnetic interference shielding and electro-photo-thermal performance. ACS Applied Materials & Interfaces *12*: 40859–69.
187. Jin, M., Chen, W., Liu, L.-X., et al. (2022). Transparent, conductive and flexible MXene grid/silver nanowire hierarchical films for high-performance electromagnetic interference shielding. Journal of Materials Chemistry A *10*: 14364–73.
188. Hu, P., Lyu, J., Fu, C., et al. (2019). Multifunctional aramid nanofiber/carbon nanotube hybrid aerogel films. ACS Nano *14*: 688–97.

189. Song, W.-L., Guan, X.-T., Fan, L.-Z., et al. (2015). Tuning three-dimensional textures with graphene aerogels for ultra-light flexible graphene/texture composites of effective electromagnetic shielding. *Carbon 93*: 151–60.

190. An, Z., Ye, C., Zhang, R., et al. (2019). Multifunctional C/SiO2/SiC-based aerogels and composites for thermal insulators and electromagnetic interference shielding. *Journal of Sol-Gel Science and Technology 89*: 623–33.

191. Zhang, Y.S., and Khademhosseini, A. (2017). Advances in engineering hydrogels. *Science 356*: eaaf3627.

192. Zhu, L., Zong, L., Wu, X., et al. (2018). Shapeable fibrous aerogels of metal–organic-frameworks templated with nanocellulose for rapid and large-capacity adsorption. *ACS Nano 12*: 4462–68.

193. Zou, Y., Zhao, J., Zhu, J., et al. (2021). A Mussel-inspired polydopamine-filled cellulose aerogel for solar-enabled water remediation. *ACS Applied Materials & Interfaces 13*: 7617–24.

194. Zeng, Z., Wang, C., Zhang, Y., et al. (2018). Ultralight and highly elastic graphene/lignin-derived carbon nanocomposite aerogels with ultrahigh electromagnetic interference shielding performance. *ACS Applied Materials & Interfaces 10*: 8205–13.

195. Wu, S., Zou, M., Li, Z., et al. (2018). Robust and stable Cu nanowire@ graphene Core–Shell aerogels for ultraeffective electromagnetic interference shielding. *Small 14*: 1800634.

196. Gao, W., Zhao, N., Yu, T., et al. (2020). High-efficiency electromagnetic interference shielding realized in nacre-mimetic graphene/polymer composite with extremely low graphene loading. *Carbon 157*: 570–77.

197. Wan, Y.-J., Zhu, P.-L., Yu, S.-H., et al. (2017). Ultralight, super-elastic and volume-preserving cellulose fiber/graphene aerogel for high-performance electromagnetic interference shielding. *Carbon 115*: 629–39.

198. Huangfu, Y., Ruan, K., Qiu, H., et al. (2019). Fabrication and investigation on the PANI/MWCNT/thermally annealed graphene aerogel/epoxy electromagnetic interference shielding nanocomposites. *Composites Part A: Applied Science and Manufacturing 121*: 265–72.

199. Zhao, S., Zhang, H.B., Luo, J.Q., et al. (2018). Highly electrically conductive three-dimensional Ti(3)C(2)T (x) MXene/reduced graphene oxide hybrid aerogels with excellent electromagnetic interference shielding performances. *ACS Nano 12*: 11193–202.

200. Wan, C., Jiao, Y., Wei, S., et al. (2019). Functional nanocomposites from sustainable regenerated cellulose aerogels: a review. *Chemical Engineering Journal 359*: 459–75.

201. De France, K.J., Hoare, T., and Cranston, E.D. (2017). Review of hydrogels and aerogels containing nanocellulose. *Chemistry of Materials 29*: 4609–31.

202. Chen, Y.-J., Li, Y., Chu, B.T.T., et al. (2015). Porous composites coated with hybrid nano carbon materials perform excellent electromagnetic interference shielding. Composites Part B: Engineering *70*: 231–37.

203. Singh, S., Tripathi, P., Bhatnagar, A., et al. (2015). A highly porous, light weight 3D sponge like graphene aerogel for electromagnetic interference shielding applications. RSC Advances *5*: 107083–87.

204. Zaman, A., Huang, F., Jiang, M., et al. (2020). Preparation, properties, and applications of natural cellulosic aerogels: a review. Energy and Built Environment *1*: 60–76.

205. Ye, C., An, Z., and Zhang, R. (2019). Super-elastic carbon-bonded carbon fibre composites impregnated with carbon aerogel for high-temperature thermal insulation. Advances in Applied Ceramics *118*: 292–99.

206. Chen, Y., Zhang, L., Mei, C., et al. (2020). Wood-inspired anisotropic cellulose nanofibril composite sponges for multifunctional applications. ACS Applied Materials & Interfaces *12*: 35513–22.

207. Han, M., Yin, X., Hantanasirisakul, K., et al. (2019). Anisotropic MXene aerogels with a mechanically tunable ratio of electromagnetic wave reflection to absorption. Advanced Optical Materials *7*: 1900267.

208. Weng, C., Wang, G., Dai, Z., et al. (2019). Buckled AgNW/MXene hybrid hierarchical sponges for high-performance electromagnetic interference shielding. Nanoscale *11*: 22804–12.

209. Zhou, Z.-H., Li, M.-Z., Huang, H.-D., et al. (2020). Structuring hierarchically porous architecture in biomass-derived carbon aerogels for simultaneously achieving high electromagnetic interference shielding effectiveness and high absorption coefficient. ACS Applied Materials & Interfaces *12*: 18840–49.

210. Yang, Q., Yang, J., Gao, Z., et al. (2019). Carbonized cellulose nanofibril/graphene oxide composite aerogels for high-performance supercapacitors. ACS Applied Energy Materials *3*: 1145–51.

211. Bi, S., Zhang, L., Mu, C., et al. (2017). A comparative study on electromagnetic interference shielding behaviors of chemically reduced and thermally reduced graphene aerogels. Journal of Colloid and Interface Science *492*: 112–18.

212. Wan, Y.J., Zhu, P.L., Yu, S.H., et al. (2018). Anticorrosive, ultralight, and flexible carbon-wrapped metallic nanowire hybrid sponges for highly efficient electromagnetic interference shielding. Small *14*: 1800534.

213. Zhou, Z.-H., Liang, Y., Huang, H.-D., et al. (2019). Structuring dense three-dimensional sheet-like skeleton networks in biomass-derived carbon aerogels for efficient electromagnetic interference shielding. Carbon *152*: 316–24.

214. Guo, T., Chen, X., Su, L., et al. (2019). Stretched graphene nanosheets formed the "obstacle walls" in melamine sponge towards effective electromagnetic interference shielding applications. Materials & Design *182*: 108029.

215. Liang, C., Qiu, H., Song, P., et al. (2020). Ultra-light MXene aerogel/wood-derived porous carbon composites with wall-like "mortar/brick" structures for electromagnetic interference shielding. Science Bulletin *65*: 616–22.

216. Chen, Y., Zhang, H.B., Yang, Y., et al. (2015). High-performance epoxy nanocomposites reinforced with three-dimensional carbon nanotube sponge for electromagnetic interference shielding. Advanced Functional Materials *26*: 447–55.

217. Mei, H., Zhao, X., Gui, X., et al. (2019). SiC encapsulated Fe@CNT ultra-high absorptive shielding material for high temperature resistant EMI shielding. Ceramics International *45*: 17144–51.

218. Bystrzejewski, M., Huczko, A., Lange, H., et al. (2010). Dispersion and diameter separation of multi-wall carbon nanotubes in aqueous solutions. Journal of Colloid and Interface Science *345*: 138–42.

219. Mei, H., Zhao, X., Xia, J., et al. (2018). Compacting CNT sponge to achieve larger electromagnetic interference shielding performance. Materials & Design *144*: 323–30.

220. Huang, H.-D., Liu, C.-Y., Zhou, D., et al. (2015). Cellulose composite aerogel for highly efficient electromagnetic interference shielding. Journal of Materials Chemistry A *3*: 4983–91.

221. Hu, P., Lyu, J., Fu, C., et al. (2020). Multifunctional aramid nanofiber/carbon nanotube hybrid aerogel films. ACS Nano *14*: 688–97.

222. Xu, W., Wang, G.-S., and Yin, P.-G. (2018). Designed fabrication of reduced graphene oxides/Ni hybrids for effective electromagnetic absorption and shielding. Carbon *139*: 759–67.

223. Wang, Z., Shen, X., Akbari Garakani, M., et al. (2015). Graphene aerogel/epoxy composites with exceptional anisotropic structure and properties. ACS Applied Materials & Interfaces *7*: 5538–49.

224. Xu, Y., Sheng, K., Li, C., et al. (2010). Self-assembled graphene hydrogel via a one-step hydrothermal process. ACS Nano *4*: 4324–30.

225. Wu, Y., Yi, N., Huang, L., et al. (2015). Three-dimensionally bonded spongy graphene material with super compressive elasticity and near-zero Poisson's ratio. Nature Communications *6*: 6141.

226. Prasad, J., Singh, A.K., Haldar, K.K., et al. (2019). CoFe2O4 nanoparticles decorated MoS2-reduced graphene oxide nanocomposite for improved microwave absorption and shielding performance. RSC Advances *9*: 21881–92.

227. Sushmita, K., Menon, A.V., Sharma, S., et al. (2019). Mechanistic insight into the nature of dopants in graphene derivatives influencing electromagnetic interference shielding properties in hybrid polymer nanocomposites. The Journal of Physical Chemistry C 123: 2579–90.

228. Pu, L., Liu, Y., Li, L., et al. (2021). Polyimide nanofiber-reinforced Ti3C2Tx aerogel with "Lamella-Pillar" microporosity for high-performance piezoresistive strain sensing and electromagnetic wave absorption. ACS Applied Materials & Interfaces 13: 47134–46.

229. Lu, Z., Jia, F., Zhuo, L., et al. (2021). Micro-porous MXene/Aramid nanofibers hybrid aerogel with reversible compression and efficient EMI shielding performance. Composites Part B: Engineering 217: 108853.

230. Liang, L., Li, Q., Yan, X., et al. (2021). Multifunctional magnetic Ti3C2Tx MXene/graphene aerogel with superior electromagnetic wave absorption performance. ACS Nano 15: 6622–32.

231. Li, Y., Meng, F., Mei, Y., et al. (2020). Electrospun generation of Ti3C2Tx MXene@graphene oxide hybrid aerogel microspheres for tunable high-performance microwave absorption. Chemical Engineering Journal 391: 123512.

232. Li, X., Yin, X., Song, C., et al. (2018). Self-assembly Core–Shell graphene-bridged hollow MXenes spheres 3D foam with ultrahigh specific EM absorption performance. Advanced Functional Materials 28: 1803938.

233. Ma, W., Cai, W., Chen, W., et al. (2021). A novel structural design of shielding capsule to prepare high-performance and self-healing MXene-based sponge for ultra-efficient electromagnetic interference shielding. Chemical Engineering Journal 426: 130729.

234. Shui, W., Li, J., Wang, H., et al. (2020). Ti3C2Tx MXene sponge composite as broadband terahertz absorber. Advanced Optical Materials 8: 2001120.

235. Chen, Y., Zhang, L., Mei, C., et al. (2020). Wood-inspired anisotropic cellulose nanofibril composite sponges for multifunctional applications. ACS Applied Materials & Interfaces 12: 35513–22.

236. Liu, X., Chen, T., Liang, H., et al. (2019). Facile approach for a robust graphene/silver nanowires aerogel with high-performance electromagnetic interference shielding. RSC Advances 9: 27–33.

237. Zeng, Z., Li, W., Wu, N., et al. (2020). Polymer-assisted fabrication of silver nanowire cellular monoliths: toward hydrophobic and ultraflexible high-performance electromagnetic interference shielding materials. ACS Applied Materials & Interfaces 12: 38584–92.

238. Gupta, S., Sharma, S.K., Pradhan, D., et al. (2019). Ultra-light 3D reduced graphene oxide aerogels decorated with cobalt ferrite and zinc oxide perform excellent electromagnetic interference shielding effectiveness. Composites Part A: Applied Science and Manufacturing 123: 232–41.

239. Huangfu, Y., Liang, C., Han, Y., et al. (2019). Fabrication and investigation on the Fe3O4/thermally annealed graphene aerogel/epoxy electromagnetic interference shielding nanocomposites. Composites Science and Technology *169*: 70–75.

240. Chen, Y., Pötschke, P., Pionteck, J., et al. (2020). Multifunctional cellulose/rGO/Fe3O4 composite aerogels for electromagnetic interference shielding. ACS Applied Materials & Interfaces *12*: 22088–98.

241. Segmehl, J.S., Laromaine, A., Keplinger, T., et al. (2018). Magnetic wood by in situ synthesis of iron oxide nanoparticles via a microwave-assisted route. Journal of Materials Chemistry C *6*: 3395–402.

242. Sudagar, J., Lian, J., and Sha, W. (2013). Electroless nickel, alloy, composite and nano coatings – A critical review. Journal of Alloys and Compounds *571*: 183–204.

243. Wang, H., Yao, Q., Wang, C., et al. (2017). Hydrothermal synthesis of nanooctahedra MnFe2O4 onto the wood surface with soft magnetism, fire resistance and electromagnetic wave absorption. Nanomaterials *7*: 118.

244. Wang, S., Shi, J., Liu, C., et al. (2011). Fabrication of a superhydrophobic surface on a wood substrate. Applied Surface Science *257*: 9362–65.

245. Liu, H., Li, J., and Wang, L. (2010). Electroless nickel plating on APTHS modified wood veneer for EMI shielding. Applied Surface Science *257*: 1325–30.

246. Xing, Y., Xue, Y., Song, J., et al. (2018). Superhydrophobic coatings on wood substrate for self-cleaning and EMI shielding. Applied Surface Science *436*: 865–72.

247. Xia, C., Zhang, S., Ren, H., et al. (2015). Scalable fabrication of natural-fiber reinforced composites with electromagnetic interference shielding properties by incorporating powdered activated carbon. Materials *9*: 10.

248. Ding, Z., Shi, S.Q., Zhang, H., et al. (2015). Electromagnetic shielding properties of iron oxide impregnated kenaf bast fiberboard. Composites Part B: Engineering *78*: 266–71.

249. Dang, B., Chen, Y., Wang, H., et al. (2018). Preparation of high mechanical performance nano-Fe3O4/wood fiber binderless composite boards for electromagnetic absorption via a Facile and Green method. Nanomaterials *8*: 52.

250. Dang, B., Chen, Y., Shen, X., et al. (2017). Fabrication of a nano-ZnO/polyethylene/wood-fiber composite with enhanced microwave absorption and photocatalytic activity via a Facile Hot-Press method. Materials *10*: 1267.

251. Lou, Z., Zhang, Y., Zhou, M., et al. (2018). Synthesis of magnetic wood fiber board and corresponding multi-layer magnetic composite board, with electromagnetic wave absorbing properties. Nanomaterials *8*: 441.

252. Zhao, B., Fan, B., Xu, Y., et al. (2015). Preparation of Honeycomb SnO2 foams and configuration-dependent microwave absorption features. ACS Applied Materials & Interfaces *7*: 26217–25.

253. Xu, H., Yin, X., Zhu, M., et al. (2017). Carbon hollow microspheres with a designable mesoporous shell for high-performance electromagnetic wave absorption. ACS Applied Materials & Interfaces *9*: 6332–41.

254. Han, J., Li, Y., You, Y., et al. (2022). Autosomal dominant optic atrophy caused by six novel pathogenic OPA1 variants and genotype–phenotype correlation analysis. BMC Ophthalmology *22*: 322.

255. Thomas, B., Sain, M., and Oksman, K. (2022). Sustainable carbon derived from sulfur-free lignins for functional electrical and electrochemical devices. Nanomaterials *12*: 3630.

256. Zeng, Z., Zhang, Y., Ma, X.Y.D., et al. (2018). Biomass-based honeycomb-like architectures for preparation of robust carbon foams with high electromagnetic interference shielding performance. Carbon *140*: 227–36.

257. Li, K., Jin, S., Jiang, S., et al. (2022). Bioinspired mineral–organic strategy for fabricating a high-strength, antibacterial, flame-retardant soy protein bioplastic via internal boron–nitrogen coordination. Chemical Engineering Journal *428*: 132616.

258. Zhao, B., Bai, P., Wang, S., et al. (2021). High-performance Joule heating and electromagnetic shielding properties of anisotropic carbon scaffolds. ACS Applied Materials & Interfaces *13*: 29101–12.

259. Shen, Z., and Feng, J. (2019). Preparation of thermally conductive polymer composites with good electromagnetic interference shielding efficiency based on natural wood-derived carbon scaffolds. ACS Sustainable Chemistry & Engineering *7*: 6259–66.

260. Ma, X., Guo, H., Zhang, C., et al. (2022). ZIF-67/wood derived self-supported carbon composites for electromagnetic interference shielding and sound and heat insulation. Inorganic Chemistry Frontiers *9*: 6305–16.

261. Li, Y., Yan, S., Zhang, Z., et al. (2022). Wood-derived porous carbon/iron oxide nanoparticle composites for enhanced electromagnetic interference shielding. ACS Applied Nano Materials *5*: 8537–45.

262. Liu, X., Liu, H., Xu, H., et al. (2022). Natural wood templated hierarchically cellular NbC/Pyrolytic carbon foams as Stiff, lightweight and High-Performance electromagnetic shielding materials. Journal of Colloid and Interface Science *606*: 1543–53.

263. Zhao, B., Bai, P., Yuan, M., et al. (2022). Recyclable magnetic carbon foams possessing voltage-controllable electromagnetic shielding and oil/water separation. Carbon *197*: 570–78.

264. Yuan, Y., Sun, X., Yang, M., et al. (2017). Stiff, thermally stable and highly anisotropic wood-derived carbon composite monoliths for electromagnetic interference shielding. ACS Applied Materials & Interfaces 9: 21371–81.

265. Liu, S., Sheng, M., Wu, H., et al. (2022). Biological porous carbon encapsulated polyethylene glycol-based phase change composites for integrated electromagnetic interference shielding and thermal management capabilities. Journal of Materials Science & Technology 113: 147–57.

266. Jia, X., Shen, B., Zhang, L., et al. (2021). Construction of shape-memory carbon foam composites for adjustable EMI shielding under self-fixable mechanical deformation. Chemical Engineering Journal 405: 126927.

267. Wang, Z.-x., Han, X.-s., Zhou, Z.-j., et al. (2021). Lightweight and elastic wood-derived composites for pressure sensing and electromagnetic interference shielding. Composites Science and Technology 213: 108931.

268. Pei, X., Zhao, M., Li, R., et al. (2021). Porous network carbon nanotubes/chitosan 3D printed composites based on ball milling for electromagnetic shielding. Composites Part A: Applied Science and Manufacturing 145: 106363.

CHAPTER 7

Multi-functional electromagnetic interference (EMI) shielding materials

INTRODUCTION

In recent years, the increased use of electronic devices in both civil and military fields has led to a growing concern over electromagnetic interference (EMI). This interference not only hampers the normal functioning of electronic devices but also poses risks to human health. Consequently, there is an urgent need to develop advanced materials that can effectively absorb and shield electromagnetic (EM) waves[1]. Traditional materials like metals and ferrites have been widely used for EM wave absorption and shielding (EMAS), but they have significant drawbacks including high weight and poor chemical stability. As a result, alternative lightweight materials have been extensively studied for EMAS applications[2]. These materials include one-dimensional (1D) carbon nanotubes (CNTs)[3] and carbon nanofibers (CNFs)[4], silicon carbide nanowires (SiC NWs)[5], two-dimensional (2D) graphene (GE) and its derivatives such as graphene oxide (GO) and reduced graphene oxide (rGO)[6], as well as transition metal carbides/carbon nitrides (MXenes)[7] and conductive polymers[8].

DOI: 10.1201/9781003490098-7

The majority of the reported powder EMAS materials require mixing with suitable matrices, such as epoxy resin, silicone rubber or Al_2O_3, to form composite coatings. However, achieving homogeneous dispersion of these powders in the matrix is challenging and often results in the formation of discontinuous networks, leading to poor conductivity distribution in the composite[9]. Additionally, the tendency of carbon materials to aggregate due to van der Waals forces hinders their absorption, shielding, and mechanical properties[10]. Moreover, these composite layers display low cohesiveness, making them susceptible to failure in harsh environments. These complex preparation processes and the resulting composite's poor durability limit the application of powder EMAS materials.

To address these limitations, researchers have started developing bulk macrostructural EMAS materials, which refer to solid materials such as films, foams and aerogels. These materials can be directly used as standalone EM wave absorbers or shields without requiring additional steps. They also overcome the cohesion issues observed in conventional composite materials, potentially enhancing durability. Therefore, incorporating macrostructure design in EMAS materials shows promise in overcoming these limitations. Furthermore, there is a significant opportunity for the development of multifunctional EMAS materials. For instance, EMAS materials with excellent flexibility are suitable for wearable and portable devices[11]. Enhancing hydrophobicity can enable self-cleaning and corrosion resistance, making them ideal for humid environments[12]. Additionally, materials with high thermal insulation capability can be employed in high-temperature settings. These desirable properties highlight the significance of multifunctional EMAS materials in expanding their applications. While powder EMAS materials only possess EM wave radiation suppression capability, macrostructured EMAS materials offer the advantage of integrating mechanical properties and other functions. However, there is a limited number of reviews focusing on multifunctional EMAS materials with macrostructures. This work aims to showcase the potential of macrostructured EMAS materials, which can more effectively fulfill these functions.

7.1 THERMAL CONDUCTIVE (TC) EMI SHIELDING MATERIALS

With the advent of the 5G era, miniaturized and high-frequency electronic devices have become highly integrated, offering a wide range of functions enabled by advanced chips. However, the operation of these devices generates continuous heat due to their increased power density. When exposed to various EM radiations, conventional EMI shields accumulate excessive heat. This overheating affects the normal functioning of the devices, necessitating effective thermal management strategies to maintain performance and reliability. In recent years, there has been significant interest in enhancing thermal conductivity by introducing electrically conductive (EC) fillers alone or in combination with TC fillers to create hybrid materials[13]. This approach aims to establish a dense thermal conduction network. However, simply blending the fillers with the matrix can lead to a decline in mechanical properties while providing limited improvement in thermal conductivity[14]. Therefore, it is preferable to engineer hierarchical structures and utilize highly crystallized monolithic carbon films to prepare high-performance EMI shields with the desired thermal conductivity[15].

7.1.1 Materials with segregated structure

Polymer composites with a segregated structure can be prepared using different methods. One approach involves mixing conductive fillers with polymer emulsion and then film-casting the mixture, while another method involves hot-pressing polymer granules with a coated conductive layer[16]. These processes result in a random dispersion of functional fillers within the polymer matrix for homogeneous polymer composites. In contrast, the formation of a segregated structure occurs when the polymer granules push the conductive fillers toward the boundaries of polymeric microdomains, creating localized filler networks. This increases the probability of interlap between the fillers, facilitating the formation of perfect filler networks even with low filler loading[17]. For instance, Kim et al. prepared segregated EMI shields by assembling polymethyl methacrylate (PMMA) beads with a hybrid shell consisting of rGO and MXene[18]. The hot-pressing process significantly enhanced the probability of interlap between the functional fillers, resulting in a high thermal conductivity of 2.75 W m^{-1} K^{-1} and an EMI shielding effectiveness (EMI SE) of 51 dB. Another approach to achieve a TC functional layer is through an in-situ growth strategy. Ren et al. developed EMI shields composed of a poly(phenylene sulfide) (PPS) matrix with selectively distributed silver (Ag) networks using electroless plating and hot molding (Fig. 7.1a)[19]. The formation of a perfect three-dimensional (3D) Ag interconnected network created efficient paths for phonon and electron conduction, leading to excellent thermal conductivity of 1.15 W m^{-1} K^{-1} with a significantly reduced Ag loading of only 1.48 vol%. Zheng et al. achieved an even higher thermal conductivity of 5.22 W m^{-1} K^{-1} using a microwave-assisted sintering method to fabricate segregated polymer composites with an EMI SE of 60 dB (Fig. 7.1b)[20]. While the formation of segregated filler networks can enhance EMI SE and thermal conductivity at low filler loading, it often results in isolated polymer domains covered by the fillers. This compromises the integrity of the polymer matrix and leads to poor mechanical properties in the resulting composites. Thus, designing a segregated polymer matrix without sacrificing mechanical properties remains a critical challenge for large-scale applications.

7.1.2 Materials with multilayered structure

Candidate TC EMI shields with multi-layered structures have been studied to concentrate functional nanofillers in specific layers, effectively increasing the concentration of nanofillers[21]. This approach maintains the integrity of the substrates and allows the composites to retain good mechanical properties. It also improves the EMI SE by enabling multiple reflections of electromagnetic waves (EMWs) within the composites, leading to significant attenuation[22]. For instance, Hu et al. used commercially available filter paper as a substrate and fabricated a two-layer TC MXene/cellulose composite using a simple dip-coating method. The composite exhibited excellent EMI SE (> 43 dB) due to the 3D interconnected MXene network. Additionally, it achieved an in-plane thermal conductivity of 3.89 W m^{-1} K^{-1}. Another type of layered composite is the sandwiched composite, where each layer's components can be tailored accordingly[23]. A composite composed of a boron nitride (BN)-embedded polymer medium layer and multiwall carbon nanotubes (MWCNTs)-loaded polymer outer layers demonstrated high EMI SE (41.4 dB) and TC properties (0.76 W m^{-1} K^{-1})[24]. However, this dielectric-conductive-dielectric design has limitations. The conductive component in the outer layers enhances the reflection loss of EMWs rather than absorption, while the BN-filled medium layer fails to effectively collect heat for optimal thermal conduction.

To address these limitations, a reasonable layer component design focuses on achieving absorption-dominated EMI shielding behavior. In this design, EC fillers are concentrated in the middle layer, while magnetic materials are concentrated in the outer layers[25]. By incorporating multiple "absorption-reflection-reabsorption" processes of EMWs, the performance can be further enhanced with an increased dielectric-conductive-dielectric

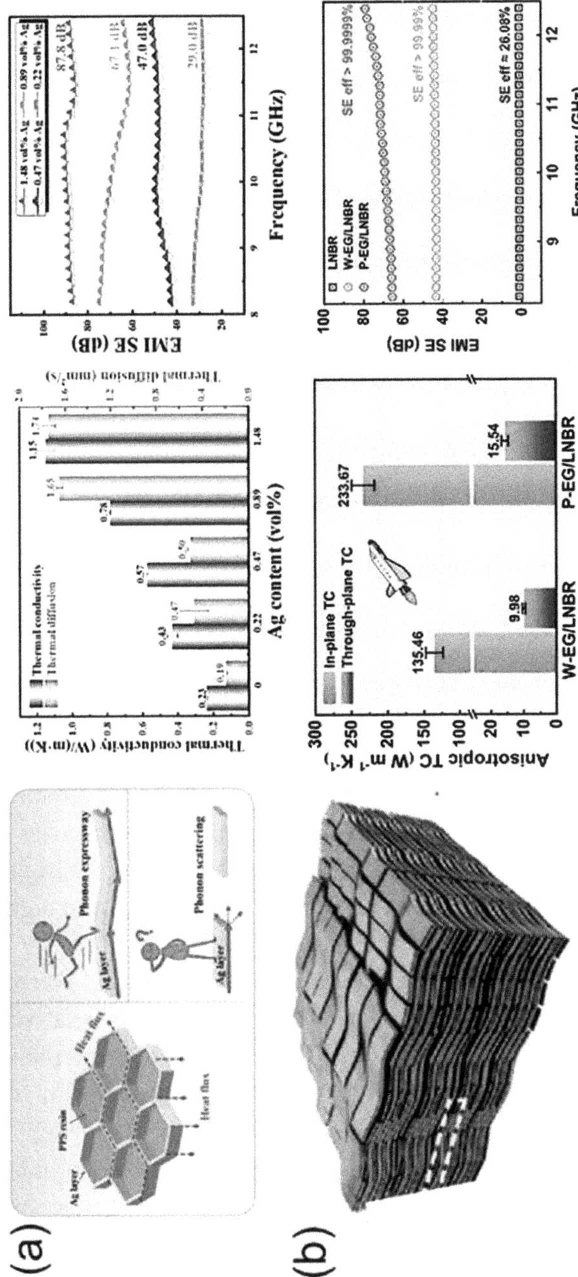

FIG. 7.1   Structures, thermal conductivities and EMI shielding properties of thermal conductive EMI shielding materials with (a) segregated structure[19]. (Reproduced with permission from Ref[19]. Copyright 2021, Elsevier.) (b) Multilayered structure[20]. (Reproduced with permission from Ref[20]. Copyright 2022, Wiley-VCH.)

alternative multi-layer structure. The introduction of multiple layers promotes interfacial polarization between the impedance-matching (dielectric) layer and the highly EC layer, ultimately leading to better EMI shielding.

7.1.3 Monolithic carbon-based films

The development of highly crystallized monolithic carbon films with excellent ECs and TCs has been considered a promising foundation for the next generation of flexible heat dissipation EMI shields[26]. Carbon nanomaterials, including expanded graphite (EG)[27] and its derivatives such as graphene nanosheets (GNS)[28], GO, and rGO[29], have been extensively researched for their exceptional thermal conductivity in EMI shielding applications. These materials possess a 2D planar structure, high aspect ratio, and conjugated π–π electronic conjugation, which contribute to superior electron and phonon transport. GE, for instance, is a 2D sheet material with a crystal structure that exhibits an electric transport speed of 1500 cm^2 V^{-1}, thermal conductivity exceeding 5000 W m^{-1} K^{-1}, and electrical conductivity of 6000 S cm^{-1}. Therefore, the production of monolithic films using 2D carbon-based nanomaterials is an effective approach to enhance thermal conductivity. In this context, Shen et al. achieved the fabrication of an ultrathin GE film (approximately 8.4 μm) through direct evaporation of GO suspension followed by graphitization at 2000 °C[30]. The resulting GE film demonstrated an impressive in-plane thermal conductivity of up to 1100 W m^{-1} K^{-1}, but exhibited a low EMI SE of 20 dB. To enhance the shielding capability, thick GE films were also produced. A GE paper consisting primarily of single or few-layer GE sheets exhibited a thermal conductivity as high as 1103 W m^{-1} K^{-1}, electrical conductivity up to 1.1×10^5 S m^{-1}, and significantly improved EMI SE of 77.2 dB[31].

The thermal conductivity of a GE film, consisting of 2D layered GE, is predominantly high in the in-plane direction, but remains relatively low in the out-of-plane direction. This is mainly due to the inherent planar structure of GE, which facilitates efficient phonon transfer within the plane, while hindering transfer through the plane due to scattering. To overcome this limitation, 1D carbon nanomaterials with exceptional thermal conductivity have been combined with GE to form hybrids. In a study, a thick film of GE/CNT hybrids was produced using a vacuum-assisted self-assembly process, followed by carbonization under hot-pressing and subsequent graphitization[32]. The resulting film exhibited a significantly enhanced out-of-plane thermal conductivity of up to 6.27 W m^{-1} K^{-1}, which was more than four times higher compared to a film without CNTs. Moreover, the hybrid film demonstrated stable EMI shielding performance across the Ku-band, achieving a maximum EMI SE of 75 dB.

7.2 ELECTROTHERMAL EMI SHIELDING MATERIALS

In recent years, the demand for heaters has significantly increased, driven by the advancements in wearable electronics and the growing need for practical medical care[33]. The integration of electro-heating (EH) devices with EMI shielding functionality opens up numerous possibilities such as human thermal physiotherapy, anti-fogging, anti-icing, and deicing for outdoor EMI shielding systems[34]. The principle behind this is that when an electric current flows through a conductive material, heat is generated as a result of the Joule effect, first demonstrated by James Prescott Joule[35]. It is understood that the heat generated per unit time, in a uniform conducting material with resistance R, is equal to I^{2}R, where I represents the current. Joule heating and heat loss eventually reach an equilibrium, affecting the power dissipation in the material and ultimately reaching a steady temperature. High-performance EMI shielding materials, which are good conductors with decent electrical conductivity, can also be utilized as EH devices[36].

When designing heating devices, it is crucial to consider the appropriate structural design that combines electrical and thermal performance with low power consumption, desired equilibrium temperature and mechanical properties. For practical applications like human thermal physiotherapy, the most suitable form is electrothermal fabric due to its easy integration with clothing and comfortable wearing experience[37]. In terms of structural design, mainstream EH-EMI shielding composites can be categorized into multi-layered structures, modified fabrics, and porous structures[38]. In this part, different structure designs are presented with the purpose of guiding the fabrication of EH EMI shielding materials.

7.2.1 Materials with multilayer structure

The multilayer structure is widely regarded as a practical and simple design that can be created using various methods including vacuum filtration, hot compression, printing, and coating. This structure overcomes the limitations of composites such as low electrical conductivity and single functionality. In a two-layer composite, the polymer layer maintains its mechanical properties while the conductive layer, with a high-density conductive network, provides high electrical conductivity. Thin polymer films like polyethylene terephthalate (PET), polycarbonate (PC), aramid nanofiber (ANF), and polyimide (PI) are used as supporting substrates for heaters.

One technique involves depositing MXene on a PC substrate using spraying coupled with a spinning technique. The resulting PC/MXene film demonstrates effective EMI shielding (>20 dB) and Joule heating performance (approximately 100 °C at 13 V). Another method is the production of double-layered conductive nanocomposites through vacuum filtration and hot-pressing[39]. These nanocomposite papers exhibit a superior EMI SE of 48.1 dB. Because of the highly conductive network, the surface temperature of the nanocomposite paper quickly exceeds 115 °C within 15 seconds at 2.5 V. Another important strategy is the use of a sandwich structure to achieve EH-EMI shielding composites. The advantage of this structure is that the conductive layer is protected from air and corrosive solvents that may otherwise degrade the conductive network and lead to failure. For example, MXene is deposited on a poly(vinylidene fluoride) (PVDF) supporter, followed by the addition of a protective PI layer to create a sandwiched PVDF/MXene/PI composite. This composite exhibits significant Joule heating effects (over 80 °C within 300 seconds at 6.5 V). Similarly, a sandwiched polytetrafluoroethylene (PTFE)/MXene/PI composite (44.0 dB) demonstrates EH conversion performance. Recently, a sandwich-structured poly(vinyl alcohol) (PVA)-based hydrogel was fabricated using electrospinning and vacuum-assisted filtration (Fig. 7.2a)[40]. These highly conductive hydrogels (1.66×10^4 S m^{-1}) display excellent EMI shielding performance (52 dB) with a low filler loading (0.23 vol%). In electro-thermal property tests, the hydrogel heats up from 20 °C to a steady-state temperature of approximately 55 °C in less than 10 seconds at a low driving voltage of 1.5 V.

Meanwhile, there have been studies focusing on the multi-layered structure of heaters. Utilizing multiple layers can effectively address the inherent shortcomings of a single layer. For instance, a four-layer protein-based integrated electronic skin has been developed to overcome the fragility and poor thermal stability of silk proteins. Additionally, multi-layered structures such as MXene/PLA[41] and natural rubber (NR)/CNTs/BN composites[42] have been reported, exhibiting both EMI shielding capabilities and heating properties. Zhou et al. have presented multilayered films created through

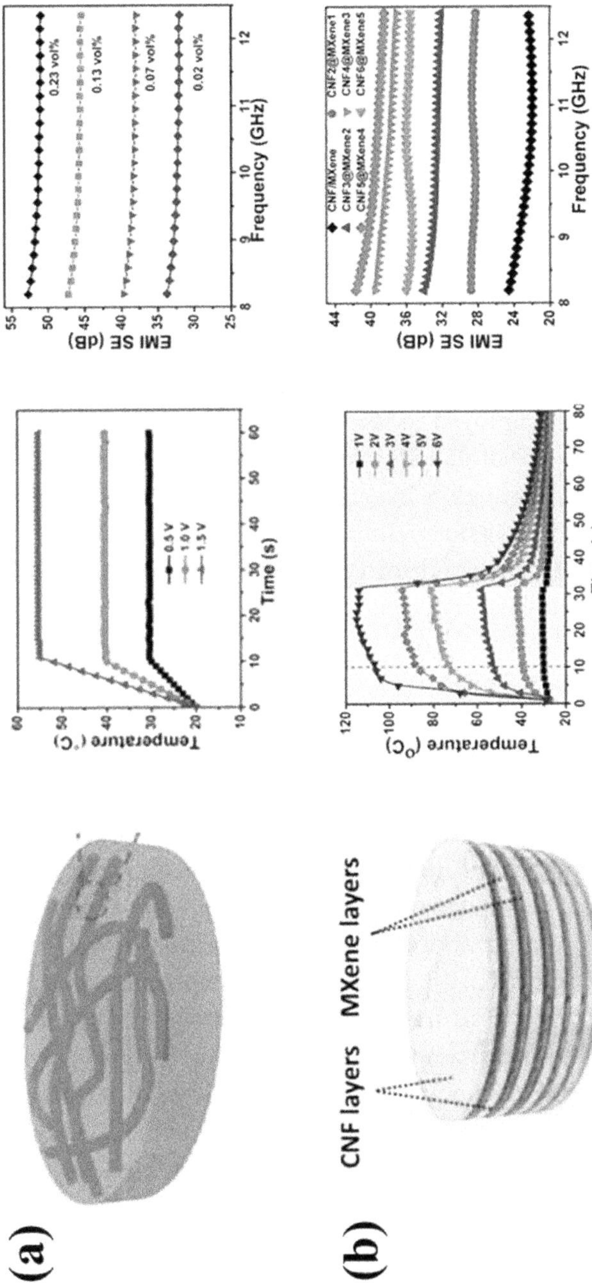

FIG. 7.2 Multilayered EMI shielding composites with electrothermal properties. (a) 3-layer PVA/Ag NWs composite with sandwiched structure. (Reproduced with permission from Ref.[40] Copyright 2021, Wiley-VCH.) (b) Alternating CNF and MXene multilayers. (Reproduced with permission from Ref.[43] Copyright 2020, American Chemical Society.)

the alternating deposition of cellulose nanofiber (CNF) layers and MXene layers using the vacuum filtration technique (Fig. 7.2b)[43]. The resulting CNF/MXene films demonstrate steady Joule heating performance, capable of reaching temperatures exceeding 100 °C within just 10 seconds with an input bias of only 6 V.

7.2.2 Conductive fabrics

Fabric, a flexible material formed by interlacing fibers into yarns through knitting or weaving processes, is an essential part of our daily lives. To impart electrical conductivity to fabrics, there are two approaches: (1) applying a conductive layer to pristine fibers before knitting them into fabrics, and (2) directly modifying the raw fabric with conductive coatings. To make pristine fibers conductive, hydrothermal and electroless plating technologies can be employed to form a conductive skin around the fiber core[44]. For example, Li et al. achieved in-situ growth of $NiCo_2O_4$ around carbon fibers (CFs) through a hydrothermal reaction, which was subsequently annealed to transform into NiCo alloy nanoparticles (NPs)[45]. Encapsulated with PI, the resulting composite film exhibits excellent EMI shielding (87 dB) and heating performance (approximately 100 °C at 5 V). Moreover, the EH element shows a uniform temperature distribution even under intense bending angles.

On the other hand, the second approach, involving the modification of raw fabrics, is more versatile and preferable for imparting conductivity, thanks to its ability to construct a well-connected conductive network and its scalability. Various techniques such as vacuum filtration[46], electroless deposition[47], and dip-coating[48] have been employed to create conductive fabrics. For instance, Jia et al. decorated a wood-pulp fabric grid with MXene coating networks using the vacuum filtration method, achieving an EMI SE of 90.2 dB. The fabric also demonstrates Joule heating performance, reaching a saturated temperature of 95 °C at 4 V within approximately 60 seconds. Another example involves poly(m-phenylene isophthalamide) (PMIA) nonwoven fabric coated with silver nanowires (Ag NWs)/PEDOT:PSS conductive coating, which exhibits a high EMI SE of up to 56.6 dB. The derived electrical heaters can reach a saturation temperature of around 110 °C at a low voltage of 2 V within 20 seconds[49].

Apart from their excellent electro-thermal conversion properties, EMI shields with EC Ag components also possess antibacterial properties. Since fabrics often come into contact with the skin during practical usage, it is important for the devices to possess antibacterial activity to prevent potential health concerns. Recently, an EMI shielding cellulosic film has been fabricated by coating cellulose scaffold (CS) fibers with conductive Ag NWs using hot-pressing and polydimethylsiloxane (PDMS) dipping processes. The composite gradually releases Ag+ ions from the Ag NWs, effectively preventing bacterial infections during long-term wearing[50]. Additionally, Zhang et al. proposed a polyvinylpyrrolidone (PVP)-assisted preparation method for highly EC and durable nanofiber composites containing Ag, which exhibit high EMI SE (approximately 96.9 dB). With favorable electrical conductivity reaching up to 245.7 S cm^{-1}, the composite demonstrates excellent breathability, antibacterial properties, and Joule heating performance, making it suitable for wearable devices or on-skin electronics[51].

7.2.3 Materials with porous structure

Compared to other structures, porous EMI shields provide improved SE due to the abundant interfaces between air and the composite material. Additionally, the low-density nature of porous materials results in a more prominent specific SE (SSE) when compared to solid materials. Porous polymers are known for their thermal insulation properties, as they have low heat diffusivity[52]. In terms of electrothermal behavior, a heating device with low thermal diffusivity helps to minimize heat loss during the heating process and requires lower electrical power input, resulting in higher saturation temperatures with reduced energy loss. Therefore, porous polymer composites show promise in electrothermal EMI shielding applications.

Various methods have been employed to create polymer composites with hierarchical porous structures such as freeze-drying, pore-foaming, and others[53]. For example, Papanastasiou et al. developed a multifunctional aerogel film composed of aligned nanofibers (ANFs) with uniformly distributed CNTs through freeze-drying[54]. The high electrical conductivity (230 S m^{-1}) of the composite enables excellent EMI shielding (54.4 dB) and low-voltage-driven Joule heating properties. The temperature of the composite increases linearly with a voltage increase of 2 V every 2 minutes, indicating a fast-heating rate and stable performance. Compared to directly foamed polymer composites, foaming the pristine polymer matrix into a porous insulated framework and subsequently treating it with a conductive coating can lead to better performance. For instance, composites prepared using this approach exhibit enhanced heating properties (~120 °C, 9 V) compared to polyethylenimine (PEI)/CNTs composites prepared by direct pore-foaming technique (53 °C, 10 V)[55]. Another method involves carbonizing porous polymers to produce conductive foam, which typically results in high conductivity, especially at elevated carbonization temperatures[56]. Jia et al. compared the performance of cotton-derived 3D carbon networks produced at different pyrolysis temperatures and found that higher pyrolysis temperatures (2000–2800 °C) led to significantly enhanced graphitization degree, resulting in higher conductivity and better EMI shielding and heating performances[57].

In summary, conductive polymer composites (CPCs) with different structural designs exhibit suitable EMI shielding properties and superior electrothermal functionality. These composites are particularly well-suited for use as resistance heating elements compared to traditional metal heating devices because their resistance is orders of magnitude lower than that of metals. Metal-based heating devices require more complex structures, often in the form of coiled wires, to increase resistance by prolonging electron transport distance and narrowing the cross-sectional area[58]. In contrast, CPCs can be used directly in a bulk form without the need for coiled structures due to their medium conductivity. To fully explore the potential of CPCs in electrothermal devices, it is important to consider their deformability and stretchability. Some studies have demonstrated encouraging results in achieving uniformly distributed temperature in deformed states[44]. However, the development of stretchable EMI shields with heating capabilities is still an ongoing challenge. The key issue lies in finding a balance between stretchability and a stable conductive network. Stretchable heaters with appropriate structures such as kirigami, wave patterns, and serpentine shapes are more suitable as they exhibit significantly reduced relative changes in resistance under strain. These materials may also have potential as effective EMI shielding materials but have not yet been fully evaluated[59].

7.3 SENSING EMI SHIELDING MATERIALS

Efforts have been made to develop multifunctional composites with excellent EMI shielding and sensing capabilities in order to ensure the normal operation of electronic instruments and enable the detection of deformation, temperature, chemical agents, and more. EMI shields, known for their conductivity, can also function as piezoresistive strain sensors when designed with robust mechanical deformation abilities[60]. The change in electrical resistivity of these materials upon mechanical deformation is attributed to variations in contact area or carrier transport barriers between adjacent conductive fillers[61]. These materials have widespread applications in electronic pressure measurement, personal health monitoring, human motion tracking, electronic skin, and other fields[62]. Different parameters have been proposed to evaluate the performance of these materials, with sensitivity being a crucial measure quantified by the gauge factor (GF). The GF has two definitions depending on the type of deformation. For sensors responding to tensile strain, $(R - R_0)/(R_0 \times \varepsilon)$ is used, where ε is the applied strain, R_0 is the initial resistance, and R is the resistance at ε strain. For pressure sensors, $(I - I_0)/(I_0 \times P)$ is used, where P is the applied stress, I_0 is the initial electrical current, and I is the current at stress P.

To overcome the limitations of conventional piezoresistive sensors, such as unstretchability, incompressibility and low GF, electroactive nanomaterials such as CNTs, rGO, and AgNWs have been introduced into the composites[63]. These nanocomposites exhibit superior piezosensitive performance due to their large surface area. The dominant sensing principle in strain sensors based on CPCs is the change in interfiller overlap and/or tunneling resistance. In the case of pressure-sensitive CPCs, introducing a porous structure enhances properties such as stretchability, elasticity, reversibility, and sensitivity to minimal strain or pressure. This section will discuss EMI shields with sensing functions, such as strain and pressure sensing, focusing on their sensing ability, working range, applications, and more. Additionally, some recently reported EMI shields with other types of sensing capabilities will be briefly introduced.

7.3.1 Strain sensing

In order to create stretchable materials, the use of stretchable substrates has become necessary. Materials such as NR, polyurethane (PU) and hydrogel have gained considerable attention in the design of strain-sensing devices. Additionally, combining conductive fillers and magnetic particles is a preferred method for achieving enhanced EMI shielding[64]. For instance, Wang et al. developed stretchable EMI shielding devices using a combination of magnetic iron oxide (Fe_3O_4), rGO, and NR with a segregated conductive network[65]. These composites demonstrated excellent EMI SE of 42.6 dB at 8.5 GHz, and a GF of 14 at 10% tensile strain. The devices also showed the ability to reliably capture resistance signals generated by various human motions, such as phonation, facial expressions, and finger and wrist movements, indicating great potential for wearable electronic devices. Another promising material is a multifunctional PVA-based hydrogel containing Fe_3O_4 clusters and poly(3,4-ethylenedioxythiophene): poly(styrene sulfonate) (PEDOT:PSS), which was produced using a freeze-casting method[66]. This hydrogel exhibits a high EMI SE value of 46 dB and possesses significant potential as a stretchable strain sensor. It demonstrates excellent overall performance, including high stretchability (904.5%), fast response time (9 ms), recovery time (12 ms), durability, and sensitivity (GF up to 7.2).

Porous conductive scaffolds have also emerged as a good option for strain-sensing EMI shields. The introduction of porous architectures provides several advantages, such as low density, enhanced multiple reflection loss, and better accommodation to tensile strain compared to solid, bulky counterparts. For example, a 3D CS impregnated with CNT/MXene hybrids demonstrates EMI shielding (29.3 dB) and strain-sensing properties. It achieves a high GF of 3.9 and exhibits good cyclic stability over 1000 cycles, making it suitable for real-time monitoring of various human motions in wearable strain sensors. Another notable development is a strain-sensing GE film (45 dB) with ultra-low detection ability, reported by Han et al. They fabricated honeycomb porous GE using a laser scribing technology (Figs. 7.3a–b)[67]. This cellular GE film exhibits high sensitivity to strain within a small range. For instance, at 1% tensile strain, the film reaches a value of $R/R_0 \sim 6$, corresponding to a GF of approximately 600. Therefore, it can be utilized to monitor weak physiological signals such as pulse, respiration and laryngeal movements of humans (Fig. 7.3c).

7.3.2 Pressure sensing

It has been widely demonstrated that incorporating rigid nanomaterials into a polymer matrix can increase the elastic modulus of the final composites. Consequently, composites with higher filler loading exhibit smaller strain variation under a given stress. However, directly compounding the polymer matrix with conductive fillers often leads to limited pressure-sensitive behavior. To tackle this issue, researchers have focused on designing conductive composites with a porous architecture.

Introducing a cellular structure in the composites offers several advantages. Firstly, it improves the glass transition temperature (GF) due to the reduced modulus. Secondly, it facilitates better EMI shielding because of the presence of abundant material-air interfaces. While solid

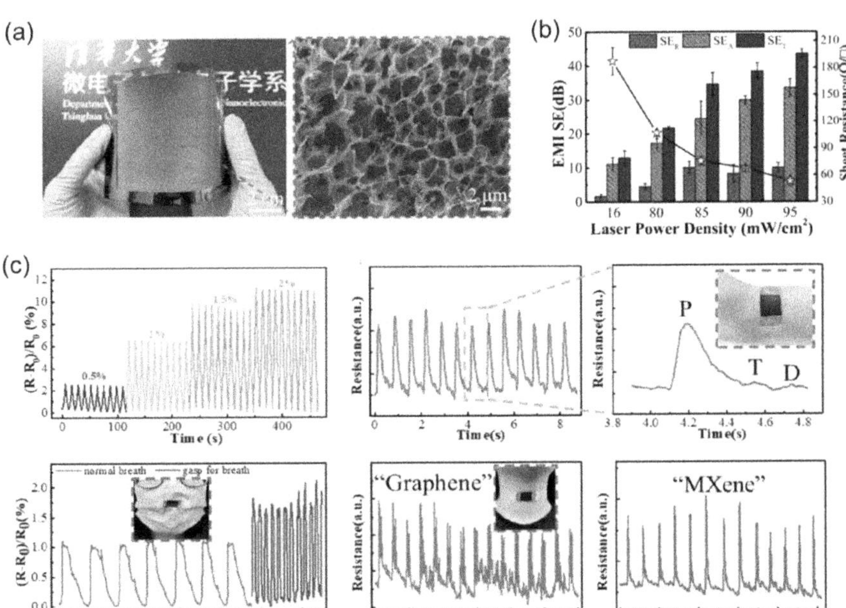

FIG. 7.3 Honeycomb porous graphene (HPG)-based EMI shield with strain sensing function. (a) Photograph and SEM image of the HPG surface morphology, and (b) EMI SE of the HPG at 10 GHz. (c) Strain sensing behaviors and multifunctional applications of the HPG sensor used to monitor pulse signal, respiration signal, and sounds. (Reproduced with permission from Ref.[67] Copyright 2021, American Chemical Society.)

composites with uniformly distributed conductive fillers have already been reported to possess EMI shielding and pressure sensing functions[68], recent developments in advanced composites with enhanced overall performance have consistently employed a porous structure design. Porous scaffolds such as PU sponge, cotton and fabric have been utilized and further treated with a conductive coating. Additionally, bottom-up pore-forming techniques, including the use of solid templates, emulsion templating, phase separation and supercritical carbon dioxide (scCO$_2$) techniques, have been developed to generate porous composites with desired cellular structures.

Currently, the most effective strategy to achieve a low percolation threshold in bulky composites is by forming a segregated structure[69]. Interestingly, the percolation threshold in porous composites can be exceptionally low by arranging conductive fillers in the wall-layer of the cellular structure. This arrangement provides a uniform and high-density conductive network surrounding a foam media[70]. For instance, a multifunctional composite composed of a carbon fiber felt (CFF) framework decorated with Ag NWs and thermoplastic polyurethane (TPU) demonstrates a remarkably high average EMI SE of up to 108.8 dB with only 7.85 wt% Ag NWs[21]. This is due to the constructed dense conductive network. When pressure is applied, the "contact effect" between adjacent Ag NWs layers forms a new conductive network, giving the composites their pressure-sensing characteristics. Similar procedures have yielded composites such as Ag NPs modified PU sponge, GE, Fe_3O_4 and MXene co-decorated hollow PDMS sponge, which exhibit good EMI shielding and pressure sensing behaviors. Recently, Ma et al. fabricated hierarchical structures of AgNWs decorated leather (leather/AgNWs) nanocomposites[71]. The micro- and nano-porous structures in the corium side of the leather, with penetrated Ag NWs, construct highly efficient conductive networks (Fig. 4a). This offers the composite an excellent EMI SE of approximately 55 dB. Upon pressing, the connection between Ag NWs improves, resulting in a tighter conductive network. This composite displays a three-stage piezoresistive behavior in the pressure range of 0–17.5 kPa (Fig. 4b) and exhibits a stable current response to fingertip press, joint bending and human speaking, among others (Figs. 4c–e).

Porous polymer composites can also be achieved through bottom-up pore-forming techniques. One well-established technique is $scCO_2$ foaming, which produces composite foam with tunable cellular structures. For example, Xu et al. used $scCO_2$ as a physical blowing agent to introduce a cellular microstructure into poly(ethylene-co-octene) (POE)/CNT composites[69]. By simply adjusting the expansion ratio, the resulting cellular POE/CNT foam demonstrates adjustable EMI shielding performance. The foam also shows stable resistance signals under multiple compression cycles. Another example is a CNT/PEI foam, which has an ultralow percolation threshold (0.029 vol%) and satisfactory EMI shielding performance (30.3 dB), as well as pressure sensing functionality. Chen et al. presented a bio-foam made from wheat flour (WF) and CNTs, forming an environmentally friendly WF/CNT composite foam (40.1 dB)[72]. This porous biomass foam possesses multifunctional sensing capabilities, including pressure sensing and acoustic sensing, and can be used to monitor the airflow frequency from the mouth, and capture, record and distinguish different signals of human motion.

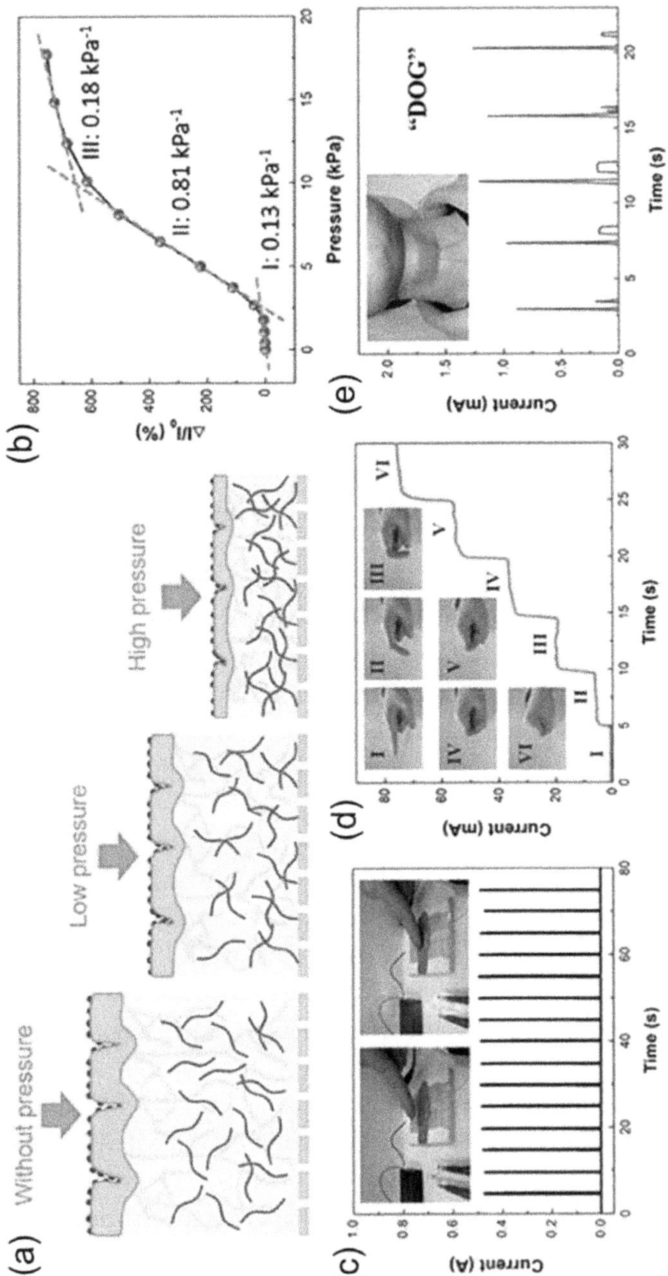

FIG. 7.4 (a) Schematic representation of the sensing mechanism of the leather/Ag NWs nanocomposite upon pressing and (b) relative current changes and corresponding GFs of the leather/Ag NWs sensor responds to various pressures. Current responses upon (c) the repeated pressure of fingertip, (d) finger bending with gradually increased angles and (e) repeatedly speaking "DOG". (Reproduced with permission from Ref.[71] Copyright 2022, Wiley-VCH.)

7.3.3 Other sensing functions

Composites with EMI shielding capabilities that respond to physical or chemical stimuli have been extensively studied. For instance, a wearable MXene/Ag NWs silk textile (MAF) has been developed, which demonstrates EMI blocking and moisture response properties[73]. As the moisture content increases, the resistance of the MAF silk increases consistently. The MAF silk also exhibits stable and reproducible resistance responses during cyclic testing. The moisture sensing ability of the MAF silk is attributed to the insertion of water molecules between MXene nanosheets, which alters the bandgap of MXene and leads to changes in electrical resistance upon water adsorption and desorption. EMI shielding textiles with acid/alkali-responsive characteristics have been achieved using polyanipne (PANI) and MXene decorated cotton fabrics (PMCFs)[74]. The conductivity of PANI is sensitive to the doping level, making the PMCF suitable for acid and alkali sensing. The PMCF shows a remarkable response to ammonia, as well as to HCl and volatile organic compounds (VOCs), indicating high selective sensitivities in complex gas environments. Composites with multiple sensing functions, including EMI shielding, have also been developed. For example, a PDMS/MWCNTs composite with thermo-expandable microspheres (EMs) exhibits temperature and pressure sensing capabilities in addition to enhanced EMI shielding[75]. MXene-based inks have been used to create composites with EMI shielding and multifunctional sensing abilities, responding to stress/strain, blowing, humidity and temperature.

Besides resistive-type sensing principles, such as those mentioned above, EMI shields based on self-powered devices[76], piezoelectric mechanosensing and wireless EM sensing have also emerged. While significant progress has been made in developing EMI shields with sensitive functions, there are still several aspects that require further consideration. For example, strain sensors often exhibit a negative piezoconductive effect, where electrical conductivity decreases under strain due to the deterioration of the conductive network. When designing strain-sensing EMI shields, finding a balance between EMI SE and mechanical durability is crucial. Additionally, when EMI shields are used as on-skin sensing devices, their conformability, biocompatibility and breathability need to be explored to ensure a comfortable wearing experience[77].

7.4 FLAME-RETARDANT EMI SHIELDING MATERIALS

Polymeric materials have certain disadvantages compared to metallic and inorganic materials, such as poor fire resistance, low thermal decomposition temperature and the release of flammable gases when burning[78]. Combustion of polymeric materials generates a significant amount of heat and hazardous substances, posing a serious threat to the environment, human health and property safety. With the increasing use of polymer composites in various applications, the occurrence of accidental fires caused by these materials has become a growing concern in society. Therefore, ensuring fire safety has become a crucial criterion for evaluating the flame-retardant ability of polymer-based EMI shields in both academic and industrial sectors.

When exposed to heat, polymeric materials undergo thermal decomposition, leading to the continuous generation of flammable gases. In the presence of heat and oxygen, these fuel vapors ignite on the surface of the polymeric materials, generating a large amount of heat. Some of this heat is conducted back to the polymer itself, further promoting the decomposition of the polymer matrix. Flame retardation can be achieved through gas phase or condensed phase mechanisms, or a combination of both. Various approaches have been developed to enhance the flame retardancy of polymers[79]. These approaches involve modifying the surface of the polymer or incorporating flame retardants (FRs) into the bulk polymer matrix. FRs can be directly mixed into the polymer matrix, applied as FR layers on the surface of the polymer, or chemically synthesized to form polymer materials with embedded flame-retardant monomeric units.

7.4.1 Materials with uniformly dispersed FRs

To enhance the flame retardancy of polymer composites, the physical blending of FRs with the polymer matrix is widely favored in industrial applications due to its simplicity. Over the past few decades, a variety of FRs have been synthesized and are commercially available, based on different mechanisms and elements for flame retardancy. The earliest flame-retardant materials were typically produced by directly compounding conductive nanofillers and FRs with the polymer matrix. For example, polystyrene (PS)-based composites containing intumescent flame retardant (IFR) and MWCNTs were fabricated through melt mixing. These composites achieved an EMI SE of 38.7 dB with the addition of 8 wt% MWCNTs and 10 wt% IFRs[80]. The synergistic effect of MWCNTs and IFRs led to a significantly improved limiting oxygen index (LOI), with a composite containing 1 wt% MWCNTs and 10 wt% IFRs exhibiting an LOI of approximately 30 vol%, meeting the requirements for flame safety.

Similar synergy effects were also observed in a TPU/CNTs/IFRs system, which achieved V-0 grade flame retardancy and an EMI shielding property of approximately 20 dB[81]. To maximize the flame-retardant effect, it is important to choose appropriate FRs[82]. For instance, a comparative study by Kim et al. found that aluminum diethylphosphinate (ADEP) outperformed bisphenol diphosphate (BDP) and resorcinol-di(bis-2,6-dimethylphenyl) phosphate (RDP) in terms of flame-retardant properties. ADEP releases diethylphosphinic acid during combustion and can produce aluminum phosphate char as a flame inhibitor by reacting with the PBT matrix. For a composite containing 8 vol% CF and 2 vol% MWCNTs, an EMI SE of 33.7 dB was observed.

However, directly blended bulk composites often exhibit low EMI SE and require high FR loading to achieve satisfactory flame retardancy, which can lead to a deterioration in mechanical properties. To overcome these limitations, FR EMI shielding materials based on hierarchical structures have been developed. For example, a flame-retardant epoxy resin back-infiltrated carbon foam exhibited both desirable flame retardancy (V-0 grade) and EMI shielding performance (33 dB)[83]. Similarly, CNF/ammonium polyphosphate (APP)/MXene porous aerogel (55 dB) and multilayered PBS/CNTs/IFRs composites (> 30 dB) demonstrated improved comprehensive performance at lower overall filler loading compared to homogeneous systems. Coextrusion technology was used to fabricate a TPU-based composite with a multilayered structure, resulting in an average EMI SE of 32.7 dB in the X band with less than 4 wt% CNTs. Additionally, this composite exhibited superior flame retardancy rated as V-0, along with low heat release and quick self-extinguishment, thanks to the production of a continuous carbonaceous structure among layers[84]. Interestingly, the authors also discovered that the EMI SE significantly improved from 32.7 dB to over 60 dB after combustion, attributed to the well-maintained multilayered architecture and the generation of atmospheric cavities.

7.4.2 Materials with flame-retardant coatings

Despite making significant progress, the addition of large amounts of FRs to bulk polymers has led to issues such as reduced mechanical properties and compatibility between the matrix and FRs. In light of this, there has been an increasing focus on surface treatment techniques to enable flame retardancy. This approach targets the interface between materials and combustion, where active flame-retardant ingredients cover the surface of a 3D skeleton to extinguish fire at an early stage. Surface treatment can therefore provide flame retardancy to the polymer matrix with minimal impact on bulk performance. Several decades ago, halogen-containing compounds like tris-(2,3-dibromopropyl) phosphonate, decabromodiphenyl ether and hexabromocyclododecane were extensively studied for high-efficiency flame-retardant coatings[85]. However, in recent years, researchers have shifted their focus towards the development of halogen-free flame-retardant systems due to the serious adverse health effects posed by halogen compounds. Nitrogen (N), phosphorus (P) and phosphorus-nitrogen (N-P) synergistic systems, such as phosphorus-containing polymer derivatives, hypophosphorous acid and cyclotriphosphazene derivatives, are now preferred for flame retardancy instead of halogen FRs.

Dip coating, a classic surface treatment method, has been widely used to impart flame retardancy to high-surface-area materials. For example, cotton textiles are commonly used in our daily lives, and ensuring their fire safety requires consideration of flame-retardant properties. A conductive fabric with an alternative MXene/PEI coating was fabricated using a dip-coating process. This multilayered coating with a nanoscale thickness (~500 nm) exhibited a 139% enhancement in EMI SE compared to the corresponding pure MXene coating. The fabric also demonstrated good flame retardancy resulting from a nitrogen-based flame-retardant mechanism[86]. Other examples include the fabrication of flame-retardant cotton fabric (33.0 dB) by dip-coating it alternatively with PEI, phytic acid (PA), and dispersion of AgNWs[87]. The synergistic effect of phosphorus and nitrogen endowed the cotton fabric with effective self-extinguishing properties and reduced peak heat release rate (PHRR) by approximately 58.6% compared to raw cotton fabric. An MXene-decorated flame-retardant cotton fabric was also reported, where alternative PEI/APP acted as a flame-retardant layer and MXene served as a shielding layer (31 dB)[88]. Another example involved P, N co-doped cellulose nanocrystals treated conductive cotton fabric (21 dB), where a 20-bilayer coating achieved a high LOI of 32 vol% and significantly reduced total smoke production (by 98.3%)[89].

7.4.3 Others

In addition to the strategies mentioned above, FRs can also be incorporated into segregated CPCs. This approach enhances the flame-retardant effect through the synergistic combination of inherent flame retardancy and the physical barrier provided by the FRs[90]. Another option is to use flame-retardant polymer matrices to create EMI shields with flame-retardant properties[91]. For example, a conductive leather solid waste (LSW)/PVA/PANI aerogel with EMI shielding capabilities (40 dB) also exhibits flame-retardant properties. The LSW component acts as an N-based FR[92].

Many CPCs demonstrate flame-retardant properties even without the inclusion of FRs[93]. Inorganic conductive nanomaterials can offer flame-retardant abilities to the resulting composites due to their inherent fire-resistant characteristics and can act as physical barriers to inhibit the decomposition of the polymer matrix during burning. For instance, Yang et al. fabricated EMI shields (47.8 dB) composed of NR and MXene and found that a higher fraction of MXene contributes to superior flame retardancy[94]. MXene provides flame retardancy to the matrix by acting as a barrier that prevents heat and oxygen from directly interacting with the polymer matrix. Similar results have been observed in Fu's research group, where ANF/MXene composite films with high specific tensile strength (143.1 MPa cm^3 g^{-1}) and EMI SE (> 40 dB) were prepared via a vacuum filtration process (Fig. 7.5)[95]. These films exhibited excellent flame retardancy and could not be ignited even when heated over a burning alcohol lamp for 60 seconds, thanks to the inherent components of the composites.

Applying the same principle, carbonaceous materials and carbon-containing hybrids can impart flame-retardant properties to the final articles[96]. Monolithic carbon composites can be used as flame-retardant EMI shields. Additionally, 3D GE/Ag NWs bi-continuous skeletons with a back-filled PDMS layer demonstrate both EMI shielding (34.1 dB) and flame-retardant properties[97]. The excellent thermostability of these composites is attributed to the intact GE/Ag NWs conductive network, which enhances the barrier property and prevents oxidation by outside oxygen, as well as the dense conductive path that acts as an efficient thermal conductor, dissipating heat and avoiding the formation of hot spots. Other examples include wood/polyethylene composites with segregated carbon black (CB) (20.2 dB), porous PU/GE (30.0 dB) and MXene/carbon foam, all of which exhibit both shielding capacity and flame retardancy.

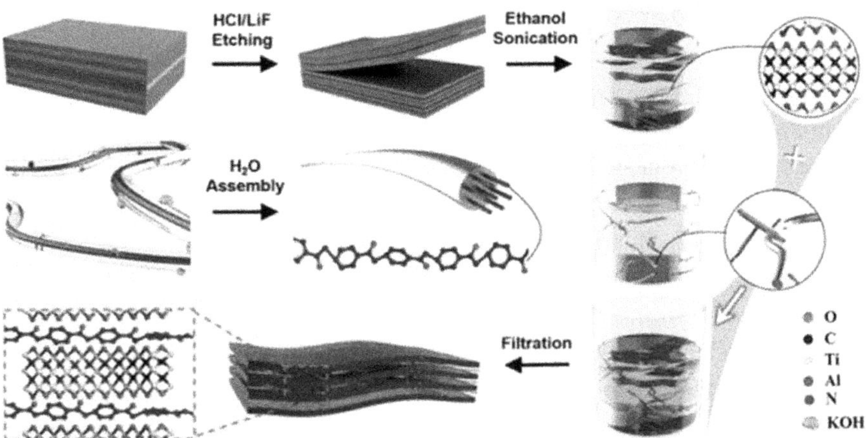

FIG. 7.5 Illustration of the preparation process of ANF/MXene composite film with inherent flame retardancy. (Reproduced with permission from Ref.[95] Copyright 2020, American Chemical Society.)

7.5 TRANSPARENT EMI SHIELDING MATERIALS

The development of transparent EMI shields is highly valuable for applications that require visual observation, such as observation windows, flexible displays, shielding cabinets and mobile communication devices. The primary requirement for these shields is to have sufficient transparency in the visible range while also achieving satisfactory EMI shielding properties. Additionally, it is important for the proposed EMI shield to maintain its shielding ability over long-term bending, torsion and folding, especially when used to cover dynamically changing surfaces.

Indium tin oxide (ITO), a well-established transparent conductor, has been widely used as a transparent EMI shield for a long time. Despite its favorable transmittance and EMI SE (80% @ 30 dB), ITO faces challenges in terms of indium scarcity and brittleness[98]. As a result, carbonaceous and metal nanomaterials have emerged as promising candidates for transparent EMI shielding and have gained increasing attention. The combination of transparent substrates with conductive nanomaterials enables the production of films with high transmittance that are suitable for EMI shielding devices. In this section, we will discuss the main conductive nanomaterials used to fabricate transparent EMI shields, considering their fabrication methods, transparency and EMI SE.

7.5.1 Materials with carbon materials

CNTs and GE are two representative carbon materials that possess both high optical transparency and electrical conductivity[99]. In addition, they exhibit mechanical flexibility, high-temperature resistance, low density and cost-effectiveness, making them highly promising for transparent EMI shields. There are two main methods for fabricating transparent conductors derived from CNTs and GE: dry process and wet process. The key difference between these approaches lies in whether they require solvent dispersion prior to film formation.

Various techniques have been developed to prepare carbon films with high transmittance. For example, Anlage et al. used transfer printing to fabricate single-walled carbon nanotube (SWCNT) films on PET substrate, which achieved an EMI SE of 28 dB at 10 GHz with 90% optical transmittance. Ha et al. employed a wet process, spraying a Ni-Pd-modified CNT solution onto sapphire wafers to create films. The transmittance of the resulting samples with thicknesses of 10 nm, 50 nm, and 100 nm was measured as 96.7%, 81.7% and 71.4%, respectively, corresponding to average EMI SE values of approximately 1.6 dB, 14.3 dB and 21.4 dB [100].

It is worth noting that transparent conductive films (TCFs) based on carbonaceous materials have been studied for decades[101]. These films serve as transparent electrodes and are essential components in various optoelectronic devices such as organic photovoltaic (OPV) cells, organic light-emitting diodes (OLEDs) and touch panels[102]. However, their potential for application in transparent EMI shields still needs to be further explored. Several reviews have summarized the fabrication techniques and applications of TCFs in flexible electronics[103], and we encourage readers to refer to these articles in order to better understand the development and potential applications of TCFs in transparent EMI shields.

GE, as the first 2D atomic crystal, is a remarkable nanomaterial with a light transmittance of up to 97%, indicating its almost complete transparency in the visible range[104]. However, the EMI SE of a monolayer GE sheet grown using chemical vapor deposition (CVD) methods is reported to be only 2.27 dB[105], which falls short of the requirements in most scenarios. Various efforts have been made to enhance the EMI SE of GE. One effective approach is to embed GE into a polymer matrix, but this often results in opaque or semi-transparent composites[106]. Another method involves constructing sandwich structures or using cascading GE films laminated on transparent substrates. For instance, Yang et al. demonstrated a flexible ultrathin large-area single-crystal graphene (SCG) grown on single crystal-line Cu substrates using CVD. The sandwich-structured film, with a MgF middle layer, exhibited an enhanced EMI SE of 8.1 dB and approximately 90% transmittance at 550 nm. To further optimize the SE value, a multi-layer structured film was constructed using pure GE sheets separated by transparent PET interlayers[107]. The PET/GE film, with a total GE thickness of only 4 nm, achieved an improved SE of 19.1 dB but compromised trans-mittance (80.5%). It is evident that the transmittance of most carbon-based materials in the visible range is lower than expected. Although interleaved and multi-layered structures can be used to improve EMI SE, they often result in significantly reduced transparency. It is clear that there is a trade-off between transmittance and EMI SE when using carbonaceous materials alone as conductive components. Currently, it remains challenging to fabricate efficient EMI shields with high transmittance based on carbon materials.

7.5.2 Materials with metals and their oxides

Compared to carbon materials, metals and their oxides are considered more suitable for transparent EMI shields due to their high inherent elec-trical conductivity. Metal materials have long been dominant in the EMI shielding market. However, bulky metal films that meet the required level of EMI SE often have limited transmittance. In light of this, researchers have started seeking transparent solutions. With advancements in nano-technology and advanced manufacturing techniques, conductive films incorporating thin metal oxides and metallic patterns have shown improved performance. So far, MXene nanosheets, metallic nanowires/nanoparticles and metal meshes have been explored for flexible and trans-parent EMI shields. Transparent films with laminated metallic coatings/patterns, such as metallic nanowires and metal meshes, have gained sig-nificant attention due to their excellent electrical conductivity and optical transparency. Transparent substrates with randomly distributed networks of nanowires have demonstrated some of the best transparent metal-based EMI shields. Examples include PU/Ag nanowires (31.3 dB @ 81%) on transparent substrates, sandwich-structured PET/Ag nanowires/PMMA (21.3 dB @ 95.6%), PDDA/Ag nanowires (31.3 dB @ 86.8%) and PI/Ag nanowires (55 dB @ 58%).

Furthermore, there are other methods for fabricating transparent EMI shields that offer a better balance between optical transmittance and EMI

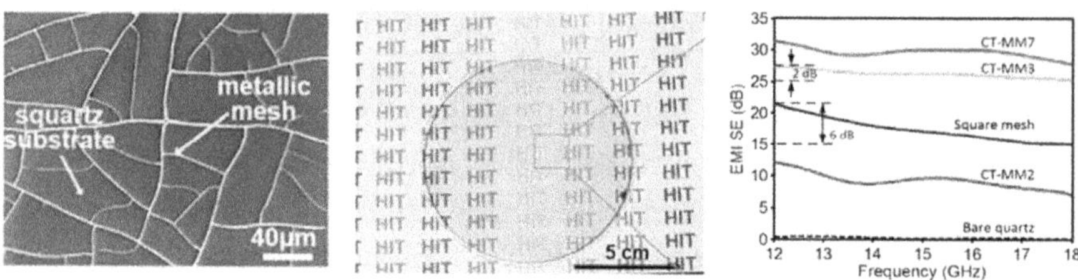

FIG. 7.6 Structures, photographs of transmittance demonstration and EMI shielding properties of transparent EMI shielding film derived from different nanomaterials. Crackle templated Ag mesh. (Reproduced with permission from Ref.[109] Copyright 2016, Springer Nature.)

shielding capability. One approach is the use of ring-shaped metal mesh, which achieves an EMI SE of 17 dB at 95% transmittance[108]. Another method involves a nature-inspired crackle-template-based metal mesh, which demonstrates an EMI SE of 26.0 dB at 91% transmittance. For example, a transparent EMI shielding film with a crack-template-based Ag mesh achieves a high EMI SE of approximately 26 dB and superior optical transmittance of approximately 91% (Fig. 7.6)[109]. However, these fabrication techniques involve CVD and/or ultraviolet photolithography, which increases cost and complexity.

Another strategy for fabricating transparent EMI shields involves depositing metal NPs on a polymer substrate, which offers a better compromise between transparency and EMI SE. An ultrathin (8 nm) and continuous doped Ag film, containing a small amount of copper (Cu), is obtained through co-deposition techniques[110]. The EMI shielding film is then constructed using a conductive-dielectric-metal-dielectric design. This film exhibits an average relative transmittance of 96.5% and excellent EMI SE of approximately 26 dB. An alternative approach is to sandwich an ultrathin Ag layer between layers of zinc oxide (ZnO) on a PET substrate[111]. This film demonstrates both high EMI SE (70 dB) and transmittance (approximately 90%), making it one of the best results reported so far.

7.5.3 Materials with nanohybrids

In addition, the use of hybrid materials has gained attention as they can offer comprehensive properties to the film. Yu et al. combined Ag NWs with MXene to fabricate a multiscale structured MXene-decorated Ag NWs film[112]. By wet welding adjacent nanowire junctions, the integrity and connection of the Ag NWs network were significantly improved, enhancing conductance and therefore EMI SE. The layered structure design resulted in a high EMI SE of 49.2 dB and a light transmittance of approximately 83%.

Another example is the transparent EMI shield composed of rGO conformably wrapped Ag NWs networks, which exhibits good EMI SE (35.5 dB) and transparency (91.1%)[113]. Ding et al. fabricated metallic-mesh/GE hybrids, an important building block of transparent EMI shielding materials. These hybrids achieved an EMI SE exceeding 47.8 dB in the Ku-band and exceeding 32.1 dB in the Ka-band, while maintaining a normalized visible transmittance of approximately 85% at 700 nm[114].

The development of flexible and transparent EMI shielding materials is crucial, and as technology advances, this need becomes even more urgent. Flexibility and transparency are largely determined by EMI SE, so compromises must be made during fabrication to achieve a satisfactory balance. The choice of suitable conductive materials is essential to achieve the desired EMI SE. Carbon-based materials, although they possess excellent chemical stability and mechanical properties, have limited application in transparent EMI shields due to their light absorption in the visible light range. On the other hand, Ag has shown promising results with high potential. However, the fabrication process of Ag-based materials often involves expensive costs and complicated procedures. Additionally, for EMI shielding research to be applicable on a large scale, the scalability of materials needs to be considered, while their overall performance in diverse environments requires further investigation. Flexibility and transparency are fundamental requirements that must be fulfilled.

7.6 HYDROPHOBIC EMI SHIELDING MATERIALS

When using CPCs as EMI shields, it is crucial to emphasize the stability of electrical conductivity in order to ensure the reliability of EMI shielding. These shields may be utilized in outdoor, underwater or contaminative environments, and under such conditions, the naked conductive nanomaterials used in the composites may undergo oxidation or corrosion[115], leading to a deterioration in the conductivity of the composites and consequently compromising their EMI shielding performance[116]. These concerns have prompted the demand for additional properties in the shield materials, including strong corrosion resistance, water repellency and even self-cleaning characteristics[117]. The high hygroscopicity of the materials may result in the absorption of water vapor from the environment, which can disrupt their properties. Therefore, it is beneficial to develop composites with a hydrophobic surface[118], as a hydrophobic surface can repel water droplets and prevent the penetration of aqueous solutions or moisture into the material[119]. The design of new multifunctional EMI shields with hydrophobic properties is thus of significant practical importance for their safety and long-term performance.

The hydrophobic/hydrophilic characteristics of materials are inherent features that depend on surface energy and surface roughness[120]. It is therefore possible to improve the water-resistance of materials by eliminating or reducing hydrophilic groups or enhancing surface roughness. The contact angle (CA), which represents the angle between the material surface and liquid droplets, is a critical parameter for evaluating hydrophobic ability. A material is considered hydrophobic when the static water contact angle (WCA) exceeds 90°. Furthermore, a material with a CA greater than 150° and a rolling/sliding angle less than 10° is commonly referred to as superhydrophobic. Over the past few decades, researchers have focused on developing hydrophobic EMI shields through various effective approaches, such as modifying the surface chemical composition from hydrophilic to hydrophobic, constructing micro/nano architectures and applying low-surface-energy coatings, among others.

7.6.1 Chemical groups enabled hydrophobicity

The introduction of low-surface-energy chemical groups has emerged as a viable strategy to confer hydrophobic properties to EMI shields. For example, MXene, which contains polar terminating groups like −OH and −O on its surface, is inherently hydrophilic. To modify its wet behavior, Liu et al. utilized a fluorocarbon agent called 1H, 1H, 2H, 2H-perfluorooctyltriethoxysilane (POTS) in conjunction with conductive silk textile containing Ag NWs and MXene (54 dB)[121]. By coating the composite with a POTS layer, they achieved a high static WCA of 145.8°. Interestingly, they observed that after five months of aging, the material exhibited self-derived hydrophobic behavior, attributed to the breakdown of unstable Ti–F bonds and the generation of hydrophobic C–F bonds. Another technique, proposed by Liu et al., involved the fabrication of a hydrophobic and flexible MXene foam through the assembly of MXene nanosheets into a film followed by a foaming process induced by hydrazine[122]. The procedure is illustrated in Fig. 7.7(a). During the foaming process, hydrazine reacted

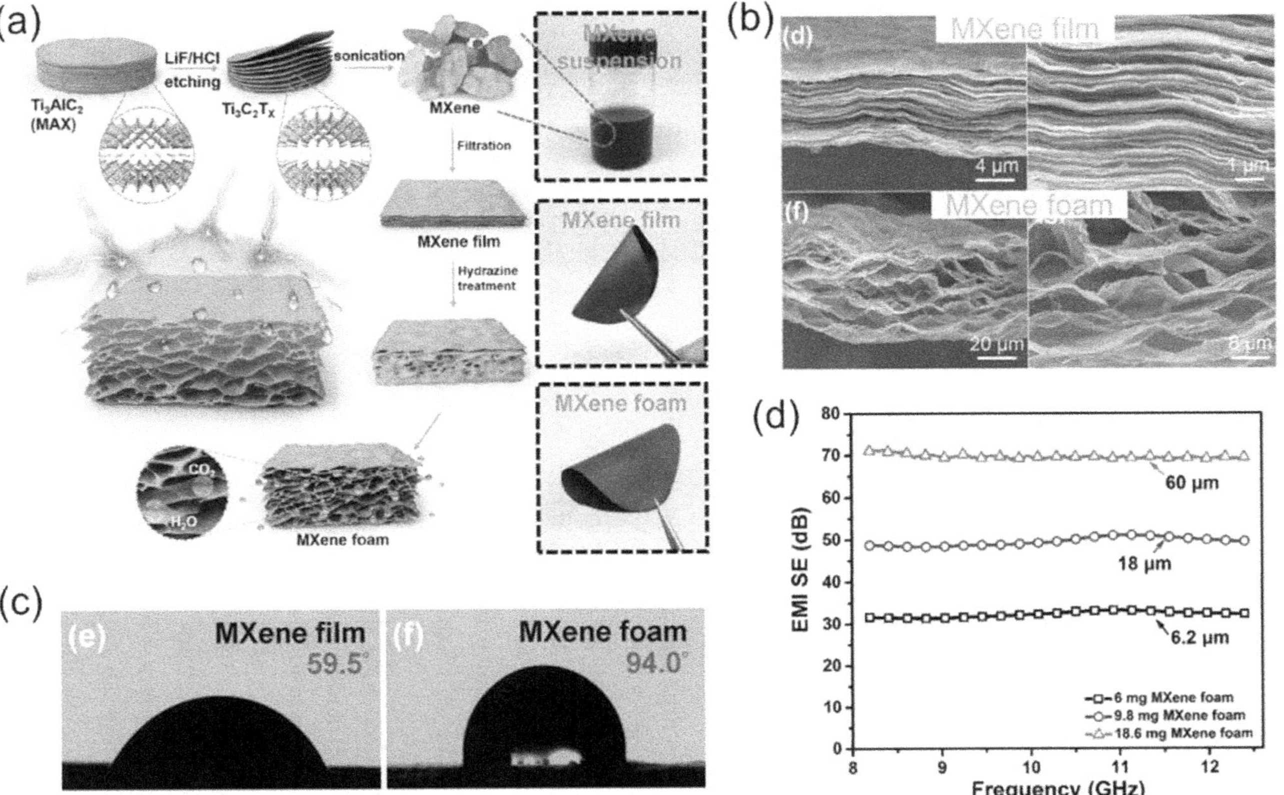

FIG. 7.7 (a) Schematic illustration of the fabrication procedures of hydrophobic MXene foam. (b) Cross-sectional SEM images of the MXene film and the MXene foam, and (c) the corresponding results of WCA test. (d) EMI SE of the MXene foam at different thicknesses. (Reproduced with permission from Ref.[122] Copyright 2017, Wiley-VCH.)

with the oxygen-containing groups of MXene, resulting in the production of H_2O and CO_2. This reaction led to the formation of cellular structures and a weakening of the hydrophilic nature of MXene, as depicted in Figs. 7.7(b)–(d).

7.6.2 Micro/nano-structure enabled hydrophobicity

Inspired by natural organisms such as lotus leaves, butterfly wings and water strider legs, researchers have explored the construction of rough micro- and nanostructures on surfaces to control wettability[123]. These surfaces with porous architectures can reduce the surface tension of liquids, preventing wetting by minimizing the contact area with falling liquids[120]. By introducing nanomaterials, the hydrophobic behavior of these porous architectures can be enhanced through the addition of nanoscale roughness, creating a micro-nano hybridized surface. For instance, Cheng et al. developed a supercritical-ethanol-dried graphene aerogel (GA-S) with SiC NWs grown on the outer surface[124]. The nanoscale roughness provided by each SiC NW increased the WCA from 53° to 134°. Similarly, a wood-derived porous composite demonstrated superhydrophobic behavior after the deposition of Ni NPs[125]. The high concentration of Ni NPs contributed to significant nanoscale surface roughness, while microscopic wrinkles in highly graphitized cellular structures added to increased microscale surface roughness. As a result, the composite exhibited hydrophobic characteristics with a WCA of 152.1° and excellent EMI shielding ability (50.8 dB). Hydrophobic porous composites are also found in Ni NPs/CNTs decorated carbonized wood (118.6°, 73.7 dB)[126] and GE-hybridized Cu NWs core-shell aerogel (138.8°, 52.5 dB), etc.

Film-based EMI shielding materials require more than just the introduction of nanomaterials to create a synergistic micro-nano structured surface. Therefore, alternative effective strategies have been proposed. For example, Park et al.[127] utilized an imprinting technique to create a surface hole pattern on a PDMS/MWCNTs nanocomposite film (80 dB). The dual-patterned composites displayed superhydrophobic properties with a high-WCA of 167° and a low sliding angle of 7.5°. This film can be applied in anti-icing and de-icing applications due to its ability to inhibit ice formation and its fast Joule heating effect. Another approach involved using a low-surface-energy material PVDF as the matrix and incorporating CNT and GE. Micro-nano structures were imparted by replica molding against a non-woven fabric, followed by a phase separation process (Fig. 7.8a)[116].

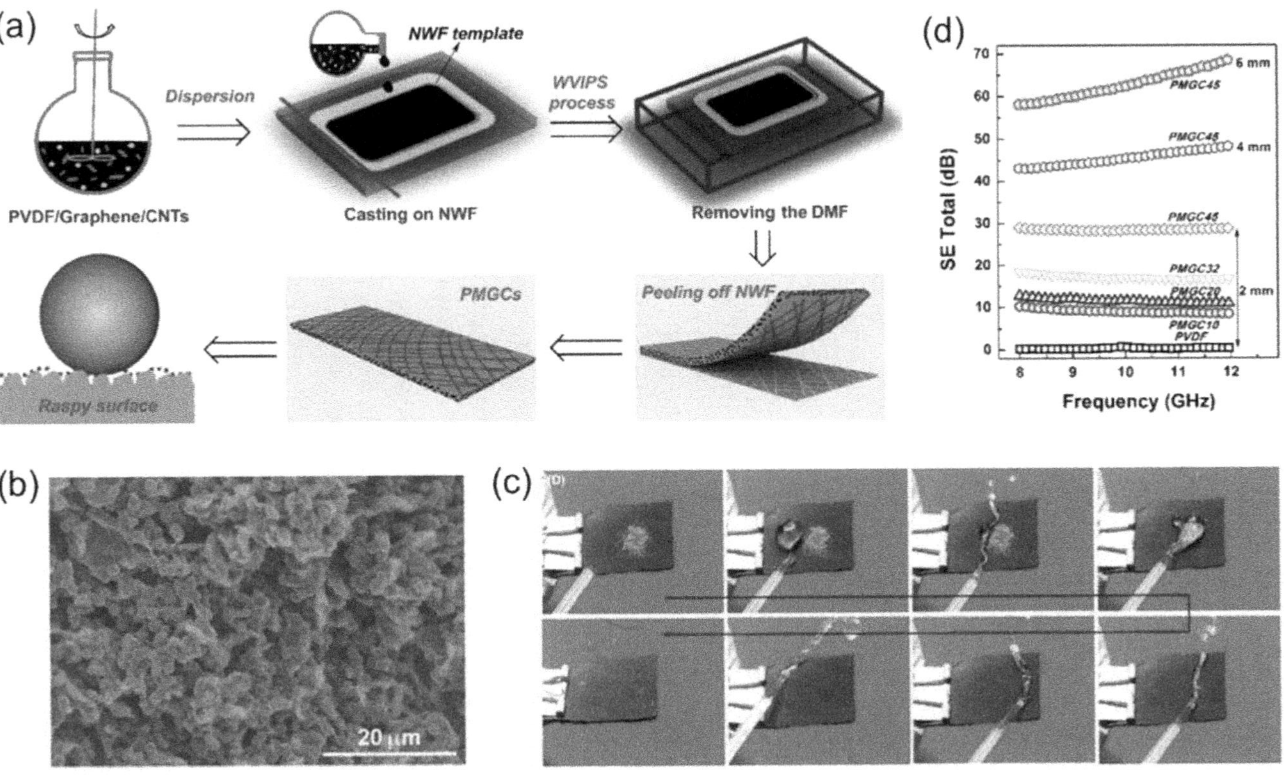

FIG. 7.8 (a) Schematic of the fabrication of the hydrophobic PVDF/GE/CNTs film with raspy surface. (b) SEM image of the imparted hierarchical surface and (c) photographs display the self-cleaning ability of the hydrophobic composite film. (d) Curves of SET for the composite film[116]. (Copyright 2018, Elsevier.)

The resulting composites, consisting of assembled spherulites and multi-scaled raspy surface morphology, exhibited a large WCA of 155.4 ± 2.7°, self-cleaning properties and high EMI SE (Figs. 7.8b–d).

7.6.3 Coating enabled hydrophobicity

Importing hydrophobic coatings onto surfaces is the most versatile strategy for enhancing hydrophobicity. Typically, these coatings contain low-surface-energy ingredients such as fluorine (F) or Si, along with active chemical groups that bond firmly to the target surface through self-polymerization[128]. PDMS, a Si-containing material with low surface energy, is widely used to impart hydrophobic properties to various EMI shielding materials[129].

For example, Yu's group synthesized a highly conductive textile decorated with polypyrrole (PPy)/MXene, which exhibited a high EMI SE of 90 dB. The textile was made waterproof by coating it with a PDMS layer, resulting in a WCA of 126°[130]. Wang et al. applied PDMS coating to a conductive porous CS filled with CNTs and MXene. The composite exhibited an EMI SE of 29.3 dB, along with an enhanced WCA of 135.7°. Lee et al. used triethoxy (4-chlorophenethyl) silane as a Si coating to control the hydrophobic properties of the surface. The resulting EMI shield (~40 dB) displayed a WCA of approximately 115.2°[131].

While Si-containing coatings can adjust the hydrophobic behavior of various surfaces to some extent, they have limited effectiveness. In contrast, F-containing coatings seem to offer better hydrophobic ability[132]. For instance, Gao et al. treated PP/Ag NPs fabric with a coating of 1H, 1H, 2H, 2H-perfluorodecanethiol (PFDT), resulting in a superhydrophobic fabric with a high WCA of 155°. This coating maintained its performance even when in direct contact with acid/alkali/salt aqueous solutions, making it suitable for harsh conditions[133]. Even better results were achieved with fluoroalkylsilane-treated conductive wood, which achieved an ultrahigh WCA of 160° and an extremely low roll-off angle of 3°. The superhydrophobic conductive wood also exhibited excellent self-cleaning performance and EMI shielding ability[134].

In addition to coatings containing Si and F, Teh et al. utilized electroplating deposition of cerium myristate on fabric surfaces. This resulted in a fabric with a high WCA of 156.4° and a remarkable EMI SE of 101.3 dB in the X band. This promising achievement offers advantages for extending the fabric's lifespan in wet or corrosive conditions[135]. Jia et al.

proposed a strategy combining surface structure and low-energy coating. They formulated a superhydrophobic layer using CNTs, PTFE NPs and fluoroacrylic polymer. The CNTs and PTFE NPs provided sufficient surface roughness, while the low surface energy was achieved through the use of fluoride. The resulting composite exhibited a high WCA of up to 160.8°, a low sliding angle of 2.9°, and a remarkable EMI shielding ability of 51.5 dB[136].

The hydrophobic EMI shielding materials mentioned above possess water-repellent properties, and some also demonstrate self-cleaning and/or anticorrosion characteristics, making them suitable for harsh environments. Despite the significant progress and efficient approaches developed to enhance water repellency, these methods have limitations in practical applications. Covalent bonding techniques that convert surface polar groups into benign groups through chemical reactions can reduce material surface energy. However, these methods are often complex, time-consuming and offer limited improvements in WCA, making it challenging to achieve superhydrophobicity. Composites with patterned micro- and nano-structures demonstrate enhanced water repellency, but they tend to have poor durability and are easily damaged under harsh conditions. The use of hydrophobic coatings appears to be a promising approach to modify materials with various architectures. However, F- or Si-based coatings have drawbacks such as organic pollution and high cost. Furthermore, the hydrophobic performance of these materials under dynamic deformation has not been comprehensively investigated. This limits their potential applications in situations where low deformation is required, as the applied coatings or patterned structures may be compromised, for example, during stretching.

7.7 SELF-HEALING EMI SHIELDING MATERIALS

Although considerable progress has been made in achieving superior EMI shielding performance, polymer-based EMI shields are susceptible to mechanical damage in practical applications. This includes issues such as cracking, scratching and mechanical cutting, which can limit their durability and cause overall malfunction of electronic equipment[137]. To address these challenges, self-healing EMI shields have been developed. These shields can restore both EMI shielding and mechanical performance after suffering mechanical damage, thereby extending their service lifespan and ensuring stable and durable performance even in harsh environments[138]. For instance, cracked GO/Ag NW composites exhibit a decrease in EMI SE from 72 dB to 56 dB. However, through self-healing, the EMI SE value can recover to 71 dB due to reversible hydrogen bonding under humid conditions[139]. Another example is the fabrication of a conductive poly-acrylamide/CNF/CNT hydrogel by Yang et al. This composite hydrogel can effectively restore its EMI SE after mechanical damage, achieving a high EMI SE retention of approximately 96%. However, the healing process takes a relatively long time (around a week)[140]. To shorten the healing time, a microwave-assisted near-instantaneous self-healing method has been proposed for PPy/polyoctylthiophene sulfonate (POTS) decorated fabric. This fabric demonstrates an EMI SE of 26.4 dB and negligible efficiency loss. It can achieve a healing efficiency of up to 99% after only 5 seconds of healing[141].

The key step in creating polymer composites with self-healing ability is the synthesis of a self-healing polymer matrix. Liu et al. achieved a self-healing conductive sponge by dip-coating it into a commercial self-healing PU solution[138]. The resulting composite exhibits a high EMI SE of 90.5 dB, and the shielding performance is retained at 59.2 dB even after two cutting/healing cycles. Lu et al. synthesized a self-healable waterborne polyurethane (ADWPU) matrix, which was blended with MXene to produce a range of EMI shielding materials with self-healing function[142]. The reversible dynamic polymer networks provided by the chain extender 2-aminophenyl disulfide contribute to the self-healing ability of these materials, allowing for healing at 60 °C for 5 minutes.

Li's research group conducted a series of studies on the development of self-healing polymer composites for EMI shielding. They utilized a dynamic crosslinking structure enabled by the reversible Diels-Alder (DA) mechanism. In one of their experiments, they synthesized a polymer matrix composed of dynamically crosslinked PU with DA bonds, and dispersed conductive CNTs homogeneously within the PU matrix. The self-healing properties of this PU-DA matrix allowed the EMI SE to recover from 16.8 dB in the damaged state to 29.8 dB after three cutting/healing cycles, demonstrating an outstanding shielding retention of 97.1%. Furthermore, Li's group also explored the use of segregated conductive networks in PU-DA/CNT composite materials[143]. After three repeated cutting/healing cycles, the EMI SE of these composites increased from 17.1 dB to 33.8 dB, with a retention rate of 96%. However, these segregated composites exhibited undesirable mechanical properties, including low failure stress (15.6 MPa) and low elongation at break (340%). To overcome these limitations, the researchers recently developed cationic waterborne polyurethane microspheres (CPA microspheres) that can electrostatically interact with anionic CNT/GO hybrids. By incorporating these CPA microspheres into the composite, they were able to construct a CNT/GO/CPA composite with a segregated conductive network. This improved the elongation at break (400%) and failure stress (>32 MPa), as well as enhanced the processing performance of the composites. The self-healing mechanism of these composites is illustrated schematically. Initially, during heat treatment at 130 °C for 5 minutes, the molecular chains of the CPA matrix temporarily split into lower molecular weight substances through DA reactions. This viscous movement allowed the broken sections to rejoin and heal. In the second thermal treatment, the DA bonds were reconstructed through further DA reactions. Ultimately, the cleaved molecular chains were recrosslinked, and the continuous conductive networks were restored. This resulted in a high retention rate of EMI SE (90%) even after three cutting/healing cycles.

CONCLUSION

After years of extensive research, there has been significant progress in enhancing the properties of single-functional EMI shielding materials and devices. However, the increasing demands in practical applications have surpassed the capabilities of materials solely designed to enhance SE value. This has led to the rapid development of multifunctional EMI shielding materials, as there are specific scenarios where EMI SE alone is not the sole acceptance criteria. Significantly, EMI shields with additional dual functionalities, such as thermal conduction and electrothermal capabilities, electrothermal and sensing features, transparency and hydrophobic properties, and more, have been developed. This approach allows for the fulfillment of performance requirements under various service conditions and further broadens the application scope of EMI shields. While several encouraging outcomes have been achieved, there are still forthcoming challenges that need to be addressed.

(1) TC EMI shields utilize a segregated conductive network to ensure isotropic thermal conductivity. However, they often exhibit weak mechanical properties and are prone to layer separation. Additionally,

monolithic carbon materials offer improved thermal conductivity but suffer from limitations such as low through-plane thermal conductivity. Therefore, there is a need for effective strategies to mitigate interface phonon scattering. Furthermore, a comprehensive understanding of the microstructure-property relationship is crucial for the development of more optimized structural designs.

(2) In order to adapt to various surfaces and environments, electrothermal devices have transitioned from being based on traditional metallic materials to using CPCs-based elements. Through well-designed structures and advanced manufacturing techniques, flexible EMI shields have been developed, demonstrating uniform temperature distribution even under twisting and bending conditions, and driven by safe voltages. However, a challenge still remains in maintaining stable temperatures after repeated heating cycles, requiring future research on materials with strain-resistant conductive networks.

(3) An important focus for sensitive EMI shielding materials is their potential use as sensors for recording human activity and changes in the surrounding environment (such as temperature and humidity). In such cases, the evaluation of sensing performance should consider factors beyond just the working range and GF. Compatibility between the sensors and the human body, as well as the accuracy and reliability of the signals, must also be taken into account.

(4) To achieve desired flame retardancy, a significant amount of FRs needs to be embedded in the polymer matrix. However, high filler content can make the resulting composites brittle and pose challenges in composite processing. While applying FR coatings can partially or completely address these issues, it introduces new problems such as complex procedures and poor water resistance, which hinder practical applications. Therefore, further research is necessary to enhance the overall performance in flame retardancy.

(5) Transparent EMI shielding films usually achieve a significant portion of EMI attenuation through reflection loss. Unfortunately, this also leads to the generation of secondary EMW pollution. Therefore, there is a need for further attention in developing transparent EMI shielding films that exhibit absorption-dominated shielding behavior.

(6) Several researchers have successfully developed materials with both excellent shielding characteristics and hydrophobic properties. Achieving superhydrophobicity requires the incorporation of surface micro- or nano-structures and/or hydrophobic coatings. However, it should be noted that surface micro- or nano-structures may suffer from durability issues and can be easily damaged in harsh environments. Additionally, organic hydrophobic coatings typically contain Si or F ingredients with low surface energy, which may introduce organic pollution. Further research is necessary to address these challenges in this field.

REFERENCES

1. Zhao, B., Shao, G., Fan, B., et al. (2014). ZnS nanowall coated Ni composites: facile preparation and enhanced electromagnetic wave absorption. RSC Advance 4: 61219–25.
2. Zhou, M., Zhang, X., Wei, J., et al. (2010). Morphology-controlled synthesis and novel microwave absorption properties of hollow Urchinlike α-MnO2 nanostructures. The Journal of Physical Chemistry C 115: 1398–402.
3. Hu, J., Zhao, T., Peng, X., et al. (2018). Growth of coiled amorphous carbon nanotube array forest and its electromagnetic wave absorbing properties. Composites Part B: Engineering 134: 91–97.
4. Chen, J., Wu, J., Ge, H., et al. (2016). Reduced graphene oxide deposited carbon fiber reinforced polymer composites for electromagnetic interference shielding. Composites Part A: Applied Science and Manufacturing 82: 141–50.
5. Cheng, Y., Hu, P., Zhou, S., et al. (2018). Achieving tunability of effective electromagnetic wave absorption between the whole X-band and Ku-band via adjusting PPy loading in SiC nanowires/graphene hybrid foam. Carbon 132: 430–43.
6. Han, Y., Liu, Y., Han, L., et al. (2017). High-performance hierarchical graphene/metal-mesh film for optically transparent electromagnetic interference shielding. Carbon 115: 34–42.
7. Yun, T., Kim, H., Iqbal, A., et al. (2020). Electromagnetic shielding of monolayer MXene assemblies. Advanced Materials 32: 1906769.
8. Zhang, Z., Tan, J., Gu, W., et al. (2020). Cellulose-chitosan framework/polyailine hybrid aerogel toward thermal insulation and microwave absorbing application. Chemical Engineering Journal 395: 125190.
9. Xu, H., Wu, J., Luo, W., et al. (2020). Dendritic cell-inspired designed architectures toward highly efficient electrocatalysts for nitrate reduction reaction. Small 16: 2001775.

10. Li, H., Yuan, D., Li, P., et al. (2019). High conductive and mechanical robust carbon nanotubes/waterborne polyurethane composite films for efficient electromagnetic interference shielding. Composites Part A: Applied Science and Manufacturing *121*: 411–17.

11. Jia, L.-C., Li, M.-Z., Yan, D.-X., et al. (2017). A strong and tough polymer–carbon nanotube film for flexible and efficient electromagnetic interference shielding. Journal of Materials Chemistry C *5*: 8944–51.

12. Li, T.-T., Wang, Y., Peng, H.-K., et al. (2020). Lightweight, flexible and superhydrophobic composite nanofiber films inspired by nacre for highly electromagnetic interference shielding. Composites Part A: Applied Science and Manufacturing *128*: 105685.

13. Wang, Y., Xin, Z., Shen, J., et al. (2022). Acrylonitrile-butadiene-styrene-based composites derived from "fish-net"-inspired Pickering emulsion for high-performance electromagnetic interference shielding and thermal management. Composites Communications *30*: 101085.

14. Zhao, B., Wang, S., Zhao, C., et al. (2018). Synergism between carbon materials and Ni chains in flexible poly(vinylidene fluoride) composite films with high heat dissipation to improve electromagnetic shielding properties. Carbon *127*: 469–78.

15. Zhou, B., Li, Q., Xu, P., et al. (2021). An asymmetric sandwich structural cellulose-based film with self-supported MXene and AgNW layers for flexible electromagnetic interference shielding and thermal management. Nanoscale *13*: 2378–88.

16. Pang, H., Xu, L., Yan, D.-X., et al. (2014). Conductive polymer composites with segregated structures. Progress in Polymer Science *39*: 1908–33.

17. Iqbal, A., Sambyal, P., and Koo, C.M. (2020). 2D MXenes for electromagnetic shielding: a review. Advanced Functional Materials *30*: 2000883.

18. Vu, M.C., Mani, D., Kim, J.-B., et al. (2021). Hybrid shell of MXene and reduced graphene oxide assembled on PMMA bead core towards tunable thermoconductive and EMI shielding nanocomposites. Composites Part A: Applied Science and Manufacturing *149*: 106574.

19. Ren, W., Yang, Y., Yang, J., et al. (2021). Multifunctional and corrosion resistant poly(phenylene sulfide)/Ag composites for electromagnetic interference shielding. Chemical Engineering Journal *415*: 129052.

20. Xie, Y., Li, P., Tang, J., et al. (2021). Highly thermally conductive and superior electromagnetic interference shielding composites via in situ microwave-assisted reduction/exfoliation of expandable graphite. Composites Part A: Applied Science and Manufacturing *149*: 106517.

21. Guo, Z., Ren, P., Zhang, Z., et al. (2021). Simultaneous realization of highly efficient electromagnetic interference shielding and human motion

detection in carbon fiber felt decorated with silver nanowires and thermoplastic polyurethane. Journal of Materials Chemistry C *9*: 6894–903.

22. Kamkar, M., Ghaffarkhah, A., Hosseini, E., et al. (2021). Multilayer polymeric nanocomposites for electromagnetic interference shielding: fabrication, mechanisms, and prospects. New Journal of Chemistry *45*: 21488–507.

23. Guo, Y., Qiu, H., Ruan, K., et al. (2022). Flexible and insulating silicone rubber composites with sandwich structure for thermal management and electromagnetic interference shielding. Composites Science and Technology *219*: 109253.

24. Wang, G., Liao, X., Zou, F., et al. (2021). Flexible TPU/MWCNTs/BN composites for frequency-selective electromagnetic shielding and enhanced thermal conductivity. Composites Communications *28*: 100953.

25. Zhang, Y., Ruan, K., and Gu, J. (2021). Flexible sandwich-structured electromagnetic interference shielding nanocomposite films with excellent thermal conductivities. Small *17*: e2101951.

26. Lewis, J.S., Perrier, T., Barani, Z., et al. (2021). Thermal interface materials with graphene fillers: review of the state of the art and outlook for future applications. Nanotechnology *32*: 142003.

27. Liu, Y., Qu, B., Wu, X., et al. (2019). Utilizing ammonium persulfate assisted expansion to fabricate flexible expanded graphite films with excellent thermal conductivity by introducing wrinkles. Carbon *153*: 565–74.

28. Li, L., Ma, Z., Xu, P., et al. (2020). Flexible and alternant-layered cellulose nanofiber/graphene film with superior thermal conductivity and efficient electromagnetic interference shielding. Composites Part A: Applied Science and Manufacturing *139*: 106134.

29. Liu, Y., Lu, M., Wu, K., et al. (2019). Anisotropic thermal conductivity and electromagnetic interference shielding of epoxy nanocomposites based on magnetic driving reduced graphene oxide@Fe3O4. Composites Science and Technology *174*: 1–10.

30. Shen, B., Zhai, W., and Zheng, W. (2014). Ultrathin flexible graphene film: an excellent thermal conducting material with efficient EMI shielding. Advanced Functional Materials *24*: 4542–48.

31. Xu, F., Chen, R., Lin, Z., et al. (2018). Variable densification of reduced graphene oxide foam into multifunctional high-performance graphene paper. Journal of Materials Chemistry C *6*: 12321–28.

32. Jia, H., Kong, Q.-Q., Yang, X., et al. (2021). Dual-functional graphene/carbon nanotubes thick film: bidirectional thermal dissipation and electromagnetic shielding. Carbon *171*: 329–40.

33. Lee, J., Sul, H., Lee, W., et al. (2020). Stretchable skin-like cooling/heating device for reconstruction of artificial thermal sensation in virtual reality. Advanced Functional Materials *30*: 1909171.

34. Zhou, Z., Song, Q., Huang, B., et al. (2021). Facile fabrication of densely packed Ti(3)C(2) MXene/nanocellulose composite films for enhancing electromagnetic interference shielding and electro-/photothermal performance. ACS Nano *15*: 12405–17.

35. Papanastasiou, D.T., Schultheiss, A., Muñoz-Rojas, D., et al. (2020). Transparent heaters: a review. Advanced Functional Materials *30*(21): 1910225.

36. Han, Y., Ruan, K., and Gu, J. (2022). Janus (BNNS/ANF)-(AgNWs/ANF) thermal conductivity composite films with superior electromagnetic interference shielding and Joule heating performances. Nano Research *15*: 4747–55.

37. Kim, T., Park, C., Samuel, E.P., et al. (2021). Supersonically sprayed washable, wearable, stretchable, hydrophobic, and antibacterial rGO/AgNW fabric for multifunctional sensors and supercapacitors. ACS Applied Materials & Interfaces *13*: 10013–25.

38. Lu, H., Xia, Z., Mi, Q., et al. (2022). Cellulose-based conductive films with superior joule heating performance, electromagnetic shielding efficiency, and high stability by in situ welding to construct a segregated MWCNT conductive network. Industrial & Engineering Chemistry Research *61*: 1773–85.

39. Ma, Z., Kang, S., Ma, J., et al. (2020). Ultraflexible mechanically strong double-layered aramid nanofiber and–Ti3C2Tx MXene/silver nanowire nanocomposite papers for high-performance electromagnetic interference shielding. ACS Nano *14*: 8368–82.

40. Zhou, Q., Lyu, J., Wang, G., et al. (2021). Mechanically strong and multifunctional hybrid hydrogels with ultrahigh electrical conductivity. Advanced Functional Materials *31*: 2104536.

41. Du, Z., Chen, K., Zhang, Y., et al. (2021). Engineering multilayered MXene/electrospun poly(lactic acid) membrane with increscent electromagnetic interference (EMI) shielding for integrated Joule heating and energy generating. Composites Communications *26*: 100770.

42. Zhan, Y., Lago, E., Santillo, C., et al. (2020). An anisotropic layer-by-layer carbon nanotube/boron nitride/rubber composite and its application in electromagnetic shielding. Nanoscale *12*: 7782–91.

43. Zhou, B., Zhang, Z., Li, Y., et al. (2020). Flexible, robust, and multifunctional electromagnetic interference shielding film with alternating cellulose nanofiber and MXene layers. ACS Applied Material Interfaces *12*: 4895–905.

44. Ying, J., Tan, X., Lv, L., et al. (2021). Tailoring highly ordered graphene framework in Epoxy for high-performance polymer-based heat dissipation plates. ACS Nano *15*: 12922–34.

45. Li, J., Zhang, X., Ding, Y., et al. (2022). Multifunctional carbon fiber@ NiCo/polyimide films with outstanding electromagnetic interference shielding performance. Chemical Engineering Journal *427*: 131937.

46. Jia, X., Shen, B., Zhang, L., et al. (2020). Waterproof MXene-decorated wood-pulp fabrics for high-efficiency electromagnetic interference shielding and Joule heating. Composites Part B: Engineering *198*: 108250.

47. Bai, Y., Qin, F., and Lu, Y. (2020). Multifunctional electromagnetic interference shielding ternary alloy (Ni–W–P) decorated fabric with wide-operating-range Joule heating performances. ACS Applied Materials & Interfaces *12*: 48016–26.

48. Li, J., Li, Y., Yang, L., et al. (2022). Ti3C2Tx/PANI/liquid metal composite microspheres with 3D nanoflower structure: preparation, characterization, and applications in EMI shielding. Advanced Materials Interfaces *9*: 2102266.

49. Liang, Y., Tong, Y., Tao, Z., et al. (2021). Multifunctional shape-stabilized phase change materials with enhanced thermal conductivity and electromagnetic interference shielding effectiveness for electronic devices. Macromolecular Materials and Engineering *306*: 2100055.

50. Zhu, M., Yan, X., Lei, Y., et al. (2022). An ultrastrong and antibacterial silver nanowire/aligned cellulose scaffold composite film for electromagnetic interference shielding. ACS Applied Materials & Interfaces *14*: 14520–31.

51. Zhang, S., Huang, X., Xiao, W., et al. (2021). Polyvinylpyrrolidone assisted preparation of highly conductive, antioxidation, and durable nanofiber composite with an extremely high electromagnetic interference shielding effectiveness. ACS Applied Materials & Interfaces *13*: 21865–75.

52. Xiang, Z., Zhu, X., Dong, Y., et al. (2021). Enhanced electromagnetic wave absorption of magnetic Co nanoparticles/CNTs/EG porous composites with waterproof, flame-retardant and thermal management functions. Journal of Materials Chemistry A *9*: 17538–52.

53. Feng, D., Liu, P., and Wang, Q. (2020). Selective microwave sintering to prepare multifunctional poly(ether imide) bead foams based on segregated carbon nanotube conductive network. Industrial & Engineering Chemistry Research *59*: 5838–47.

54. Papanastasiou, D.T., Schultheiss, A., Muñoz-Rojas, D., et al. (2020). Transparent heaters: a review. Advanced Functional Materials *30*: 1910225.

55. Sun, Z., Chen, J., Jia, X., et al. (2021). Humidification of high-performance and multifunctional polyimide/carbon nanotube composite foams for enhanced electromagnetic shielding. Materials Today Physics *21*: 100521.

56. Zhao, B., Bai, P., Wang, S., et al. (2021). High-performance Joule heating and electromagnetic shielding properties of anisotropic carbon scaffolds. ACS Applied Materials & Interfaces *13*: 29101–12.

57. Jia, X., Shen, B., Chen, J., et al. (2022). Multifunctional TPU composite foams with embedded biomass-derived carbon networks for electromagnetic interference shielding. Composites Communications *30*: 101062.

58. Rao, K.D.M., and Kulkarni, G.U. (2014). A highly crystalline single Au wire network as a high temperature transparent heater. Nanoscale *6*: 5645–51.

59. Choi, S., Park, J., Hyun, W., et al. (2015). Stretchable heater using ligand-exchanged silver nanowire nanocomposite for wearable articular thermotherapy. ACS Nano *9*: 6626–33.

60. Amjadi, M., Kyung, K.-U., Park, I., et al. (2016). Stretchable, skin-mountable, and wearable strain sensors and their potential applications: a review. Advanced Functional Materials *26*: 1678–98.

61. Jason, N.N., Ho, M.D., and Cheng, W. (2017). Resistive electronic skin. Journal of Materials Chemistry C *5*: 5845–66.

62. Choi, S., Han, S.I., Kim, D., et al. (2019). High-performance stretchable conductive nanocomposites: materials, processes, and device applications. Chemical Society Reviews *48*: 1566–95.

63. Sanli, A., Benchirouf, A., Müller, C., et al. (2017). Piezoresistive performance characterization of strain sensitive multi-walled carbon nanotube-epoxy nanocomposites. Sensors and Actuators A: Physical *254*: 61–68.

64. Pasha, A., Khasim, S., Darwish, A.A.A., et al. (2022). Flexible, stretchable and electrically conductive PDMS decorated with polypyrrole/manganese-iron oxide nanocomposite as a multifunctional material for high performance EMI shielding applications. Synthetic Metals *283*: 116984.

65. Wang, J., Liu, B., Cheng, Y., et al. (2021). Constructing a segregated magnetic graphene network in rubber composites for integrating electromagnetic interference shielding stability and multi-sensing performance. Polymers *13*: 3277. doi: 10.3390/polym13193277.

66. Hao, M., Wang, Y., Li, L., et al. (2021). Stretchable multifunctional hydrogels for sensing electronics with effective EMI shielding properties. Soft Matter *17*: 9057–65.

67. Han, L., Song, Q., Li, K., et al. (2021). Hierarchical, seamless, edge-rich nanocarbon hybrid foams for highly efficient electromagnetic-interference shielding. Journal of Materials Science & Technology *72*: 154–61.

68. Tung, T.T., Karunagaran, R., Tran, D.N.H., et al. (2016). Engineering of graphene/epoxy nanocomposites with improved distribution of graphene nanosheets for advanced piezo-resistive mechanical sensing. Journal of Materials Chemistry C *4*: 3422–30.

69. Xu, D., Wang, Q., Feng, D., et al. (2020). Facile fabrication of multifunctional poly(ethylene-co-octene)/carbon nanotube foams based on tunable conductive network. Industrial & Engineering Chemistry Research *59*: 1934–43.

70. Zheng, X., Wang, P., Zhang, X., et al. (2022). Breathable, durable and bark-shaped MXene/textiles for high-performance wearable pressure sensors, EMI shielding and heat physiotherapy. Composites Part A: Applied Science and Manufacturing *152*: 106700.

71. Ma, Z., Xiang, X., Shao, L., et al. (2022). Multifunctional wearable silver nanowire decorated leather nanocomposites for Joule heating, electromagnetic interference shielding and piezoresistive sensing. Angewandte Chemie International Edition *61*: e202200705.

72. Chen, Y., Liu, Y., Li, Y., et al. (2021). Highly sensitive, flexible, stable, and hydrophobic biofoam based on wheat flour for multifunctional sensor and adjustable EMI shielding applications. ACS Applied Materials & Interfaces *13*: 30020–29.

73. Liu, L.X., Chen, W., Zhang, H.B., et al. (2019). Flexible and multifunctional silk textiles with biomimetic leaf-like MXene/silver nanowire nanostructures for electromagnetic interference shielding, humidity monitoring, and self-derived hydrophobicity. Advanced Functional Materials *29*: 1905197.

74. Li, D.-Y., Liu, L.-X., Wang, Q.-W., et al. (2022). Functional polyaniline/MXene/cotton fabrics with acid/alkali-responsive and tunable electromagnetic interference shielding performances. ACS Applied Materials & Interfaces *14*: 12703–12.

75. Cai, J.H., Li, J., Chen, X.D., et al. (2020). Multifunctional polydimethylsiloxane foam with multi-walled carbon nanotube and thermo-expandable microsphere for temperature sensing, microwave shielding and piezoresistive sensor. Chemical Engineering Journal *393*: 124805.

76. Du, Y., Wang, X., Dai, X., et al. (2022). Ultraflexible, highly efficient electromagnetic interference shielding, and self-healable triboelectric nanogenerator based on Ti3C2Tx MXene for self-powered wearable electronics. Journal of Materials Science & Technology *100*: 1–11.

77. Chen, F., Huang, Q., and Zheng, Z. (2022). Permeable conductors for wearable and on-skin electronics. Small Structures *3*.

78. Song, P., Qiu, H., Wang, L., et al. (2020). Honeycomb structural rGO-MXene/epoxy nanocomposites for superior electromagnetic interference shielding performance. Sustainable Materials and Technologies *24*: e00153.

79. Camino, G., Costa, L., and Luda di Cortemiglia, M.P. (1991). Overview of fire retardant mechanisms. Polymer Degradation and Stability *33*: 131–54.

80. Ma, Y., Ma, P., Ma, Y., et al. (2017). Synergistic effect of multiwalled carbon nanotubes and an intumescent flame retardant: toward an ideal electromagnetic interference shielding material with excellent flame retardancy. Journal of Applied Polymer Science *134*: 45088.

81. Ji, X., Chen, D., Wang, Q., et al. (2018). Synergistic effect of flame retardants and carbon nanotubes on flame retarding and electromagnetic shielding properties of thermoplastic polyurethane. Composites Science and Technology *163*: 49–55.

82. Kim, I.C., Kwon, K.H., and Kim, W.N. (2019). Effects of hybrid fillers on the electrical conductivity, EMI shielding effectiveness, and flame retardancy of PBT and PolyASA composites with carbon fiber and MWCNT. Journal of Applied Polymer Science *136*: 48162.

83. Guo, W., Zhao, Y., Wang, X., et al. (2020). Multifunctional epoxy composites with highly flame retardant and effective electromagnetic interference shielding performances. Composites Part B: Engineering *192*: 107990.

84. Ji, X., Chen, D., Shen, J., et al. (2019). Flexible and flame-retarding thermoplastic polyurethane-based electromagnetic interference shielding composites. Chemical Engineering Journal *370*: 1341–49.

85. Inagaki, N., Hamajima, K., and Katsuura, K. (1978). Flame-retardant action of chlorine compounds and antimony trioxide on cellulose fabric. Journal of Applied Polymer Science *22*: 3283–91.

86. Lan, C., Jia, H., Qiu, M., et al. (2021). Ultrathin MXene/polymer coatings with an alternating structure on fabrics for enhanced electromagnetic interference shielding and fire-resistant protective performances. ACS Applied Materials & Interfaces *13*: 38761–72.

87. Zhang, Y., Tian, W., Liu, L., et al. (2019). Eco-friendly flame retardant and electromagnetic interference shielding cotton fabrics with multi-layered coatings. Chemical Engineering Journal *372*: 1077–90.

88. Cheng, W., Zhang, Y., Tian, W., et al. (2020). Highly efficient MXene-coated flame retardant cotton fabric for electromagnetic interference shielding. Industrial & Engineering Chemistry Research *59*: 14025–36.

89. Mao, Y., Wang, D., and Fu, S. (2022). Layer-by-layer self-assembled nanocoatings of Mxene and P, N-co-doped cellulose nanocrystals onto cotton fabrics for significantly reducing fire hazards and shielding electromagnetic interference. Composites Part A: Applied Science and Manufacturing *153*: 106751.

90. Gao, C., Shi, Y., Chen, Y., et al. (2022). Constructing segregated polystyrene composites for excellent fire resistance and electromagnetic wave shielding. Journal of Colloid and Interface Science *606*: 1193–204.

91. Zhang, Q., Wang, J., Yang, S., et al. (2019). Facile construction of one-component intrinsic flame-retardant epoxy resin system with fast curing ability using imidazole-blocked bismaleimide. Composites Part B: Engineering *177*: 107380.

92. Zhang, T., Zeng, S., Jiang, H., et al. (2021). Leather solid waste/poly(vinyl alcohol)/polyaniline aerogel with mechanical robustness, flame retardancy, and enhanced electromagnetic interference shielding. ACS Applied Materials & Interfaces *13*: 11332–43.

93. Liang, C., Qiu, H., Song, P., et al. (2020). Ultra-light MXene aerogel/wood-derived porous carbon composites with wall-like "mortar/brick" structures for electromagnetic interference shielding. Science Bulletin *65*: 616–22.

94. Yang, W., Liu, J.J., Wang, L.L., et al. (2020). Multifunctional MXene/natural rubber composite films with exceptional flexibility and durability. Composites Part B: Engineering *188*: 107875.

95. Lei, C., Zhang, Y., Liu, D., et al. (2020). Metal-level robust, folding endurance, and highly temperature-stable MXene-based film with engineered aramid nanofiber for extreme-condition electromagnetic interference shielding applications. ACS Applied Materials & Interfaces *12*: 26485–95.

96. Yuan, Y., Liu, L., Yang, M., et al. (2017). Lightweight, thermally insulating and stiff carbon honeycomb-induced graphene composite foams with a horizontal laminated structure for electromagnetic interference shielding. Carbon *123*: 223–32.

97. Li, Y., Li, C., Zhao, S., et al. (2019). Facile fabrication of highly conductive and robust three-dimensional graphene/silver nanowires bicontinuous skeletons for electromagnetic interference shielding silicone rubber nanocomposites. Composites Part A: Applied Science and Manufacturing *119*: 101–10.

98. Xu, H., Anlage, S.M., Hu, L., et al. (2007). Microwave shielding of transparent and conducting single-walled carbon nanotube films. Applied Physics Letters *90*: 183119

99. Du, J., Pei, S., Ma, L., et al. (2014). 25th Anniversary article: carbon nanotube- and graphene-based transparent conductive films for optoelectronic devices. Advanced Materials *26*: 1958–91.

100. Park, J.B., Rho, H., Cha, A.N., et al. (2020). Transparent carbon nanotube web structures with Ni-Pd nanoparticles for electromagnetic interference (EMI) shielding of advanced display devices. Applied Surface Science *516*: 145745.

101. Wang, L., Yue, X., Sun, Q., et al. (2022). Flexible organic electrochemical transistors for chemical and biological sensing. Nano Research *15*: 2433–64.

102. Ma, C., Liu, Y.-F., Bi, Y.-G., et al. (2021). Recent progress in post treatment of silver nanowire electrodes for optoelectronic device applications. *Nanoscale 13*: 12423–37.

103. Kim, S.-W., and Lee, S.-Y. (2020). Transparent supercapacitors: from optical theories to optoelectronics applications. *Energy & Environmental Materials 3*: 265–85.

104. Wu, J., Agrawal, M., Becerril, H.A., et al. (2010). Organic light-emitting diodes on solution-processed graphene transparent electrodes. *ACS Nano 4*: 43–48.

105. Hong, S.K., Kim, K.Y., Kim, T.Y., et al. (2012). Electromagnetic interference shielding effectiveness of monolayer graphene. *Nanotechnology 23*: 455704.

106. Yang, S.B., Kong, B.-S., Jung, D.-H., et al. (2011). Recent advances in hybrids of carbon nanotube network films and nanomaterials for their potential applications as transparent conducting films. *Nanoscale 3*: 1361–73.

107. Lu, Z., Ma, L., Tan, J., et al. (2016). Transparent multi-layer graphene/polyethylene terephthalate structures with excellent microwave absorption and electromagnetic interference shielding performance. *Nanoscale 8*: 16684–93.

108. Wang, H., Lu, Z., and Tan, J. (2016). Generation of uniform diffraction pattern and high EMI shielding performance by metallic mesh composed of ring and rotated sub-ring arrays. *Optics Express 24*: 22989–3000.

109. Han, Y., Lin, J., Liu, Y., et al. (2016). Crackle template based metallic mesh with highly homogeneous light transmission for high-performance transparent EMI shielding. *Scientific Reports 6*: 25601.

110. Wang, H., Ji, C., Zhang, C., et al. (2019). Highly transparent and broadband electromagnetic interference shielding based on ultrathin doped Ag and conducting oxides hybrid film structures. *ACS Applied Materials & Interfaces 11*: 11782–91.

111. Yuan, C., Huang, J., Dong, Y., et al. (2020). Record-high transparent electromagnetic interference shielding achieved by simultaneous microwave Fabry–Pérot interference and optical antireflection. *ACS Applied Materials & Interfaces 12*: 26659–69.

112. Chen, W., Liu, L.-X., Zhang, H.-B., et al. (2020). Flexible, transparent, and conductive Ti3C2Tx MXene–silver nanowire films with smart acoustic sensitivity for high-performance electromagnetic interference shielding. *ACS Nano 14*: 16643–53.

113. Yang, Y., Chen, S., Li, W., et al. (2020). Reduced graphene oxide conformally wrapped silver nanowire networks for flexible transparent heating and electromagnetic interference shielding. *ACS Nano 14*: 8754–65.

114. Lu, Z., Ma, L., Tan, J., et al. (2017). Graphene, microscale metallic mesh, and transparent dielectric hybrid structure for excellent transparent electromagnetic interference shielding and absorbing. 2D Materials 4: 025021.

115. Wu, M., Li, Y., An, N., et al. (2016). Applied voltage and near-infrared light enable healing of superhydrophobicity loss caused by severe scratches in conductive superhydrophobic films. Advanced Functional Materials 26: 6777–84.

116. Ma, X., Shen, B., Zhang, L., et al. (2018). Porous superhydrophobic polymer/carbon composites for lightweight and self-cleaning EMI shielding application. Composites Science and Technology 158: 86–93.

117. Zhang, L., Luo, J., Zhang, S., et al. (2022). Interface sintering engineered superhydrophobic and durable nanofiber composite for high-performance electromagnetic interference shielding. Journal of Materials Science & Technology 98: 62–71.

118. Luo, J., Huo, L., Wang, L., et al. (2020). Superhydrophobic and multi-responsive fabric composite with excellent electro-photo-thermal effect and electromagnetic interference shielding performance. Chemical Engineering Journal 391: 123537.

119. Yuan, R., Wu, S., Yu, P., et al. (2016). Superamphiphobic and electroactive nanocomposite toward self-cleaning, antiwear, and anticorrosion coatings. ACS Applied Materials & Interfaces 8: 12481–93.

120. Dixit, D., and Ghoroi, C. (2020). Role of randomly distributed nanoscale roughness for designing highly hydrophobic particle surface without using low surface energy coating. Journal of Colloid and Interface Science 564: 8–18.

121. Liu, L.X., Chen, W., Zhang, H.B., et al. (2019). Flexible and multifunctional silk textiles with biomimetic leaf-like MXene/silver nanowire nanostructures for electromagnetic interference shielding, humidity monitoring, and self-derived hydrophobicity. Advanced Functional Materials 29.

122. Liu, J., Zhang, H.B., Sun, R., et al. (2017). Hydrophobic, flexible, and lightweight MXene foams for high-performance electromagnetic-interference shielding. Advanced Materials 29: 1702367.

123. Wang, H., Tang, L., Wu, X., et al. (2007). Fabrication and anti-frosting performance of super hydrophobic coating based on modified nano-sized calcium carbonate and ordinary polyacrylate. Applied Surface Science 253: 8818–24.

124. Cheng, Y., Tan, M., Hu, P., et al. (2018). Strong and thermostable SiC nanowires/graphene aerogel with enhanced hydrophobicity and electromagnetic wave absorption property. Applied Surface Science 448: 138–44.

125. Zheng, Y., Song, Y., Gao, T., et al. (2020). Lightweight and hydrophobic three-dimensional wood-derived anisotropic magnetic porous carbon for highly efficient electromagnetic interference shielding. ACS Applied Materials & Interfaces *12*: 40802–14.

126. Cheng, M., Ren, W., Li, H., et al. (2021). Multiscale collaborative coupling of wood-derived porous carbon modified by three-dimensional conductive magnetic networks for electromagnetic interference shielding. Composites Part B: Engineering *224*: 109169.

127. Park, S.H., Cho, E.H., Sohn, J., et al. (2013). Design of multi-functional dual hole patterned carbon nanotube composites with superhydrophobicity and durability. Nano Research *6*: 389–98.

128. Kar, E., Bose, N., Dutta, B., et al. (2019). Ultraviolet- and microwave-protecting, self-cleaning e-skin for efficient energy harvesting and tactile mechanosensing. ACS Applied Materials & Interfaces *11*: 17501–12.

129. Luo, J., Wang, L., Huang, X., et al. (2019). Mechanically durable, highly conductive, and anticorrosive composite fabrics with excellent self-cleaning performance for high-efficiency electromagnetic interference shielding. ACS Applied Materials & Interfaces *11*: 10883–94.

130. Wang, Q.W., Zhang, H.B., Liu, J., et al. (2019). Multifunctional and water-resistant MXene-decorated polyester textiles with outstanding electromagnetic interference shielding and Joule heating performances. Advanced Functional Materials *29*: 1806819.

131. Lee, S., Park, J., Kim, M.C., et al. (2021). Polyvinylidene fluoride Core–Shell nanofiber membranes with highly conductive shells for electromagnetic interference shielding. ACS Applied Materials & Interfaces *13*: 25428–37.

132. Sang, M., Wang, S., Liu, S., et al. (2019). A hydrophobic, self-powered, electromagnetic shielding PVDF-based wearable device for human body monitoring and protection. ACS Applied Materials & Interfaces *11*: 47340–49.

133. Gao, J., Luo, J., Wang, L., et al. (2019). Flexible, superhydrophobic and highly conductive composite based on non-woven polypropylene fabric for electromagnetic interference shielding. Chemical Engineering Journal *364*: 493–502.

134. Xing, Y., Xue, Y., Song, J., et al. (2018). Superhydrophobic coatings on wood substrate for self-cleaning and EMI shielding. Applied Surface Science *436*: 865–72.

135. Mei, X., Lu, L., Xie, Y., et al. (2020). Preparation of flexible carbon fiber fabrics with adjustable surface wettability for high-efficiency electromagnetic interference shielding. ACS Applied Materials & Interfaces *12*: 49030–41.

136. Jia, L.-C., Zhang, G., Xu, L., et al. (2019). Robustly superhydrophobic conductive textile for efficient electromagnetic interference shielding. ACS Applied Materials & Interfaces *11*: 1680–88.

137. Menon, A.V., Choudhury, B., Madras, G., et al. (2020). 'Trigger-free' self-healable electromagnetic shielding material assisted by co-doped graphene nanostructures. Chemical Engineering Journal *382*: 122816.

138. Ma, W., Cai, W., Chen, W., et al. (2021). A novel structural design of shielding capsule to prepare high-performance and self-healing MXene-based sponge for ultra-efficient electromagnetic interference shielding. Chemical Engineering Journal *426*: 130729.

139. Sim, H.J., Lee, D.W., Kim, H., et al. (2019). Self-healing graphene oxide-based composite for electromagnetic interference shielding. Carbon *155*: 499–505.

140. Yang, W., Shao, B., Liu, T., et al. (2018). Robust and mechanically and electrically self-healing hydrogel for efficient electromagnetic interference shielding. ACS Applied Materials & Interfaces *10*: 8245–57.

141. Zou, L., Lan, C., Zhang, S., et al. (2021). Near-instantaneously self-healing coating toward stable and durable electromagnetic interference shielding. Nano-Micro Letters *13*: 190.

142. Lu, J., Zhang, Y., Tao, Y., et al. (2021). Self-healable castor oil-based waterborne polyurethane/MXene film with outstanding electromagnetic interference shielding effectiveness and excellent shape memory performance. Journal of Colloid and Interface Science *588*: 164–74.

143. Wang, T., Yu, W.C., Zhou, C.G., et al. (2020). Self-healing and flexible carbon nanotube/polyurethane composite for efficient electromagnetic interference shielding. Composites Part B: Engineering *193*: 108015.

For Product Safety Concerns and Information please contact our EU
representative GPSR@taylorandfrancis.com
Taylor & Francis Verlag GmbH, Kaufingerstraße 24, 80331 München, Germany

www.ingramcontent.com/pod-product-compliance
Ingram Content Group UK Ltd.
Pitfield, Milton Keynes, MK11 3LW, UK
UKHW022317100726
473146UK00009B/513